Gerhard Pfaff

Inorganic Pigments

Gerhard Pfaff

Inorganic Pigments

2nd, Revised and Extended Edition

DE GRUYTER

Author
Prof. Dr. Gerhard Pfaff
Technische Universität Darmstadt
Eduard-Zintl-Institut
Petersenstr. 18
64287 Darmstadt
Germany
pfaff.pigmente@gmx.de

ISBN 978-3-11-074391-3
e-ISBN (PDF) 978-3-11-074392-0
e-ISBN (EPUB) 978-3-11-074399-9

Library of Congress Control Number: 2023934970

Bibliographic information published by the Deutsche Nationalbibliothek
The Deutsche Nationalbibliothek lists this publication in the Deutsche Nationalbibliografie;
detailed bibliographic data are available on the Internet at http://dnb.dnb.de.

© 2023 Walter de Gruyter GmbH, Berlin/Boston
Cover image: Variation of special effect pigments, Merck KGaA
Typesetting: Integra Software Services Pvt. Ltd.
Printing and binding: CPI books GmbH, Leck

www.degruyter.com

Preface to the second edition

Six years have passed since the first edition of *Inorganic Pigments* was published. There was a lot of movement in pigments during these years. Several new developments were advanced and some pigment producers have merged or varnished. Progress was made with regard to the application systems for pigments. Environmental aspects are now being taken into account more and more, from pigment development to application in the various systems. Issues relating to the provision of suitable raw materials for pigment production are also becoming more important because of the increased efforts to protect environment and climate. By including all these topics, the second edition gives the up-to-date status of the field. The author has taken the opportunity to thoroughly review, revise, and update all chapters. A large amount of new information has been included and informative figures have been added. Chapters such as "Fundamentals", "Colored pigments", or "Application systems for pigments" are covered much more extensively than in the first edition.

The objective of the book is still to provide comprehensive knowledge about the field of inorganic pigments in an understandable way. Objectivity and thoroughness have a high priority in all included topics. The structure of the chapters has been retained, as it has proven its worth. Several new subchapters have been added. The number of figures has been greatly expanded. For example, many photographs of pigment samples have now been included to give a better idea of their appearance. The bibliography for each chapter has been thoroughly revised and supplemented with the latest literature citations. As already in the first edition, study questions are designed to test the acquired knowledge in each section.

I would like to thank all my colleagues and friends who supported me in the completion of the book and contributed to its improvement with helpful discussions. Special thanks go to Karin Sora and her team from De Gruyter for their continued support of this book project, without which the second edition would not have been possible.

Berlin, March 2023   Gerhard Pfaff

https://doi.org/10.1515/9783110743920-202

Preface to the first edition

The present textbook has its origins in the long term employment of the author with inorganic pigments in teaching, research, and application. The objective was to provide a technically correct and up-to-date introduction to these pigments with regard to their properties, manufacture, and application.

Care has been taken to give the reader an overview of basic principles, synthesis, chemical and physical properties, the application technical behavior in various systems, and the production of inorganic pigments. A special section for each pigment type deals with questions of toxicology and occupational health. The book is addressed not only to students with main focus in chemistry, material science, paint and color technology, color design, physical optics, but also to pigment and color specialists in industry and academia and to pigment users in the application areas such as coatings, paints, plastics, printing inks, cosmetics, and ceramics.

Inorganic pigments have been very important substances for human life for thousands of years. They have contributed, and still contribute, to the beautification of daily life and habitat; to accentuation and differentiation of objects; and to influence thoughts, moods, and feelings. The industrial manufacture of inorganic pigments based on grown chemical and physical knowledge started only around 1800. Before that, pigments based on natural materials or manufactured by mostly empirical processes were used. Most of the technically relevant inorganic pigments were developed and produced in industrial scale in the nineteenth and twentieth centuries. There are ongoing developments not only for the improvement of existing conventional pigments but also for the synthesis of new innovative colorants, especially in the class of effect pigments.

The present book covers a wide range of aspects of inorganic pigments from fundamentals up to application. The individual classes, such as white, colored, black, transparent, and effect pigments are discussed. In addition, ceramic colors, magnetic, anticorrosive, and luminescent pigments are covered. A further chapter deals with general chemical and physical properties of inorganic pigments, color properties including the origin of color and color vision, and application technical behavior. Additional focal points are general methods of manufacture and application systems for inorganic pigments. Finally, fillers have been included because they are very often used in combination with pigments.

The objective of the book is to provide comprehensive knowledge about inorganic pigments in an understandable way without neglecting scientific objectivity and thoroughness. Particular emphasis has been laid on a clear structure of the single topics to quickly disseminate the desired information. Figures and tables inserted illustrate fundamental aspects, pigment properties, and manufacturing processes. For readers interested in further information and details, literature and patent references are included at the end of each chapter. Study questions are designed to test the acquired knowledge in each section.

https://doi.org/10.1515/9783110743920-203

At this point, I wish to thank a number of friends and colleagues for their support in the preparation of this book: Anja Kilian (Merck KGaA) and Dr. Adalbert Huber (Carl Schlenk AG) for a variety of electron-microscopic images; Dr. Carsten Plüg and Dr. Michael Rösler (both Merck KGaA) for helpful discussions; and Werner Rudolf Cramer, Hagen Ißbrücker, Volker Ißbrücker, and Byk-Gardner GmbH for the provision of illustrations. Finally, I want to express special thanks to Karin Sora and her team from De Gruyter for their patience and continued support of this book project.

Berlin, May 2017   Gerhard Pfaff

Contents

1 Fundamentals, general aspects, color, application

1.1 Definitions and classification

Pigments are defined in the modern context as substances consisting of small particles that are practically insoluble in an application system and that are used as colorants or because of their anticorrosive or magnetic properties [1]. They differ from dyes, which also belong to the class of colorants, primarily due to the fact that dyes are practically completely soluble in the medium of application. The most important application media for pigments are automotive and industrial coatings, paints, plastics, printing inks, cosmetic formulations, and building materials. Other uses of pigments are in paper, rubber, glass, porcelain, glazes, and artists' colors.

The term "pigment" descends from the Latin word "pigmentum". It was originally used in the sense of a coloring matter. The use of the word was later extended to indicate colored decoration. The term "pigment" was also used in the late Middle Ages for plant and vegetable extracts, particularly for those with coloring properties. In biological terminology, it is still used to indicate vegetable and animal colorants that are present in "solved" form as extremely small particles in cells or cell membranes, as deposits in tissues, or suspended in bodily fluids. In all these cases, the term "pigment" is misleading and would better be replaced by the more suitable term "dye" or "dyestuff".

The term "colorant" covers all colored compounds regardless of their origin and utility for coloration or other purposes. Colorants are divided not only into pigments and dyes but also into natural and synthetic compounds. Some pigments and some dyes exist as natural and synthetic variants. Pigment particles, when applied, have to be attached to surfaces (substrates) by additional materials, such as binder systems (paints, coatings, printing inks, cosmetics), plastics, or glazes. Dyes are applied to various substrates, such as textiles, leather, paper, or hair, using a liquid in which they are dissolved. In contrast to pigments, dyes must have an affinity to the substrates on which they are fixed.

Pigments are differentiated according to their chemical composition and with respect to their optical and technical properties. A fundamental distinction is that between inorganic and organic pigments.

Figure 1.1 shows a rough classification of pigments and dyes within the category of colorants (coloring materials). Fillers as a substance class closely related to pigments are involved here. It can be clearly seen that dyes are based only on organic compounds. It also becomes clear that organic white pigments do not exist.

Figure 1.2 contains a detailed classification of inorganic pigments. White, colored, black, and special pigments exist. The most important representatives for the different pigment categories are shown. White pigments are represented by titanium dioxide (rutile and anatase), zinc sulfide including lithopone, and zinc oxide. Colored pigments show the broadest variation, ranging from blue (mixed metal oxides, ultramarine, iron blue) via green (chromium oxide, mixed metal oxides) and yellow (iron oxide hydroxide, mixed metal oxides, lead chromate, bismuth vanadate, cadmium sulfide) up to red (iron oxide, cadmium selenide, lead molybdate, cerium sulfide, oxonitrides). The main representative for black pigments is carbon black. Based on this variety of colors, whites and blacks, with inorganic pigments it is possible to design

https://doi.org/10.1515/9783110743920-001

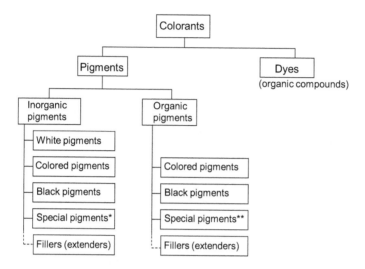

Fig. 1.1: Classification of colorants (* effect pigments, transparent pigments, luminescent pigments, functional pigments; ** effect pigments, luminescent pigments).

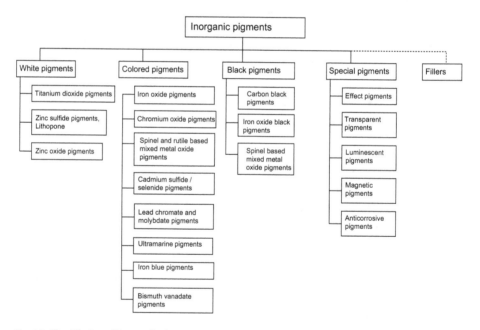

Fig. 1.2: Classification of inorganic pigments.

nearly all thinkable colors and noncolors (white, black, gray), including bright and dark color shades by the use of pure single pigments or pigment mixtures.

Special pigments are subdivided into the classes of effect pigments (luster pigments) with the two subclasses special effect pigments (pearlescent pigments, interference pigments) and metal effect pigments, transparent pigments, luminescent pigments with the two subclasses fluorescent pigments and phosphorescent pigments, magnetic pigments, and anticorrosive pigments. The last two classes, as well as a part of the transparent pigments belong to the category functional pigments. Other materials that count as functional pigments are electrically conductive, IR-reflective, UV-absorbing, and laser-marking pigments.

The optical behavior of effect pigments is based either on the directional reflection of visible light from predominantly two-dimensional and aligned metallic (metal effect pigments) or highly refractive transparent pigment particles (pearlescent pigments) or on the phenomenon of interference (interference pigments).

Transparent pigments are characterized by very small particles with sizes in the range below 100 nm and large specific surface areas. Most of the technically relevant pigments consist of particles which are even smaller than 30 nm. They are classified as nanomaterials. Pigmentation with these pigments leads to a transparent appearance of the application systems.

Luminescent pigments (luminescent materials, luminophores, phosphors) show optical effects based on the ability to absorb radiation and emit it as light of longer wavelength with a time delay (phosphorescence) or without a time delay (fluorescence). Light emission often occurs in the visible spectral range. External energy is necessary to enable luminescent materials to generate light.

Functional pigments are not about color but about different physical properties that cannot be derived from the original term "pigmentum". Such materials, which exhibit magnetic, anticorrosive, electrically conductive, IR-reflective, UV-absorbing, and other physical properties can be appropriately described by the term "functional pigments". Some of the transparent pigments and the luminescent pigments can also be classified as functional pigments. The reason why functional pigments are classified as pigments is because of their similar morphological and application technical properties.

Fillers (extenders) are powdery substances that, like pigments, are practically insoluble in the application system. They are typically white and are used because of their chemical and physical properties. The distinction between pigments and fillers is made based on the specific application. Another criterion is the refractive index, which for fillers is usually below 1.7 and above this value for pigments. There is, however, no fixed definition of the value for the refractive index to distinguish both product classes. A filler is not a colorant in the proper sense, but a substance that modifies the application medium in order to improve its technical characteristics, to influence optical and coloristic properties, or to increase the volume. Fillers are also used to lower the consumption of more expensive binder components. Those fillers, which are used mainly for cost reduction reasons are also called extenders. Fillers are often applied together with pigments to improve the properties of the medium in which they are incorporated.

Another classification for inorganic pigments is based on the chemical composition. Table 1.1 contains relevant chemical compositions of inorganic pigments together with examples. Some of the pigments mentioned here are historically significant, but are no longer of any practical importance today. These include not only some of the sulfides (HgS, As_2S_3) but also stannate, phosphate, arsenate, and antimonate-based pigments. The main reason why these pigments are no longer used is their content of toxic components. The formerly important lead oxides PbO and Pb_3O_4 are becoming less important due to the undesired lead content. The carbonate hydroxides of lead and copper have also lost the great importance they once had. Some newer developments such as γ-Ce_2S_3 or the oxonitrides have not entered the pigment market because they do not meet important application criteria or are too costly to produce.

Tab. 1.1: Classification of inorganic pigments based on the chemical composition.

Chemical composition	Pigment examples
Oxide, oxide hydroxide	TiO_2, ZnO, α-Fe_2O_3, α-FeOOH, γ-Fe_2O_3, Fe_3O_4, Cr_2O_3, CrOOH, PbO, Pb_3O_4, Mn_3O_4, α-MnOOH, Sb_2O_3
Complex oxide	$CoAl_2O_4$, $CuCr_2O_4$, Co_2TiO_4, (Ti,Ni,Sb)O_2, (Ti,Cr,Sb)O_2
Carbonate hydroxide	2 $PbCO_3 \cdot Pb(OH)_2$, 2 $CuCO_3 \cdot Cu(OH)_2$, $CuCO_3 \cdot Cu(OH)_2$
Sulfide, selenide	ZnS, CdS, Cd(S,Se), CdSe, γ-Ce_2S_3, HgS, As_2S_3
Chromate, molybdate	$PbCrO_4$, Pb(Cr,S)O_4, Pb(Cr,S,Mo)O_4, $ZnCrO_4$, $BaCrO_4$, $SrCrO_4$
Vanadate	$BiVO_4$, 4 $BiVO_4 \cdot 3$ Bi_2MoO_6
Stannate	Pb_2SnO_4, $PbSn_2SiO_7$, Co_2SnO_4, $CoSnO_3$
Phosphate	$Co_3(PO_4)_2$
Antimonate	$Pb(SbO_3)_2$
Arsenate	$Cu(AsO_3)_2$
Ultramarine	$Na_6Al_6Si_6O_{24}(NaS_n)$
Hexacyanoferrate	$K[Fe^{III}Fe^{II}(CN)_6] \cdot x$ H_2O (x = 14–16)
Oxonitride	$CaTaO_2N$, $LaTaON_2$
Element	C, Al, Cu, Cu/Zn, Au

Generally, inorganic pigments are more stable against light, weather, temperature, and chemicals than organic pigments. Another advantage of inorganic pigments is their lower manufacturing costs. Organic pigments need mostly multistage syntheses and more expensive raw materials for their production, which leads finally to higher prices for them in the market. They are, however, in various cases more color intensive and, therefore, more attractive than inorganic colored pigments. A disadvantage of organic pigments is their lower stability against the factors mentioned above. Degradation of organic colorants – pigments as well as dyes – by exposure to UV light and under atmospheric influences is an issue. The use of UV absorbers together with the organic colorants is a very helpful possibility to ensure a long shelf life also for these materials in their application.

1.2 History and economic aspects

The history of inorganic pigments goes back to prehistoric times. Human beings in primitive society were already able to color textiles, furs, and other stuff by using natural substances, predominantly from vegetables, but also from animals. The reconstruction of historical pigments, dyes, and formulations is often ambiguous from today's perspective. The earlier manufacturing processes corresponded to the level of knowledge of the artists and craftsmen of their time. A scientific approach, as we practice today, is an achievement of modern times. Various sources are used for the reconstruction of historical materials for paintings and other works of art. The most important sources of our knowledge about historical pigments, dyes, and binders in painting are, above all, preserved paintings or parts of them, but also technical writings and original recipes of artists. The earliest surviving writings in this context date back to the Middle Ages. The interest in these sources arose in the eighteenth and nineteenth centuries, at a time when classical studies were becoming very popular.

Today, modern nondestructive scientific examination methods such as X-ray diffraction, neutron diffraction, or infrared spectroscopy support the identification of historical painting materials. Even multilayer structures of artworks can be traced using these methods. The use of modern analytical methods leads to an understanding of how the artists selected, procured and processed the materials, but it is almost impossible to translate the content of the historical recipes into the language of today's chemistry or material science. The complex raw materials used in former times were often of low homogeneity. Another difficulty in understanding the materials is that some of them came from unknown natural locations. The old recipes often require interpretation from experts in this field, since the terminologies and the manufacturing processes described do not correspond to today's materials and production methods.

The first surviving artworks of human activity with color are cave drawings like those in Altamira or Lascaux (Figure 1.3). They were most probably created 15,000 to 30,000 years ago in the Ice Age. Natural manganese oxides and charcoal were used for black shades and iron oxides for yellow, orange, and red color tones. It can be assumed that natural pigments in a pulverized state have also been used for body painting since early periods of human development. The application of pigments for coloration is considered to be among the oldest cultural activities of mankind. The artists of the Ice Age used pigments without a binder. The drawings survived over thousands of years because colorless minerals contained in seepage water encased the colored pigments in transparent mineral layers [2–4].

Important inorganic colorants of the early period were red chalk (a yellowish iron oxide) and manganese dioxide. Calcium carbonate was the basis of white color shades. Mastering fire allowed early humans to make charcoal, bone black, and brick red-fired ocher [5–7].

The methods for the manufacture of inorganic materials suitable for pigment uses and the dyeing techniques were developed further in ancient times, especially in Egypt

Fig. 1.3: Grotte de Lascaux, axial gallery, right cave wall, Panneau of the Chinese horses, second Chinese horse, (source: bpk/Ministère de la Culture – Médiathèque du Patrimoine, Dist. RMN-Grand Palais).

and Babylon. The walls of temples and funerary chambers, stone constructions, wood, and ceramics were used as painting surfaces for pigments in those days (Figure 1.4). Iron oxide black, manganese oxide black, antimony sulfide, lead sulfide (galena), and carbon black were used for black shades, iron oxide red and colloidal copper (formed by reductive firing of basic copper carbonate) for red shades, and calcium carbonate and kaolinite for white tones. Arsenic sulfide and Naples yellow were the first yellow

Fig. 1.4: Bust of Queen Nefertiti, 1,351–1,334 B.C., Ägyptisches Museum und Papyrussammlung, Staatliche Museen zu Berlin (source: bpk/ Ägyptisches Museum und Papyrussammlung, SMB/S. Steiß).

pigments, and ultramarine (lapis lazuli), Egyptian blue (a copper calcium silicate, also known as synthetic lapis lazuli), cobalt aluminate, and ground cobalt glass (smalt) were the first blue pigments used. The first green pigments were green earth, malachite, and copper hydroxychloride. Calcium carbonate, calcium sulfate, and kaolinite were the mainly used white pigments in ancient times [7–9].

Some of these pigments were not of natural origin. They were synthesized by empirical chemical methods. The use of toxic substances like arsenic and antimony compounds in cosmetic formulations was quite common. Egyptian blue became the most important pigment in Egypt. Blue smalt glass was used amongst others for the depiction of eyes, hair, and decoration in the graphic representation of pharaohs. Blue was the color of the divine. Lapis lazuli was initially not used in the manufacture of pigments but in gemstones. Nobler metals such as gold, silver, and copper, however, have already found a broader application as colorants [7–9].

A thousand years later, red cinnabar was used as red pigment in Europe by the Greeks and subsequently by the Romans (Figure 1.5). The Romans mined red mercury ore in large quantities in Almadén in Spain. Ground ocher was very popular as red pigment, but it was also expensive. The wall paintings in Pompeii testify to the use and popularity of red pigments like cinnabar and realgar. The Romans also used white lead and red lead. The trade routes to the Far East also enabled the import of organic and inorganic colorants such as indigo and colored minerals from India to Europe. Both Greeks and Romans developed techniques for painting and dyeing to an advanced state [7, 10–13]. Since the Romans were interested in artistic images that were stable over time, the technique of mosaics was developed. The wall area was

Fig. 1.5: Puglia hydra, IV century B.C., Antikensammlung, Staatliche Museen zu Berlin (source: bpk/Antikensammlung, SMB/J. Laurentius).

coated with cement and colorful flat stones or ceramic plates were pressed into the still fresh cement. The mosaics produced in this way have withstood the weather until today and look like colored paintings from a distance.

Starting in the eighth century after Christ, crusaders and Moors brought the knowledge of the Orient to Europe. In the twelfth century, Venice became the main trading place for colorants in the world. The wealth of several Venetian and Florentine noble families promoted the arts. For this reason, painters and artists from the world known at that time came to the important metropolises in Italy. The church was traditionally a very good client for painters and artists. Fresco painting experienced a heyday in the early Renaissance with the painter Giotto di Bondone (ca. 1270–1337). Figure 1.6 shows an example of Giotto's fresco painting. The Florentine painter Cennino Cennini (ca. 1370–1440) was the first to publish a handbook on painting with the title, "Il libro dell'arte o trattato di pittura" [14]. This book, written around the year 1400, was the most influential textbook on painting of the late Middle Ages. It has remained of special cultural-historical and art-historical importance until today because it is the first document in which a professional artist disclosed the secrets of painting, including all relevant materials.

Fig. 1.6: Giotto di Bondone: Adoration of the Magi, 1304–1306, Capella degli Scrovegni all'Arena, Padua (source: bpk/Scala).

The lime compatibility of the pigments used was a special problem in fresco painting. Blue color tones were typically applied "a secco" because the available blue pigments were not compatible with the lime. The color palette ranged from earth colors (yellow ocher, Terra di Siena) via the common mineral colors up to synthetically manufactured pigment colors like lead tin yellow (produced by a high-temperature reaction of red lead with tin dioxide) and smalt glass. Since early medieval times, the so-called Fra Angelico blue, named after the monk and painter Fra Angelico of Florence was used besides smalt as blue pigment for fresco paintings. However, this pigment too had compatibility problems. From the fourteenth century, Fra Angelico blue was also used for book illustrations. For this purpose, it was often applied in combination with gold leaf [7].

Paint merchants, who produced pigments and supplied brushes and screens, established themselves from the middle of the seventeenth century on. The colors were produced on the basis of recipes, some of which were kept secret. Existing knowledge was passed on and used again and again, but the basic colorants and binders were always further developed and sensitively adapted to the differentiated fields of application [7, 15–17].

From the very beginning, artists' colors were among the most desirable colorants and thus contributed greatly to the development of new pigments, dyes, and binders over the centuries [7, 18–21].

Painting saw a significant development in the Renaissance, with techniques of oil painting offering a completely new quality. The range of pigments already used earlier was extended to include green earth (natural earth pigment as a weathering product of ferrous silicates, purified). Color-intensive varieties of this green pigment came from the Verona area (Veronese green earth) and from Bohemia (Bohemian green earth). Other important paint pigments of this time were Naples yellow, auric pigment, lead tin yellow, cinnabar, realgar, carmine (calcium and aluminum salts of carminic acid), malachite, white lead, and painter's charcoal [7, 22–24].

Several important principles of fine art were established in the Renaissance. These principles consisted of conventions that applied to all aspects of image-making, including subject, composition, line, and color. Color was considered secondary to overall design. This can be inferred, for example, from the fact that art students or apprentices spent most of their time learning to draw and only much later learned the art of pigmenting and coloring. After the Renaissance, this approach to fine art painting was adopted by all the major European academies and became embedded in the style of academic art. Painting was not even on the curriculum in most academies, and color continued to have a secondary function as more of a supporting element. The famous painter Peter Paul Rubens (1577–1640) was sometimes criticized in the Baroque period for his intense use of colorants, while his painter colleague Nicolas Poussin (1594–1665) was revered as an example of more balanced colorism. One of the basic principles of academic painting concerned the primacy of the naturalistic palette. Colors should reflect the natural perception of man, which means that grass had to be green and the sea had to be blue. This situation continued in the following centuries. The range of pigments used in painting and other work of art hardly increased during this period [5, 6].

In the eighteenth century, on the basis of the level of knowledge of chemistry achieved at that time, which had increasingly developed from alchemy into a science, the targeted synthesis of various inorganic pigments was achieved for the first time. Thus, in the course of this century, pigments such as Prussian blue, Schweinfurt green, or cobalt blue became available. In particular, the discovery of Prussian blue as a new blue pigment is well-known. The color maker Johann Jacob von Diesbach (lived around 1700) and the alchemist Johann Conrad Dippel (1673–1734) were, in Berlin in 1706, the first to succeed in synthesizing the pigment. It took some time until the preparation route was fully developed. The synthesis was finally made public by the chemist and pharmacist Caspar Neumann (1683–1737) from Berlin, who provided the Prussian blue recipe to the Royal Society for anonymous publication [25, 26]. Today, Prussian blue is one of the technically important blue pigments, which is mainly used in paints and printing inks. In technical applications, the pigment is usually referred to as iron blue. The production of Prussian blue and of other newly developed pigments initially took place in workshops and small companies. These were to become the nuclei of the rapid upswing in color chemistry that took place in the nineteenth century.

A specialty among the historical pigments is Maya blue. This pigment was rediscovered in 1931 at the archaeological site of Chichén Itzá [27]. This intensively colored blue pigment that was widely used in wall painting, pottery, sculptures, and luxury art in Mesoamerica and even as therapeutic agent in pre-Columbian times can be described as a hybrid organic–inorganic material resulting from the attachment of indigo, a blue dye extracted from leaves of anil (indigofera suffruticosa and other plant species), to the clay matrix of palygorskite, a fibrous phyllosilicate of the ideal composition $(Mg,Al)_4Si_8(O,OH,H_2O)_{24} \cdot n\ H_2O$ [28, 29].

Table 1.2 contains important representatives of historical pigments (period: early human history until the eighteenth century).

Tab. 1.2: Representatives of historical pigments.

Color	Representatives
White	Lead white (2 $PbCO_3 \cdot Pb(OH)_2$), kaolin
Black	Coal (charcoal), groutite (α-MnOOH), manganite (γ-MnOOH), hausmannite (Mn_3O_4)
Yellow	Yellow ochre (α-FeOOH), auric pigment (As_2S_3), lead ochre (PbO), lead tin yellow (Pb_2SnO_4, $PbSn_2SiO_7$), Naples yellow ($Pb(SbO_3)_2$), zinc yellow (Zn_2CrO_4), Indian yellow ($C_{19}H_{16}O_{10}$)
Red	Red ocher (α-Fe_2O_3), Terra di Siena (α-Fe_2O_3), vermilion (HgS), lead red (Pb_3O_4), alizarin madder varnish, alizarin red ($C_{14}H_8O_4$)
Green	Green earth (Fe silicates), Schweinfurt green ($C_4H_6As_6Cu_4O_{16}$)
Blue	Lazurite (lapis lazuli), Egyptian blue ($CaCuSi_4O_{10}$), azurite (2 $CuCO_3 \cdot Cu(OH)_2$), malachite ($CuCO_3 \cdot Cu(OH)_2$), cobalt blue ($CoAl_2O_4$)
Brown	Burnt umber ($Fe_2O_3 \cdot x\ MnO_2$), brown ocher (α-Fe_2O_3 + Mn-oxides), limonite (mixture of different Fe-oxides)

The real industrial development and production of colorants started later, around 1800. Subsequently, ultramarine pigments, Guignet's green (CrOOH), further cobalt-containing pigments, iron oxide pigments, and cadmium sulfide pigments were developed and introduced into the market. The twentieth century was a time of increased scientific investigation and understanding of color and pigments. Pigments like cadmium selenide, manganese blue, molybdenum red, and bismuth vanadate were synthesized and offered to customers of various industries. High-purity titanium dioxide and zinc oxide were produced for the first time on an industrial scale and started their triumphal procession as white pigments. Metal effect pigments and pearlescent pigments were developed, produced on a scale of tons, and introduced into the market.

Table 1.3 contains a summary of important historical data for synthetic inorganic pigments. Organic colorants have experienced a similarly impressive development since the first industrial synthesis of the red azo pigment para red (today C.I. Pigment Red 1) in the year 1885. Many other organic pigments and also dyes were developed and introduced at the end of the nineteenth century and in the twentieth century, among these copper phthalocyanine (1935) and quinacridone (1955).

Tab. 1.3: Historical data for selected inorganic pigments.

Pigment	Formula	Year of first manufacture
Iron blue	$Fe^{III}[Fe^{II}Fe^{III}(CN)_6]_3 \cdot x\, H_2O$	1704
Scheele's green	$CuHAsO_3$	1775
Cobalt green	$ZnCo_2O_4$	1780
Chrome yellow	$PbCrO_4$	1797
Cobalt blue	$CoAl_2O_4$	1799
Chromium oxide green	Cr_2O_3	1809
Cadmium yellow	CdS	1817
Zinc white	ZnO	1824
Ultramarine blue	$Na_6Al_6Si_6O_{24}(NaS_n)$	1828
Metal bronzes	Cu/Zn	1830
Iron oxide red	$\alpha\text{-}Fe_2O_3$	1878
Silver bronzes	Al	1910
Cadmium red	$Cd(S,Se)$	1910
Titanium dioxide	TiO_2	1916
Ceramic colors	ZrO_2 based	1950
Pearl luster pigments on mica	TiO_2/mica	1969
Bismuth vanadate	$BiVO_4$	1977

The availability of more and more synthetic pigments was accompanied by the development of suitable application systems. Paints, coatings, printing inks, cosmetic formulations, and other media experienced steady improvement and adaptation. With the introduction of the car, a revolutionary development of automotive coating systems began. Printing inks were adapted according to the requirements of the different printing processes and substrates used. The development of plastics and other polymer materials in the nineteenth, but especially in the twentieth century, opened a new wide application spectrum for inorganic and organic pigments.

The worldwide production and consumption of pigments in the year 2020 may be estimated to be in the range of 12 Mio tons. The most important market segment with more than 45% of this tonnage is paints and coatings. At great distance follow the segments plastics, building materials, printing inks, and paper.

The global pigment market is dominated by titanium dioxide (rutile and anatase). TiO_2 pigments account for nearly 60% of the global demand. The second largest sales market for pigments is that of iron oxide types, followed by carbon black. More than 90% of all pigments used worldwide are inorganic ones. However, organic pigments with less than 1 million tons are also an attractive class of colorants with a very broad variety of applications [30].

1.3 Uses

Pigments are used in a broad variety of application systems. The most important of these systems are coatings (varnishes), paints, plastics, printing inks, cosmetic formulations, and building materials (construction materials). Other important application areas for pigments are paper, textiles, leather, rubber, floor coverings, glass, porcelain, glazes, enamels, and artists' colors.

Coatings are divided into automotive and industrial coatings. Both coating types require specific compositions to fulfill the demands with respect to appearance, resistance, durability, and workability. In these applications, an optimal, uniform particle size is necessary to achieve the desired gloss, hiding power, tinting strength, and lightening power. Paint and coating films have to be thin, in most cases, when they are cured. Pigments with high tinting strength and good hiding power combined with suitable dispersion properties are, therefore, needed in this application segment.

Pigments also have to meet high requirements in the other applications. There are specific requirements for the use of pigments in plastics, printing inks, cosmetic formulations, and building materials. In any case, the pigments must meet the respective application, processing, and stability criteria.

The choice of a pigment for a specific application includes several criteria to achieve an optimal application result:
- color properties: color, tinting strength, lightening power, hiding power;
- general chemical and physical properties: chemical composition, particle size, particle size distribution, density, moisture and salt content, content of water-soluble and acid-soluble matter, hardness;

- stability properties: resistance against light (especially UV light), heat, humidity, and chemicals, retention of gloss, corrosion resistance;
- behavior in binders and other application systems: dispersion properties, interaction and compatibility with binder components, solidifying properties.

A pigmented system is described as a dispersion of pigment particles in a transparent or translucent binding medium, which is initially liquid, and after hardening of the binder preferably solid. In many cases, a thin film is formed, e.g., a paint layer, with thicknesses of a few to hundreds of micrometers. In some cases, such as in cosmetic formulations, the application takes the form of a liquid dispersion. In the case of plastics and building materials, pigmentation usually takes place in the entire mass (mass coloring). Incident light undergoes reflection, refraction, scattering, and absorption in interaction with the dispersion that results in the visual appearance of the pigmented medium.

1.4 General chemical and physical properties

Inorganic pigments consist of single particles of mostly uniform chemical composition and crystal structure. Exceptions are pigment mixtures (blended pigments) and layer–substrate pigments (typical for special effect pigments), which consist of chemically and crystallographically nonuniform or multicomponent particles.

Blended pigments are colorants consisting, from the outset, of homogeneously mixed or ground single pigments (where appropriate, with a suitable proportion of a filler). An example for this pigment type is lead chrome green, which consists of a mixture of chrome yellow and iron blue. If the single components are too different in particle size, particle shape, density, or surface tension, separation of the pigments during application may occur.

In the case of the layer–substrate pigments, at least one additional component is deposited as a thin layer on a substrate particle. The attraction between layer and substrate is so strong here that no separation of the two components occurs. Surface treated pigments can be regarded as a special type of layer–substrate pigments. Following the production of a base pigment, a thin inorganic and/or organic layer is deposited to stabilize the pigment and to adapt the surface to the application medium. In this way, the catalytic and photochemical reactivity of a pigment may be reduced, the dispersion behavior improved, and the hydrophilic or hydrophobic character of the surface adjusted.

Inorganic pigments can be described with respect to their optical properties, as shown in Table 1.4. The interaction of pigment particles with visible light is decisive for the appearance of pigments in their pure powdered form and in the particular application medium.

Tab. 1.4: Classification of inorganic pigments with respect to the optical effect (with examples).

Term	Definition
White pigments	Nonselective light scattering in the visible range (titanium dioxide, zinc sulfide, lithopone, zinc oxide)
Colored pigments	Selective light absorption together with light scattering in the visible range (iron oxide red, iron oxide yellow, chromium(III) oxide, cadmium sulfide, cadmium selenide, ultramarine, chrome yellow, cobalt blue, iron blue, nickel rutile yellow)
Black pigments	Nonselective light absorption in the visible range (carbon black, iron oxide black)
Effect pigments	Regular reflection or interference on parallel-oriented, metallic (metal effect) or highly refractive transparent or semi-transparent (pearlescent and interference effects), platelet-shaped particles
Metal effect pigments	Regular reflection on parallel-oriented, metallic, platelet-shaped particles (aluminum flakes, copper/zinc flakes)
Pearlescent pigments	Regular reflection on parallel-oriented, highly refractive, platelet-shaped particles (titanium dioxide on mica, bismuth oxychloride, basic lead carbonate)
Interference pigments	Optical interference by superposition of reflected light waves on parallel-oriented, highly refractive, platelet-shaped particles (titanium dioxide on mica, iron oxide on mica)
Luminescent pigments	Light absorption followed by emission of radiation as light of a longer wavelength (fluorescence, phosphorescence)
Fluorescent pigments	The light of longer wavelength is emitted after absorption and excitation without a delay (silver-doped zinc sulfide)
Phosphorescent pigments	The light of longer wavelength is emitted after absorption and excitation within several hours (copper-doped zinc sulfide)

The optical characteristics of pigmented systems are determined by
–   the pigment properties such as particle size, particle size distribution, particle shape, refractive index, scattering coefficient, absorption coefficient;
–   the properties of the application system (binder) such as refractive index, self-absorption;
–   the pigment volume concentration (PVC) in the pigmented system;
–   the hiding power of the pigments used, for white pigments in addition the tinting strength, for black pigments in addition the color strength.

The manufacture of inorganic pigments requires efficient chemical and physical methods of analysis and suitable determination procedures [31–34]. The investigations include characterization of the raw materials, monitoring of the manufacturing processes, compliance with the legal environmental requirements, as well as testing and control of intermediates and final products. By-products and waste products (wastewater and exhaust gas) must also be monitored. Definitions and generally accepted test and determination methods are laid down in national (DIN, AFNOR), European (EN), and international (ISO) standards. Pigments are essentially characterized with regard to chemical and crystallographic composition, specific chemical properties (solubility in various solvents, pH value of aqueous suspensions), particle geometry, optical properties, behavior in the application medium, and weather resistance. Quality control is carried out in specially equipped laboratories.

Quality control and quality assurance are two aspects of quality management. Both aspects play an important role in the production process for pigments. Quality control is the part of quality management that focuses on meeting quality requirements. While quality assurance refers to how a process is performed or how a product is manufactured, quality control is more the inspection aspect of quality management. Inspection, in this sense, means the process of measuring, examining, and testing of one or more characteristics of a product or service. This includes the comparison of these characteristics with specific product and service requirements in order to determine the conformity. Products, processes, and various other outputs can be inspected to ensure that the product coming from a manufacturing line or the service provided is correct and meets specifications. A comprehensive process of quality management as is typical of pigment-related approaches places emphasis on elements of management (defined and well-managed processes, monitoring, performance criteria, job management, identification of records), competence (knowledge, skills, experience, qualification), and soft elements (personnel, integrity, confidence, organizational culture, motivation, team spirit). It starts from the product idea and ends with the sale of a product, a product group, or a service. All functions from development to production and marketing are included. In addition, functions such as information and patent management and others can play an important role.

The technical delivery conditions for pigments and fillers, including the analytical methods to be applied (e.g., for the exact composition, the required crystalline phase or for the determination of impurities), are written down in numerous standards. Admixtures in the trace range may have a considerable influence on the pigment properties. They may cause a color cast or a dirty appearance. They may also change the anticorrosive, magnetic, or electrical properties of functional pigments and limit the usability in cases where maximum levels for impurities are defined (e.g., for toxicological reasons). Some of the analytical methods enter, if required, concentrations up to the µg/kg range. X-ray diffraction provides information on the crystallographic characteristics of a pigment.

Producers and users utilize a series of measures that are easy to execute and that record characteristic parameters for the production and the application technology. Aqueous suspensions of a pigment or filler mostly contain dissolved substances. It is often sufficient to determine the sum of these substances by simple measurement procedures (e.g., determination of the salt amount soluble in water, of the electrical conductivity, or of the pH value). Extraction with hydrochloric acid or other acids is a typical method used for pigments and fillers. The content of certain elements in the extract must thereby not exceed an upper limit.

Geometric parameters such as particle size (particle diameter), particle size distribution, and particle shape are some of the most important characteristics of pigments. Particles are definable, individual units of a pigment. The structure, size, and shape of particles can vary widely. Typical particle units in pigments and fillers are primary particles, agglomerates, and aggregates (Figure 1.7). Flocculates are a special form of agglomerates, which can occur in suspensions and liquid application media.

– Primary particles (individual particles): particles recognizable as such by appropriate imaging techniques (optical microscopy, electron microscopy).

– Agglomerates: assemblies of primary particles and/or aggregates typically adhering to one another at corners and edges; the total surface area does not differ significantly from the sum of the surface areas of the individual particles and/or aggregates; agglomerates can be destroyed to a certain extent by shear forces.

– Aggregates: assemblies of primary particles typically adhering to one another side by side; the total surface area is smaller than the sum of the surface areas of the primary particles involved; aggregates can only be conditionally or not at all destroyed by strong shear forces.

– Flocculates: agglomerates occurring in suspensions or in liquid application media (paints, coatings, printing inks, nail lacquers, etc.), which can be destroyed already by weak sheer forces.

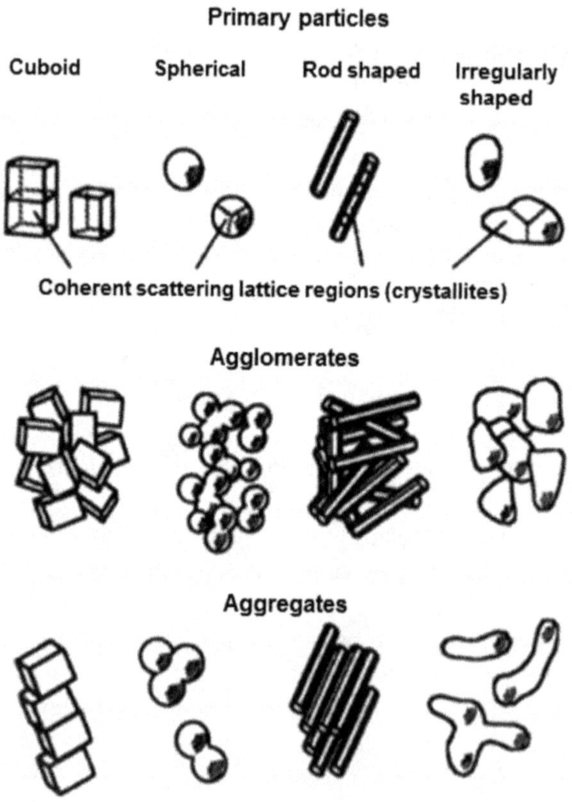

Fig. 1.7: Primary particles, agglomerates, and aggregates.

The term "particle size" has to be used carefully to avoid misunderstandings. Table 1.5 contains a summary of frequently used terms in connection with particle sizes, particle size distributions, and other characteristic particle properties. Particle size distributions are mostly reported in the form of tables, histograms, and curve presentations. Some of the determination procedures record only a mean particle diameter, while others allow complete information on the particle size distribution. Particle sizes of inorganic pigments range from several tens of nanometers for transparent pigments to more than 100 micrometers for effect pigments.

Tab. 1.5: Particle size, particle size distribution, and other characteristic particle properties.

Term	Definition
Particle size	Geometrical value characterizing the spatial extent of a particle
Particle diameter	Diameter of a spherical particle or characteristic diameter of a regularly shaped particle
Equivalent diameter	Diameter of an irregularly shaped particle converted into a spherical shape
Particle surface area	Surface area of a particle; a distinction is made between the internal and external surface areas
Particle volume	Volume of a particle; a distinction is made between effective volume (excluding cavities) and apparent volume (including cavities)
Particle mass	Mass of a particle
Particle density	Density of a particle
Particle size distribution	Statistical representation of the particle sizes of a material consisting of particles
Distribution density	Relative amount of a particulate material in relation to a given particle size diameter (density distribution functions must always be normalized)
Cumulative distribution	Normalized sum of particles that have a diameter less than a given particle diameter
Particle size fraction	Fraction of particles lying between two selected values for the particle size

The mean particle size or the mean particle diameter as the average value of the size of a representative quantity of particles of a specific pigment is an information of interest for the producer and the user, although it does not give any information about the distribution spread. Mean particle sizes of inorganic pigments range from 0.01 to 10 μm. They are usually between 0.1 and 1 μm. Exceptions are effect pigments with mean diameters of 5 to 50 μm.

Anisotropic particles, such as acicular or platelet-shaped ones, need specific mathematical statistics for the determination of geometrical parameters. Thus, it is possible to determine parameters like the mean particle size by using two-dimensional logarithmic normal distributions of the length and breadth of the particles [34].

Measurement methods for particle size analysis are primarily based on the different physical properties of the particle systems to be analyzed. Depending on the particle size, the prevailing continuous phase, and the particle concentration, different measurement methods are used. Important methods used in the particle size analysis of inorganic pigments are sedimentation analysis, light microscopy, electron microscopy, dynamic light scattering, laser diffraction, and sieving analysis.

Sedimentation analysis provides quantitative data on the sinking speed distribution of particles in a liquid medium. In the case of optimal dispersed pigments, the sedimentation in the Earth's gravity field is very slow because of the small size of the particles and must, therefore, be accelerated by centrifugation. Sedimentation analysis is used for the determination of grain sizes in the range of about 100 to 1 μm. The sample to be analyzed is suspended in a liquid (deionized water, cyclohexanol, ethylene glycol, isobutyl alcohol, etc.). Subsequently, the sinking speed of the particles is measured. Influencing factors for the measurement are the diameter of the particles, the viscosity) of the measuring liquid, and the density difference between the particles and the measuring liquid. Thus, if the falling distance and the falling time are measured, the size of the spherically imagined particles can be calculated in the laminar range, according to Stokes' law.

Some optical procedures are based on light scattering and laser diffraction, e.g., Fraunhofer diffraction (with additional correction due to Mie scattering), and allow the determination of the particle size in the range of 0.05 to 1,500 μm. Light scattering or sedimentation rate is measured and must now be converted to a particle size distribution based on models, assumptions (e.g., that all particles are spherical), and further input (e.g., the complex index of refraction or the density). Different measurement methods do not usually result in exactly consistent particle size distributions. The comparison with standardized samples is always important.

In addition, sieving methods are used for the determination of the particle size of pigments. Sieve analysis is a particle measurement technique for determining the particle size distribution of powdered materials. The most common method of sieve analysis is dry sieving with a sieve tower mounted on a sieving machine. For difficult screening materials, which tend to agglomerate, and finer screening materials, wet screening and air jet screening are also used. Wet sieving by hand and wet sieving by a mechanical flushing procedure are the two most important methods used [34].

Counting methods are very suitable for the particle size measurement of pigments. The most comprehensive information is accessible by counting of electron micrographs. At the same time, they give information on size, size distribution, shape, and degree of dispersion. The high number of particles necessary for the determination of the particle size distribution leads to the fact that this method is too expensive and time-consuming for rapid analysis, although the measurement can be supported by automated means. The duration for measuring makes it nearly impossible to use this method for production control. Basically, scanning electron microscopy (SEM) as well as transmission electron microscopy (TEM) are used for the optical investigation of inorganic pigments. Optical microscopy is also applied for the characterization of pigments, especially when the particles have larger dimensions.

X-ray diffraction (XRD) as a method for the determination of the crystal size is not expedient for particle size analysis of pigments because these do not usually consist of single crystals.

The specific surface area (mostly defined as the area per unit mass), mostly determined according to the BET method (nitrogen adsorption or adsorption of other gases, developed by Brunauer, Emmett, and Teller), is another important parameter for the characterization of pigments and fillers. It can also be used for the calculation of the mean particle diameter. However, in this case, it is necessary to consider the effect of the internal surface area, in addition to that of the external surface area. If this part is neglected, the specific surface area is not the correct basis for the calculation of the mean particle diameter. This problem plays a role in pigments with surface treatment because the layer with the treatment is often very porous.

Crystallographic data of inorganic pigments are obtained from XRD measurements. Appropriate examinations provide information on the crystal structure, including the distinction between different crystal modifications of a certain composition, state of stress, and lattice defects of the smallest coherent regions. Each pigment has its specific diffractogram, which is comparable with a fingerprint and which allows qualitative and semiquantitative information on the composition and the degree of crystallinity. XRD can also be used for the identification of more than one component in pigment mixtures. Laboratory XRD equipment relies on the use of X-ray tubes (sources), which are used to produce the X-rays for the measurement. The most commonly used laboratory X-ray tube is the copper anode (wavelength of Cu Kα 154.18 pm). Other used anode materials for X-ray tubes consist of cobalt, iron, or molybdenum.

The following crystal classes are of significance for inorganic pigments:
- cubic: zinc blende (sphalerite) lattice (ZnS, CdS)
- cubic: spinel lattice ($CoAl_2O_4$, Fe_3O_4, $CuCr_2O_4$, $ZnFe_2O_4$)
- hexagonal: corundum lattice (α-Fe_2O_3, Cr_2O_3)
- hexagonal: wurtzite lattice (ZnS, ZnO, CdS, $CdSe$)
- tetragonal: rutile lattice (TiO_2, SnO_2)
- tetragonal: scheelite lattice ($BiVO_4$)
- rhombic: goethite lattice (α-$FeOOH$)
- monoclinic: monazite lattice ($PbCrO_4$)

Spectroscopic methods are also used for the characterization of inorganic pigments. The most important of these are infrared (IR) and Raman spectroscopy, as well as spectroscopy in the ultraviolet (UV) and visible (VIS) range. For the investigation of the organic part of the surface treatment, nuclear magnetic resonance spectroscopy (NMR) and mass spectrometry (MS) may be applied.

Thermoanalytical measurement techniques are also used for the investigation of inorganic pigments. The behavior during the thermal treatment of the pigments can be investigated by differential thermal analysis (DTA), thermogravimetry (TG), and differential thermal gravimetry (DTG). Information about water/moisture content, solvent content, decomposition temperature, or phase transformation temperature is obtained.

Qualitative and quantitative chemical analyses play an important role in the course of the investigation of inorganic pigments. Classical wet-chemical procedures are used besides modern methods like atomic absorption spectroscopy (AAS) and X-ray fluorescence techniques (XRF).

The content of volatile matter is determined by drying a pigment sample in an oven at a defined temperature, mostly at 105 °C. Weight loss can be correlated with the moisture content. Loss on ignition is determined by weighing a pigment sample, heating it up to a defined temperature, cooling down, and reweighing. The content of matter soluble in water is determined by defined extraction conditions and is carried out in cold and hot water. A similar determination exists for the extraction in 0.1 M hydrochloric acid. Investigations in other acids and also alkalis are likewise possible.

Other investigations of inorganic pigments are performed in aqueous suspensions. These include the measurement of the pH value and of the electrical conductivity of the suspension and the determination of water soluble sulfates, chlorides, and nitrates. The acidity or alkalinity of a pigment suspension is measured as the quantity of alkali or acid, which is necessary to achieve the neutral point (typically 100 g pigment suspended under defined conditions in cold or hot water).

The electric surface charge of pigment particles can be measured in a suspension using the electrokinetic sonic amplitude (ESA). The common determination method is the measurement of the zeta potential, including the isoelectric point (IEP). The IEP is the pH value at which a specific particle carries no net electrical charge in the statistical mean.

The density of a pigment is determined by pyknometry at a standard temperature of 25 °C. The apparent density after tamping is defined as the mass of 1 cm^3 of the material after tamping in a specific tamping volumeter. The tamped volume is the volume of 1 g of the material. The apparent density and tamped volume depend on the true density of the material as well as on the size and shape of the particles. Both parameters are important for the choice of the dimensions of the packaging material used for a pigment.

A summary of the most important methods for the characterization of inorganic pigments is shown in Table 1.6. The ways and means of taking samples (sampling) for the investigations are decisive for the achievement of proper results. Most of the analytical methods use the pure pigment for the investigation. However, the procedures for the quantitative determination of the chemical composition need the dissolution of the pigment in a suitable manner, mostly in an acid.

Tab. 1.6: Characterization methods used for inorganic pigments.

Method	Information
Sedimentation analysis	Particle size distribution, mean particle size
Light scattering methods (Fraunhofer diffraction)	Particle size distribution, mean particle size
Sieving analysis	Particle size distribution, mean particle size
BET measurement	Specific surface area, mean particle size
Porosimetry (mercury porosimetry)	Total pore volume, pore diameter, surface area
Scanning and transmission electron microscopy (SEM and TEM)	Particle size, particle shape, particle size distribution, indication of inhomogeneities, elemental composition (when combined with energy dispersive X-ray spectroscopy, EDX), indication of agglomerates and aggregates
Optical microscopy	Particle size, particle shape, indication of agglomerates and aggregates
X-ray diffraction	Composition, crystal structure, degree of crystallization, state of stress, crystallite size
VIS spectroscopy	Spectrum, indication of the composition
UV spectroscopy	Spectrum, indication of the composition
IR spectroscopy	Spectrum, indication of the composition
Raman spectroscopy	Spectrum, indication of the composition
Atomic absorption spectroscopy (AAS)	Quantitative elemental composition
Nuclear magnetic resonance spectroscopy	Analysis of the organic part of the surface treatment
Mass spectrometry	Analysis of the organic part of the surface treatment
Wet-chemical analysis	Qualitative and quantitative elemental composition
X-ray fluorescence analysis (XRF)	Qualitative and quantitative elemental composition
Thermal analysis	Water/moisture content, solvent content, behavior during thermal treatment, decomposition temperature, phase transformation temperature measured by DTA, TG, and DTG
Determination of volatile matter and loss on ignition	Content of matter volatile upon heating, moisture content, loss on ignition
Determination of soluble matter in water and hydrochloric acid	Content of matter soluble in water and hydrochloric acid (other acids and also alkalis are likewise possible)
Measurement of the ESA	Electric charge at the pigment surface, zeta potential, IEP
Pyknometry	Density

1.5 Color properties of inorganic pigments

1.5.1 Fundamental aspects

When light strikes the surface of a pigmented system, e.g., a pigmented film, three different events may occur:

- light absorption by the pigment particles (in the case of a colored or black pigment);
- diffuse reflection of light at the pigment particles (scattering) or, in specific cases, (effect pigments) directed reflection;
- the light passes the pigmented medium without further interaction (no striking of pigment particles, the binder is assumed to be transparent and nonabsorbent).

Very complex relations for each pigment in the application medium result from the interaction of these events, which find an expression in reflection spectra $\rho(\lambda)$ or in spectral reflection factor curves $R(\lambda)$. Reflection spectra of a white pigment (TiO_2), of a black pigment (Fe_3O_4), and of some colored pigments in a pigmented film are demonstrated in Figure 1.8. Theoretically, the spectral curve for a pure white pigment in the visible range (400 to 700 nm) should be parallel to the baseline at a spectral reflectance of 100% and for a pure black pigment also parallel to the baseline but at a spectral reflectance of 0%. Colored pigments have very different spectra dependent on their absorption caused by the specific interaction of visible light with the valence electrons of the material.

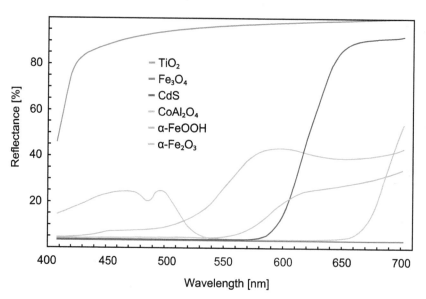

Fig. 1.8: Reflectance spectra of some inorganic pigments in paints in the visible range (white pigments represented by TiO_2, black pigments by Fe_3O_4, and colored pigments by CdS, $CoAl_2O_4$, α-FeOOH, and α-Fe_2O_3).

The reduction of a reflection spectrum and, with it, of the pigment-optical properties to the corresponding physical basic parameters is of high importance for the development and reproducible manufacture of technical inorganic pigments [35].

The fundamental relationships among the optical characteristics of pigments can be comprehensively discussed on the basis of four theoretical considerations:
- the Mie theory,
- the theory of multiple scattering,
- the Kubelka–Munk theory,
- colorimetry.

Figure 1.9 shows the relationships between the optical properties of pigments and pigmented systems and their theoretical basics. Reflectance spectra and spectral reflection

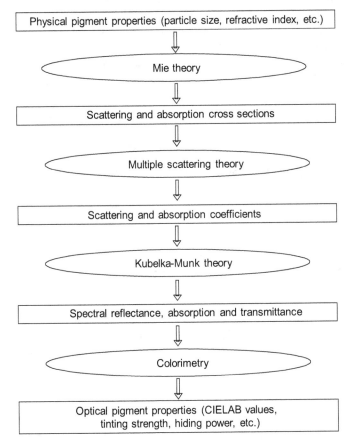

Fig. 1.9: Relationship between the optical properties of pigments and pigmented systems and their theoretical principles.

factor curves can be almost entirely derived from physical parameters and relationships by using these physical fundaments.

1.5.1.1 Mie theory

Mie scattering or Lorenz–Mie scattering refers to the elastic scattering of electromagnetic waves at spherical objects, e.g., pigment particles in an application medium, whose diameters correspond approximately to the wavelength of the radiation. This scattering can be described physically using the Lorenz–Mie theory. The Mie scattering produces the Tyndall effect [36, 37].

The scattering diameter Q_s and the absorption diameter Q_a are related to the particle size D, the wavelength λ, and the optical constants of the material (refractive index n and absorption index κ). The Mie theory shall apply, in addition to the theories of Fresnel (for very large particles) and Rayleigh (for very small particles) for medium-sized particles as they are found in the case of inorganic pigments. In general, the reflection factor (ratio of reflected and incoming light intensity, essential for the optical efficacy of a pigmented system) increases at the same particle size very strongly with the relative refractive index m ($m = n/n_B$, n_B is the refractive index of the binder) and, consequently, also with the refractive index n of the pigment. Table 1.7 contains the refractive indices of important colored and black inorganic pigments, and Table 1.8 shows the refractive indices and optimal particle sizes of inorganic white pigments [38, 39].

Tab. 1.7: Refractive indices of important colored and black inorganic pigments derived from measurements on chemically corresponding minerals, respectively, synthetically prepared single crystals [36].

Mineral	Formula	Wavelength λ, nm	Refractive index		
			n_ω / n_α	n_ε / n_β	n_γ
Bismuth vanadate	$BiVO_4$	670	2.45[a]		
Cobalt blue	$CoAl_2O_4$		1.74		
Eskolaite	Cr_2O_3	671	2.5[a]		
Greenockite	CdS	589	2.506	2.529	
Goethite	$\alpha\text{-FeOOH}$	589	2.275	2.409	2.415
Hematite	$\alpha\text{-Fe}_2O_3$	686	2.988	2.759	
Carbon[b]	C	578	1.97		
Crocoite	$PbCrO_4$	671	2.31	2.37	2.66
Magnetite	Fe_3O_4	589	2.42		
Red lead	Pb_3O_4	671	2.42		
Ultramarine	$Na_6Al_6Si_6O_{24}(NaS_n)$		1.50		

[a]mean refractive index.
[b]for carbon arc lamps.

Tab. 1.8: Refractive indices and optimal sizes of some white pigments ($\lambda = 550$ nm) derived from measurements on chemically corresponding minerals and calculated according to the van de Hulst formula [37].

Mineral	Formula	Mean refractive index		Optimal particle size (relative to binder) D_{opt}, μm
		Vacuum	Binder	
Rutile	TiO_2	2.80	1.89	0.19
Anatase	TiO_2	2.55	1.72	0.24
Zinc blende	α-ZnS	2.37	1.60	0.29
Baddeleyite	ZrO_2	2.17	1.47	0.37
Zincite	ZnO	2.01	1.36	0.48
Basic lead carbonate	$2\,PbCO_3 \cdot Pb(OH)_2$	2.01	1.36	0.48

Furthermore, the relations of scattering and absorption cross sections to the physical and application technical pigment properties are known through experimental and theoretical activities in the frame of the Mie theory. In the case of white pigments with their high scattering and low absorption proportion, primary particles are developed in view of a high tinting strength and a strong hiding power. Experimentally determined dependencies of the scattering cross section, which is of special relevance here, are shown in Figure 1.10. The introduction of the parameter a, which is proportional to D/λ and, therefore, to the mean particle diameter, shows that light scattering grows strongly with an increasing refractive index. At the same time, the optimal particle diameter for scattering is shifted to smaller values. Another fact is that the light scattering at a given particle size is higher for short-wave than for long-wave light. Based on this knowledge, it is possible for practical applications to determine the optimal particle size for scattering and thus for the hiding power for each white pigment (Tab. 1.8).

The Mie theory also has consequences for the absorption of inorganic pigments. Color properties like tinting strength are related to the absorption behavior and the particle size. The refractive index n, the absorption index κ, and again, the parameter a as a relative measure of the particle size, are used for Mie theory calculations. Resulting from this, it can be seen that in the case of very small particles, the absorption is independent of the particle size. Any further reduction of the particle size does not result in additional absorption. On the other hand, the absorption of very small particles grows with increasing absorption index κ. In the case of larger particles, the absorption is approximately the same for all refractive indices n and absorption indices κ.

The introduction of the complex refractive index $n^* = n\,(1 - i\kappa)$, which includes the refractive index n and the absorption index κ, enables the capture of further relationships. The optical properties of absorbing pigments may now be attributed to the particle size D and the complex refractive index n^*. Subsequently, the reflectance spectrum and, hence, the color properties of a pigment may be calculated if its

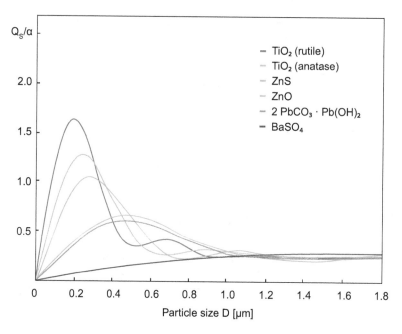

Fig. 1.10: Scattering of white pigments as a function of the particle size (λ = 550 nm). (a) TiO$_2$ (rutile), (b) TiO$_2$ (anatase), (c) ZnS, (d) ZnO, (e) 2 PbCO$_3$ · Pb(OH)$_2$, and (f) BaSO$_4$.

particle size distribution, the complex refractive index, and the pigment concentration are known [34]. In practical cases, the necessary values for the refractive index n and the absorption index κ are often not known with sufficient accuracy. Both parameters depend on the wavelength and, because of the optical anisotropy of most pigments, on the illumination and viewing direction. There are no direct methods available that allow the determination of n and κ for colored pigments. Values for both parameters can be determined by ellipsometrical measurements of larger crystals, typically of natural minerals. However, results thus obtained do not necessarily match the exact values of n and κ for a pigment. Also, the data for the refractive indices in Tables 1.7 and 1.8 are measured on large crystals and represent only approximate values, with regard to pigments.

1.5.1.2 Theory of multiple scattering

The theory of multiple scattering is the mathematical formalism that is used to describe the propagation of a wave through a collection of scatterers, e.g., a variety of pigment particles in an application system [40–43]. In detail, the theory deals with the relation of the scattering coefficient S to the pigment volume concentration σ and the scattering diameter Q_S of the individual particles. The absorption coefficient K is directly related to the absorption diameter Q_A and to the pigment volume concentration σ. K follows the Lambert–Beer's law and increases linearly with σ. Scattering is disturbed with increasing σ, when the particles are closely packed. If the distance of two primary particles falls below $\lambda/2$ of the irradiated light, the particles are no longer resolved as individuals, but are treated as larger particles. The scattering coefficient S, therefore, increases linearly (Lambert–Beer shear area) only for low pigment volume concentrations and remains behind the linearly extrapolated values at higher σ [44].

1.5.1.3 Kubelka–Munk theory

The Kubelka–Munk theory describes the light absorption and light scattering properties of pigmented systems, such as paint films or coatings [39, 45, 46]. In detail, the theory relates the reflectance spectrum $\rho(\lambda)$ to scattering and absorption, as well as to the film thickness h [39, 45]. Relevant parameters are the scattering coefficient S and the absorption coefficient K. The most common equation resulting from the Kubelka–Munk theory is that for the reflectance ρ_∞ of an opaque (infinitely thick) film:

$$K/S = (1 - \rho_\infty)^2 / (2\rho_\infty) \tag{1.1}$$

This so-called Kubelka–Munk function expresses that the reflectance ρ_∞ only depends on the ratio of the absorption coefficient to the scattering coefficient and not on the individual values of K and S. The equation is particularly suitable for reflectance measurements, which are used to obtain information on absorption and scattering (IR spectroscopy, thin layer chromatography, textile dying). The main application of the Kubelka–Munk theory is in the matching of pigmented systems based on computer programs [38, 47–49]. The calculations are often based on the additive combination of K and S using the specific coefficients of the components involved multiplied by their concentrations.

1.5.1.4 Colorimetry

Colorimetry is concerned with the quantification and physical description of human color perception. It is similar to spectrophotometry, but specifically aims to reduce spectra to the physical correlates of color perception, mostly to the tristimulus values of the Commission Internationale de l'Éclairage (CIE) 1931 XYZ color space and related parameters [50, 51]. To elaborate, colorimetry compares the color quality of a pigmented application system with the color properties of a pigment described in the reflectance spectrum (spectral reflectance) $\rho(\lambda)$ [52]. The main objective of colorimetry is the determination of a numerical value for a color valency (color stimulus specification) of a colorimetrical standard observer (theoretical colorimetrical observer whose sensitivity sensors correspond to the color-matching functions defined by the CIE. This observer is the representative for more than 95% of all people, who are normal-sighted and who assess visible radiation in the range between 380 and 780 nm, according to the same response function.

Color valency is the evaluation of color stimulus through the three sensitivity functions of the human eye to a uniform impression. One essential assumption is that there are three different types of stimulation centers in the cone system of the retina. These centers differ in their spectral sensitivity. During the visual process, the simultaneous stimulation of the three centers, which react sensitively for red, green, and blue, results in color valency and, subsequently, in color perception. Color valency is determined by three colorimetric values. Any color can be understood as an additive mixture of the color valency of all spectral radiations appearing in the color stimulus. Colorimetry includes three different procedures: the adjustment method, the tristimulus method, and the spectral method.

The adjustment method as the first colorimetrical procedure is based on the comparison of the color to be characterized with a colored reference field. The comparison should be carried out under standard illumination by an observer who perceives color normally. Systematic color reference collections can be used if the precision requirements are low. More precise results are achieved by additive color mixing of three color samples, whose mixing ratio is varied until the desired accordance with the target color is obtained. Three-color measures are used to facilitate the color adjustment.

The tristimulus method as the second colorimetrical procedure evaluates integrating a color stimulus according to three different response functions of three separate recipients. The spectral sensitivity of these recipients must comply with the set standard color matching functions.

The spectral method as the third colorimetrical procedure allows exact color measurement by using the standard color values (tristimulus values) X, Y, and Z from the convolution of the color stimulus φ_λ with the standard spectral value functions. The standard color values are converted into the chromaticity coordinates x, y, and z $(x + y + z = 1)$:

$$x = X/(X + Y + Z) \tag{1.2}$$

$$y = Y/(X + Y + Z) \tag{1.3}$$

$$z = Z/(X + Y + Z) \tag{1.4}$$

The chromaticity coordinates represent clearly the color locations of a color by their proportions with regard to the sum of the standard color values. It is sufficient to specify the two coordinates, x and y, since the third one, z, is determined by $z = 1 - x - y$). In an equi-energy spectrum, white, gray, and black ideally have the chromaticity coordinates of $x = y = 0.3333$. The chromaticity coordinates of all other colors are located between 0 and 1.0.

The conversion of the standard color values into the chromaticity coordinates is the basis for the three-dimensional representation of the colors in the CIE color system (CIE standard chromaticity diagram). The coordinates x and y define the color location in the chromatic diagram; the associated polar coordinates, which are attributed to the achromatic point, can be interpreted as saturation (radius vector) and color shade (polar angle). For the indication of the lightness, the standard color value Y is used.

The CIE standard chromaticity diagram images the color locations of the standard color values in a rectangular coordinate system with x-axis (horizontal) and y-axis (vertical) and with Z at the coordinate origin (Figure 1.11). In the diagram, the x value and the y value of a color can be used as coordinates for the graphic illustration of a chromaticity. The chromaticity coordinates derived from monochromatic color stimuli of the color-matching functions of the defined CIE standard observer limit the CIE standard chromaticity diagram. Each color location of a spectral color is, therefore, situated on the edge of a spectral curved line, which has the shape of shoe sole. This spectral curved line is closed by the so-called purple straight line (the connecting line between the shortwave and the longwave spectral ends). All physically feasible colors are located within the area. Unrealizable color locations outside of the area belong to the virtual colors [53].

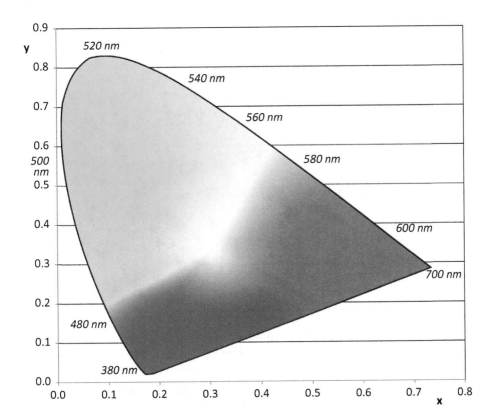

Fig. 1.11: CIE standard chromaticity diagram [53].

With the CIE system, it is possible to determine the equality of colors exactly, but in the case of color differences, the color distances in in the CIE system are not in accordance with the color perception of the eye. The determination and evaluation of color differences, however, is one of the most important tasks of colorimetry for the development and the monitoring of pigments. From the large number of conversion formulas for the transformation of standard color values into true-to-perception units, the CIELAB color difference formula has achieved the widest popularity. This formula converts the standard color values X, Y, and Z into the L^*, a^*, b^* values; L^* means lightness, a^* with $+ a^*$ and $-a^*$ is the red–green axis, and b^* with $+ b^*$ and $-b^*$ is the yellow–blue axis. The total color distance ΔE_{ab}^* in the CIELAB system (L^*, a^*, b^* color space) is the geometrical distance between the color locations of a sample and a reference pigment. It is calculated as space diagonal using the following relation:

$$\Delta E_{ab}^* = (\Delta a^*)^2 + (\Delta b^*)^2 + (\Delta L^*)^2 \tag{1.5}$$

A major advantage of the CIELAB system shown in Figure 1.12 is the fact that the color distance can be subdivided into individual amounts, namely according to lightness (ΔL), chroma (ΔC), and hue (ΔH):

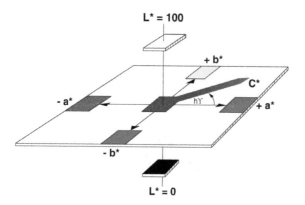

Fig. 1.12: CIELAB color space system (source: Byk-Gardner GmbH).

$$\Delta E_{ab}^* = [(\Delta L^*)^2 + (\Delta C_{ab}^*)^2 + (\Delta H_{ab}^*)^2]^{1/2} \tag{1.6}$$

The CIELAB color distance formula is particularly suitable for the evaluation of colored objects (coatings, textiles, plastics, printing inks, etc.). The result of such evaluations can be an agreement on color tolerances. The measuring instruments used for the determination of standard color values are recording spectrophotometers and tristimulus photometers. Integrated processors determine the standard color values for further transformation into the CIELAB system [54].

The CIE color system is one of several color order systems. All these systems are systematic collections of colors that should cover the complete color space. A distinction is made between empirical and colorimetrical-based color systems. The systems have been developed to achieve a better communication in the production of colored objects or in marketing (between manufacturer and customer). Ordering criteria used in all color systems are lightness, hue, and saturation. Examples for color systems besides the CIE color system are the Natural Color System (NCS), the Munsell color system, the OSA-UCS color system, the DIN color system, and the RAL design system [55].

1.5.2 Color and color vision

Color is the characteristic of human visual perception described through color categories, e.g., blue, green, yellow, or red. The perception of color is derived from the stimulation of cone cells in the human eye by visible light. Luster, which is also a perception of the eye is excluded from the term "color". The presence of natural or artificial light is a condition for the perception of color. There are objects that emit light on their own (self-luminous bodies), e.g., the sun, light bulbs, burning candles, or molten iron. However, most of the objects get their color through the incident light of a self-luminous body, which is wavelength-dependent absorbed, transmitted, or reflected. The term "color" is also used in general communication for means used for the change of the color impression of objects (painting color, printing color).

The perception of color by the human eye begins with specialized retinal cells containing photoreceptors (light sensitive molecules) with different spectral sensitivities, known as cone cells. The three types of cones in the human eye are sensitive to three different spectra, resulting in a trichromatic color vision. Cone cells are neurons, which serve as specialized sensory cells of the eye for the photopic vision in daylight. A distinction is made for human beings between three cone types: S type (blue receptor), M type (green receptor), and L type (red receptor). Their stimulus response describes the spectral absorption curve, which is the basis for the metric of colors. There are vertebrates with two, three, or four types of cones.

The spectral ranges of the cones in the human eye are as follows: S type 400 to 500 nm (peak wavelength 420 to 440 nm), M type 450 to 630 nm (peak wavelength 534 to 555 nm), and L type 500 to 700 nm (peak wavelength 564 to 580 nm). A certain range of wavelengths stimulates each of the three receptor types to a certain degree at the same time. The human brain combines the information from each receptor type to create different perceptions of different light wavelengths.

There are 6 million cones and 120 million rods in the photoreceptor layer of the retina. The percentage of the blue-sensitive cones is almost constant at about 12% for all people. The ratio between red and green cones at the retina varies very strongly within one family. The density of cones varies among the species. For humans, the density of cones on the retina is highest at the center (fovea centralis, area of greatest visual acuity) and decreases towards the periphery. Vice versa, the density of rods on the retina increases from the center towards the periphery. The differentiation in cones and rods has functional reasons. The cones only function in brightness and twilight and make color vision possible. In the dark twilight or in almost total darkness, more or less, only the rods function due to their higher light sensitivity. The rods are even able to perceive single photons in absolute darkness.

The visual process as a whole can be described as follows (Fig. 1.13): Light reaches the eye on the cornea and gets into the eye through the pupil as the central part of the iris. From there, the light reaches the ocular lens where it is directly focused on the retina. The focal point of the light is located directly on the retina in the so-called yellow spot (macula lutea) of the eye. The iris regulates the incidence of light as an aperture and ensures with the help of the elastic ocular lens a dynamic adjustment between close and distance viewing. The vitreous body stabilizes the shape of the eyeball, which is enclosed by the dermis. The light is converted on the retina by the photoreceptors (cones and rods) into electrical impulses and forwarded from there after filtering and sorting via the optic tract into the brain. Further image processing in the brain takes place in a very complex manner through a specialization of cell functions, including a separate and parallel running subsequent processing of shape, color, and movement information.

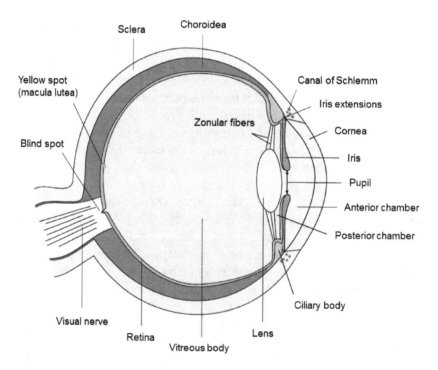

Fig. 1.13: Anatomy of the human eye (source: Talos, colorized by Jakov CC BY-SA 3.0, German Wikipedia).

1.5.3 Origin of color for colored pigments

Selective light absorption by colored pigments is based on the interaction of visible light with the valence electrons of solids. The type of elements (cations, anions), oxidation states, and electron configurations of the same, as well as the crystal lattice, determine certain electronic structures in the solid, which can lead to transitions of electrons between energetic states [56, 57]. Based on this, there are six principles for the development of color, which are relevant for inorganic pigments:

-   d–d transfers of electrons within transition metal polyhedrons;
-   charge–transfer transitions of electrons from ligands to a central atom or ion;
-   intervalence transitions of electrons between cations from one and the same element in different valence states;
-   electron transitions based on radical ions in a solid;
-   electron transitions from donor levels of "guest ions" (dopant ions of a transition metal) into the conduction band of the host lattice;
-   electron transitions from the valence band into the conduction band within a solid.

These electronic transitions result in a selective light absorption of different intensity and generate the color of inorganic colored pigments, if the absorption bands are in the visible spectral range or in the border area of the UV range or the IR range.

Transfers of electrons within d electron levels are found in transition metal compounds with a specific oxidation state, electron configuration (d^n with n from 1 to 9), and coordination sphere (octahedral, tetrahedral, or square planar; in some cases, distorted). Such transfers lead to absorption bands and determine the colors of some inorganic pigments. Examples for this principle of color generation are cobalt blue ($CoAl_2O_4$ with Co^{2+} on tetrahedral positions), cobalt green (Co_2TiO_4 respectively $[Co]_{tetr.}[Co,Ti]_{oct.}O_4$ with Co^{2+} on tetrahedral and octahedral positions) as well as Cr_2O_3 and $CrOOH$ (both with octahedrally coordinated Cr^{3+}) [56–63].

Charge–transfer transitions of electrons in connection with the generation of color can take place within transition metal polyhedrons in several ways. The typical way that plays a role for some of the inorganic pigments is the charge–transfer transition from ligands (O^{2-}, OH^-) to central atoms or ions. Intensive absorption bands and appropriate yellow to red colors are the result. Examples of this principle of color generation are yellow α-FeOOH, orange γ-FeOOH, and red α-Fe$_2$O$_3$. Basically, a charge–transfer transition is one in which a relatively large redistribution of electron density occurs. In addition to the transfer of electrons from ligands to the central atom or ion, there are also transitions in charge–transfer compounds that play no role or only a subordinate role for inorganic pigments. These include electron transfers from one central atom (ion) to another central atom (ion) or from one central atom (ion) to a ligand [56, 57, 63–66].

Intervalence transitions of electrons may occur if cations of one and the same element, mostly of a transition metal, appear in one compound in different valence states. These cations are interlinked with each other by common ligands. The electron transition from the cation with the lower valence to that with the higher valence, e.g., from

Fe^{2+} to Fe^{3+}, takes place across these ligands. This color generation principle can be found in many compounds with Fe^{2+} and Fe^{3+} ions, particularly in iron phosphates and silicates. The deep blue color of iron blue pigments can also be attributed to this principle [67, 68].

Electronic transitions based on radical ions in a solid lead to the colors of ultramarine pigments. Depending on the structure of radical S_n anions (with n-values from 2 to 4), which are placed in interstices of the ultramarine lattice, blue, green, red, or violet colors are generated. Polysulfide ions of the composition S_2^-, S_3^- and S_4^- occur in the crystal lattice of ultramarine pigments. The S_3^- unit, for example, consists of a triangle of three sulfur atoms together with one additional electron. In natural ultramarine, the color depends not only on the exact composition of the radical ions, but also on the amount of other components like calcium, chloride, and sulfate [69–74].

The insertion of transition metal ions, e.g., of Ni^{2+} or Cr^{3+} into the TiO_2 host lattice on the sites of octahedrally coordinated Ti^{4+} ions combined with the simultaneous insertion of metal ions with a higher charge, e.g., of Nb^{5+} or Sb^{5+} (charge compensation), results in yellow to orange pigments. This is due to electronic transitions from the donor levels formed by the inserted Ni^{2+} and Cr^{3+} "guest ions" into the conduction band of the TiO_2. The energetic state of the donor levels thus formed is slightly above that of the valence band. Examples of this principle of color generation are (Ti,Ni,Sb)O_2 and (Ti,Cr,Sb)O_2 [64, 75–77].

Brilliant yellow, orange, and red inorganic pigments are accessible if a band structure exists in the solid concerned, in which electronic transitions from the valence band into the conduction band are possible at excitation with visible light. Absorption spectra with step edges are the result of appropriate measurements. Examples of this principle of color generation are the cadmium sulfide/selenide pigments ranging from yellow via orange to red. The insertion of Zn^{2+} or Se^{2-} in the CdS basic lattice (formation of mixed crystals) causes a change of the band structure and thus a color shift from yellow to deep red. The color differences of chromate and molybdate pigments can be explained in a similar manner by changes of band gaps in the $PbCrO_4$ structure [64, 78–81].

1.5.4 Optical parameters

Tinting strength, lightening power, scattering power, hiding power, and transparency are the most important parameters for the efficiency of pigments in paints, coatings, and other application systems.

Tinting strength, also called coloring power, is defined as the measure of the ability of a pigment to convey its color to the surrounding medium. It gives information on the efficiency of a pigment and is used for the optical benchmarking of colorants. In practice, only the relative tinting strength is used, i.e., for a colorant to be tested, relative to an agreed standard. The relative tinting strength is determined by the weight ratio of a reference pigment and a test pigment if both have the same coloristic appearance for an observer. The term "relative tinting strength" is only applicable to colored and black pigments. In the case of white pigments, a distinction is made between the lightening power and the tinting strength. Various characteristic values are defined for the metrological registration of the tinting strength, and several procedures have been developed for the measurement. These procedures, which are an adjustment of tinting strength, are based on the Kubelka–Munk theory. Typical measurement instruments for the tinting strength are spectrophotometers [82].

Lightening power is defined as the ability of a pigment to increase the lightness of a colored, gray, or black medium. It can also be considered as the tinting strength of a white pigment. Another definition is that the lightening power is a measure of the ability of a white pigment to increase the reflectance of an absorbing (colored, gray, or black) medium by virtue of its scattering power. Analogously to the consideration for tinting strength, the lightening power is the weight ratio in which the reference pigment can be replaced by the test pigment to achieve the same lightness in a colored system. Photometric procedures are used for the determination of the lightening power besides visual adjustment methods [82]. The gray paste procedure has the advantage that the pigment concentration can be adjusted to suit the application [83].

Scattering power is the ability of an object to re-emit incident radiation, typically light. The object may consist of one single particle or of different particles or scattering centers. In practice, the relative scattering power S is often used, which is the ratio of the scattering power S_T of pigment to be tested to the scattering power of a reference pigment S_R. The determination of the relative scattering power is particularly important for white pigments. Two procedures have been established for the determination of S: the black ground method and the gray paste method [34]. In the case of the black ground method, S is determined from the tristimulus values of the pigmented medium applied in various film thicknesses on black substrates. The gray paste method determines S from the tristimulus values of gray pastes. An advantage of this procedure is that it is less time-consuming than the black ground method. Spectrophotometers are used in both cases for the determination of the relative scattering power.

Hiding power is a function of scattering and absorption. The term "hiding power" refers to the ability of a paint or a coating to cover the contrast between a black and a white background in a way that the color distance ΔE_{ab}^* between the layer over black and the layer over white is only 1 CIELAB unit. The hiding power is determined by applying pigmented paint or coating films of different thicknesses on a black-white background, measurement of ΔE_{ab}^*, and interpolation to $\Delta E_{ab}^* = 1$. This method is suitable for all pigments (white, colored, black). Instead of this procedure, the hiding power can also be determined by using the so-called principle of spectral evaluation [84]. In this case, the hiding power is measured by using computer simulation. The thickness of the hiding film is calculated in advance from the reflectance curves of a single film on a black-white background at a known film thickness. Another procedure is used for achromatic coatings. The reflectance of equally thick pigmented coatings applied on a black-white background is measured using a photometer. The calculation of the hiding power results here is based on the reflection on the two different backgrounds under involvement of the film thickness.

Transparency is a measure for the light transmission of plane–parallel colored media, e.g., of a coating film. It is high in cases where incident light is only little or not scattered. In other words, the transparency of a pigmented system describes its ability to scatter light as little as possible. The color change of a transparent pigmented film on a black substrate has to be very small. The rule is that the lower the color change, the higher the transparency. Determination of the transparency is of relevance for the assessment of transparent printing inks and varnishes.

1.6 Application technical behavior

1.6.1 Pigment–binder interaction

The dispersion behavior of a pigment in a binder system (paint, coating, printing ink, etc.) depends on the type and the number of aggregates and agglomerates present, the energy deployed for dispersing during incorporation, and the interaction between pigment surface and surrounding medium. Laws and concepts of colloid chemistry find their application in the discussion and understanding of pigment–binder interactions. The degree of dispersion affects the rheological properties, the settling behavior, the optical performance, and the photochemical stability of the organic matrix. Pigment powders are always agglomerated. Several steps have to be conducted simultaneously or successively during the process of pigment dispersion:

– wetting: removal of air from the surface of the pigment particles and formation of a solvate layer around the same [85, 86];
– disintegration: destruction of agglomerates by energy input;
– stabilization: maintenance of the dispersed state by generating repulsive forces between the pigment particles, e.g., by adsorption of macromolecules on the surface of the particles (steric stabilization) or by electrostatic repulsion of particles with identical loading; the repulsive forces must be stronger than the attracting van der Waals forces, which are responsible for the formation of flocculates [87–89].

The degree of dispersion has a significant influence on the properties of a paint or a coating system [90]. Parameters for the dispersion behavior of a pigment in a binder are determined preferably in practical paint or coating recipes. These parameters include the viscosity and color properties of the dispersion, as well as color characteristics of dried paint or coating films containing the pigments (hue, tinting strength, gloss, and haze). The gloss of dry coating layers is a function of the surface. Increasing content of coarse particles in the surface area of the layer substantially deteriorates the gloss.

In systems where different pigments are used simultaneously, e.g., colored and white pigments, segregation effects can take place. Such effects can lead to changes in optical appearance by flooding [91]. A possibility to counteract this is the targeted pre-flocculation of the pigments with suitable gelling agents or the addition of specific fillers. Color changes due to flocculation in the form of so-called rub out effects can occur in solvent-containing and sometimes also in full-shade systems containing several pigments together. Such effects occur especially if the systems are subjected to mechanical stress during application and drying [34].

Test methods used for the determination of pigment–binder interactions are oil absorption, binder absorption (smear point, yield point), viscosity, and fineness of grind [34].

The dispersion behavior of pigments in paints is measured with dependence on the specific medium. In the case of low-viscosity media, the time that is required to produce a homogeneous suspension of particles is measured using an oscillatory shaking machine equipped with several containers. The dispersion properties of high-viscosity media are determined with an automatic muller.

The fineness of grind is determined using a grindometer gauge. The quantity measured using this method is the dispersing effect needed to achieve a given fineness of grind. Samples of the pigment are measured at various stages of the dispersion process.

Further standard procedures relevant for the determination of dispersion properties are related to the change in tinting strength and the increase in strength. The context here is that the dispersibility of a pigment can be measured by means of their dispersion resistance. The determination is based on the measurement of half-life times in relation to the final tinting strength. The increase in strength is determined from the difference between the initial and the final tinting strengths [92]. Half-life times correspond to dispersion rates. A defined selection of dispersion equipment, dispersion process, and medium are necessary to achieve useful results.

The pigment volume concentration (PVC) σ has an enormous importance for the pigment–binder interaction and the dispersion behavior. It is defined as the fractional volume of a pigment in the volume of total solids of a dry paint film:

$$\sigma = \frac{V_p}{V_p + V_b} \tag{1.7}$$

V_p is the pigment volume and V_b the volume of the binder. The PVC is determined by separating the pigment fraction from a weighed paint sample.

The critical pigment volume concentration (CPVC) is defined as that PVC at which the pigment particles are packed with the maximum density. In this case, the interstices are completely filled with binder. Smaller amounts of binder lead to incompletely filled interstices. Abrupt changes in the properties of the pigmented binder system are observed after the CPVC is reached.

Pigment–binder interactions in plastics, building materials, paper, and board have to be determined by different methods. In the case of plastics, these procedures include preparation and testing of pigment–polymer basic mixtures under defined conditions. There are specific methods to determine bleeding (migration of pigment particles from the inside to the surface of the polymer), heat stability, and tinting strength (increase in strength).

Relevant tests for building materials are carried out in pigmented cement and lime bonded compositions. Important methods used in this segment are those for the determination of the relative tinting strength, the color fastness in cement, the color fastness in lime, lightfastness, heat stability, the influence of the pigment on the hardness of the building material, and the influence of the pigment on the setting properties.

Specific methods for the investigation of pigment–binder interactions for paper and board are the measurement of the reflectance (paper and board) and the determination of opacity and transparency (paper).

1.6.2 Stability against light, weather, heat, and chemicals

Color or structural changes are typical phenomena that are recognized during and after irradiation or weathering of a pigmented system. The best known are chalking, yellowing, and loss of gloss. These changes are initiated by photochemical reactions in which the pigment can act as a catalyst. The pigment itself undergoes changes during these reactions.

Inorganic pigments are very inert chemical compounds. They can be regarded as the most stable substances among the colorants. The oxides, in particular, are known to be very stable pigments, which often have a stabilizing effect for the entire system [93]. Sulfide pigments, on the other hand, are mostly less stable and tend to oxidize in air under formation of sulfates. These are soluble and can be washed out by rainwater [94].

The presence of TiO_2 in coating compositions can lead to chalking initiated by photochemical processes. Reactive radicals are formed under the influence of sunlight, rainwater, and atmospheric oxygen. The pigmented binder system is destroyed successively by the attack of these radicals. Solutions to avoid damage to or destruction of the system include an appropriate surface treatment, doping, or lattice stabilization. Rutile pigments should preferably be used in critical applications because of their higher stability compared to anatase pigments [95, 96].

The degradation of the binder takes place gradually and leads to a loss of gloss. Surface gloss, which is decisive for the attractive appearance of a coating, is initially achieved through effective dispersing of the pigment particles in the application system. The particles lose their hold in the system with progressive breakdown of the binder, and chalking takes place.

Resistance to light and weathering are crucially dependent on the particle size and the shape of the pigments used, the chemical composition, inhomogeneities in the pigment structure, the pigment volume concentration, and the quality of the binder. The experimental investigation of a pigmented system with regard to stability is carried out by outdoor exposure (open-air weathering), accelerated weathering, and chemical laboratory procedures.

Outdoor exposure as a natural weathering method is the most realistic test for a pigmented system. Weathering stations offer the possibility of testing under complex real-life conditions. They are located either in places with extreme climatic conditions, e.g., with arid climate, or with a climate as constant as possible, e.g., maritime climate. Typical sites with demanding weather conditions that are useful for outdoor exposure are located in Florida and Arizona. The exposure to sunlight is very high in both places. High air humidity and strong salt exposure caused by the marine environment are additional factors for the choice of Florida as a place for open-air weathering. Arizona, on the other hand, is characterized by its desert climate. All test samples, mostly coated panels, are exposed to the same climatic factors during the weathering procedure. The tests may typically run for several years.

Accelerated weathering tests are more reproducible because seasonal fluctuations and inter-annual fluctuations can be avoided. Stochastic fluctuations and long testing times typical for the outdoor exposure are not an issue for accelerated weathering with its shorter experimental duration and the easy possibility of repetition and reproducibility. Sunlight and rain are simulated in specific laboratory devices. The spectral composition of the light used is largely consistent with global radiation. However, accelerated weathering tests allow only limited conclusions about the stability of pigments in the application system and should be confirmed by outdoor exposure experiments. Typical devices for accelerated weathering are Weather-Ometer, the QUV accelerated weathering tester, Xenotest, and Suntest. Chemical laboratory tests deliver only limited information about the weather resistance of pigmented systems. Some of them allow at least semiquantitative statements, e.g., about the photochemical reactivity.

Determination methods for light stability include accelerated tests (daylight phase with standard illumination D 65, possibly in the presence of moisture), evaluation of color changes (CIE tristimulus values X, Y, and Z are determined using a colorimeter, CIELAB color differences are calculated from the tristimulus values before and after light exposure, measurement with a spectrophotometer), and standard depth of shade (the measure of the intensity of a color sensation, the color fastness of a colored surface is dependent on depth of shade) [34].

Determination methods for weather resistance include the assessment of gloss and haze (measurement using a reflectometer with defined incident light angles), the degree of chalking of a coating (quantity of loose pigment particles determined by the Kempf method, the adhesive tape method, or the photographic method), the effect of humid climates (test for the corrosion resistance of metals, anticorrosive coatings, and composite materials in buildings using the Kesternich condensation device), and accelerated tests for the resistance of pigmented plastics against the influence of outdoor exposure) [34].

Oxide pigments are characterized by the highest thermal stability among the inorganic pigments. They are followed by the sulfide pigments, particularly the cadmium sulfide pigments, which can still be applied in enamel varnishes and glass melts. Oxide hydroxides and carbonates possess the lowest thermal stability. The stability at higher temperatures is also of fundamental importance for modern coating technologies. The stability of a pigmented binder system is primarily dependent on the composition of the binder and the duration of thermal stress. The thermal stability is typically determined by measuring the change of the hue of a pigmented system after a defined thermal treatment.

The chemical resistance of inorganic pigments, in particular of oxide-based types, is generally very high and is, in most cases, not an application technical problem. Some specific fastness properties, which are certainly more important for organic pigments, are overspray fastness and resistance to plasticizers. Determination methods for the fastness to chemicals include resistance to water in sulfur dioxide atmosphere (exposure of samples to a specific climate with condensing water and sulfur dioxide, measurement in a condensation device), salt spray fog test with sodium chloride solutions (continuous spraying of a sample, mostly a coated panel, with a 5% aqueous NaCl solution in a spray chamber, variations are the acetic acid salt spray test and the copper chloride–acetic acid salt spray test), and the resistance to spittle and sweat (test for the usability of a pigment in colored toys by wetting a strip of filter paper with $NaHCO_3$ and NaCl solutions and pressing the wetted paper against test surfaces, visual judgement of the paper with regard to discoloration) [34].

1.7 Conclusions

Pigments are substances consisting of small particles that are practically insoluble in an application system and that are used as a colorant or because of their anticorrosive or magnetic properties. Dyes are the second class of colorants, which differ from pigments primarily by the fact that they are practically completely soluble in the medium of application. Important application media for pigments are automotive and industrial coatings, paints, plastics, printing inks, cosmetic formulations, and building materials.

Pigments are differentiated according to their chemical composition and with respect to their optical and technical properties. A fundamental distinction is that between inorganic and organic pigments. Pigments are also differentiated into synthetic and natural types. There are white, colored, black, and special pigments. White pigments are represented by titanium dioxide (rutile and anatase), zinc sulfide including lithopone, and zinc oxide. The colored pigments range from blue (mixed metal oxides, ultramarine, iron blue) via green (chromium oxide, mixed metal oxides) and yellow (iron oxide hydroxide, mixed metal oxides, chromates, bismuth vanadate, cadmium sulfide), up to red (iron oxide, cadmium selenide, molybdates, cerium sulfide, oxonitrides). The most important black pigment by far is carbon black. Based on these variety of colors, whites and blacks, with inorganic pigments, it is possible to create nearly all thinkable colors and noncolors (white, black, gray) including bright and dark color shades by the use of pure single pigments or pigment mixtures.

Special pigments are subdivided into effect pigments (luster pigments) with special effect pigments (pearlescent pigments, interference pigments) and metal effect pigments, transparent pigments, luminescent pigments with fluorescent pigments and phosphorescent pigments, magnetic pigments, and anticorrosive pigments. The last two classes as well as a part of the transparent pigments belong to the functional pigments. Further functional pigments are electrically conductive, IR-reflective, UV-absorbing, and laser-marking pigments.

Fillers (extenders) are powders that like pigments are practically insoluble in the application system. They are typically white and are used because of specific chemical and physical properties. The distinction between pigments and fillers is based on the specific application.

Generally, inorganic pigments are more stable against light, weather, temperature, and chemicals than organic pigments. Another advantage of inorganic pigments is their lower manufacturing costs. Several organic pigments are, however, more color intensive and, therefore, more attractive than inorganic colored pigments.

Inorganic pigments consist of single particles of mostly uniform chemical composition and crystal structure. Exceptions are pigment mixtures (blended pigments) and layer-substrate pigments (typical for special effect pigments), which consist of chemically and crystallographically nonuniform or multicomponent particles.

Geometric parameters such as particle size (particle diameter), particle size distribution, and particle shape are very important for the characterization of pigments. Particles are definable, individual units of a pigment. The structure, size, and shape of pigment particles vary widely. Particle units in pigments and fillers are divided into primary particles, agglomerates, and aggregates.

The interaction of light with pigmented films can take place by absorption, reflection and transmission. The fundamental relationships among the optical characteristics of pigments can be comprehensively discussed on the basis of four theoretical considerations: the Mie theory, the theory of multiple scattering, the Kubelka–Munk theory, and colorimetry.

There are several color order systems. All these systems, e.g., the CIE color system, are systematic collections of colors, which should cover the complete color space. The systems have been developed for a better communication in the production of colored objects, in the application technology, or in marketing (between manufacturer and customer).

The selective light absorption by colored pigments is based on the interaction of visible light with the valence electrons of solids. There are six principles for the development of color that are relevant for inorganic pigments: d–d transfers of electrons within transition metal polyhedrons, charge–transfer transitions of electrons from ligands to a central atom or ion, intervalence transitions of electrons between cations from one and the same element in different valence states, electron transitions based on radical ions in a solid, electron transitions from donor levels of "guest ions" (dopant ions of a transition metal) into the conduction band of the host lattice, and electron transitions from the valence band into the conduction band within a solid.

Tinting strength, lightening power, scattering power, hiding power, and transparency are the most important parameters for the efficiency of pigments in paints, coatings, and other application media. An optimal pigment–binder interaction is a decisive factor for the application of inorganic pigments in the different media. The stability against light, weather, heat and chemicals must be high enough to fulfill the requirements associated with the application of the pigmented systems.

Study questions

1.1	What are the definitions for "pigment", "dye", and "colorant"?
1.2	How are pigments classified? How are inorganic pigments classified?
1.3	What are fillers? How are they different from pigments?
1.4	What are the differences between inorganic and organic pigments?
1.5	What are the criteria for the choice of a pigment for a specific application?
1.6	How are the optical properties of pigmented systems determined?
1.7	What are primary particles, agglomerates, aggregates, and flocculates?
1.8	What are important analytical methods for the characterization of inorganic pigments?
1.9	Which optical events may happen when light strikes the surface of a pigmented system?
1.10	Which are the four theoretical considerations describing the fundamental relationships among the optical characteristics of pigments?
1.11	What is color and how is the term "color" also used?
1.12	Which are the six principles for the development of color relevant for inorganic pigments?
1.13	What are the main steps which have to be conducted simultaneously or successively during the process of pigment dispersion in an application system?
1.14	What are the most important parameters for the efficiency of pigments in paints, .media?

Bibliography

[1]   DIN EN ISO 18451: Pigmente, Farbstoffe und Füllstoffe – Begriffe – Teil 1: Allgemeine Begriffe, 2017.
[2]   Lascoux RG. Höhle der Eiszeit, Verlag von Zabern, Mainz, 1982.
[3]   Bataille G. Die vorgeschichtliche Malerei. Lascaux oder Die Geburt der Kunst, Skira-Klett-Cotta, Stuttgart, 1986.
[4]   Lorblanchet M. Höhlenmalerei, 2nd edn. Jan Thorbecke Verlag, Ostfildern, 2000.
[5]   Eastaugh N, Walsh V, Chaplin T, Siddall R. Pigment Compendium. A Dictionary of Historical Pigments, Elsevier Butterworth-Heinemann, Oxford, 2004.

[6]   Finlay V. The Brilliant History of Color in Art, Getty Publications, Los Angeles, 2014.

[7]   Gärtner M., in Encyclopedia of Color, Dyes, Pigments, vol. 2. Pfaff G, ed. Walter de Gruyter GmbH, Berlin, Boston, 2022, 691.

[8]   Davies WV. Colour and Painting in Ancient Egypt, The British Museum Press, London, 2001.

[9]   Scott DA. Conservation. 2016,61(4),185.

[10] Spiteris T., in Griechische und etruskische Malerei, Weltgeschichte der Malerei, vol. 3. Editions Rencontre, Lausanne, 1966.

[11] Scheibler I. Griechische Malerei der Antike, Verlag C.H.Beck, München, 1994.

[12] Mielsch H. Römische Wandmalerei, Wissenschaftliche Buchgesellschaft, Darmstadt, 2001.

[13] Baldassarre I, Pontrandolfo A, Rouveret A, Salvadori M. Römische Malerei, DuMont, Köln, 2003.

[14] Cennini C. Il libro dell'arte o trattato o pittura. A cura die Fernando Teimpesti (reprint of the original of Cennino Cennini), Longanesi, Milano, 1984.

[15] Hale G. The Technique of Fresco Painting, Dover Publications, New York, 1966.

[16] Bertini F Affresco e Pittura Murale. Tecnica e Materiali, Edizioni Polistampa, Firenze, 2011.

[17] Merryfield Marry P. The Art of Fresco Painting in the Middle Ages and the Renaissance, Dover Publications, Mineola, 2012.

[18] Feller RL, ed. Artists' Pigments. A Handbook of Their History and Characteristics, vol. 1. Cambridge University Press, Cambridge and National Gallery of Art, Washington, 1986.

[19] Roy A, ed. Artists' Pigments. A Handbook of Their History and Characteristics, vol. 2. National Gallery of Art, Washington and Archetype Publications, London, 1993.

[20] FitzHugh EW, ed. Artists' Pigments. A Handbook of Their History and Characteristics, vol. 3. National Gallery of Art, Washington and Archetype Publications, London, 1997.

[21] Berrie BH, ed. Artists' Pigments. A Handbook of Their History and Characteristics, vol. 4. Archetype Publications, London, 2007.

[22] Murray P, Murray L. The Art of the Renaissance (World of Art Series), Thames & Hudson Ltd, London, 1963.

[23] Baxandall M. Painting and Experience in 15th Century Italy, Oxford University Press, Oxford, 1988.

[24] Humfrey P. Painting in Renaissance Venice, Yale University Press, New Haven, 1995.

[25] Kraft A. Bull. Hist. Chem. 2008,33,61.

[26] Kraft A. Chem. Texts. 2018,4,16.

[27] Morris AA, in The Temple of the Warriors at Chichen-Itzá, Yucatán, vol. 1. Morris EH, Charlotand J, Morris AA, eds. Carnegie Inst., Washington, 1931, 355.

[28] Gómez-Romero P, Sanchez C. New J. Chem. 2005,29,57.

[29] Doménech A, Doménech-Carbo MT, Sánchez Del Rio M, Vázquez de Agredos Pascual ML, Lima E. New J. Chem. 2009,33,2371.

[30] Pfaff G. Estimation based on data from professional journals, reports, and conference information, 2022.

[31] Gall L. Farbe + Lack. 1975,81,1015.

[32] Rechmann H, Sutter G. Farbe + Lack 1976,82,793.

[33] Pfaff G., in Winnacker–Küchler: Chemische Technik, Prozesse und Produkte, vol. 7, Industrieprodukte, 5th edn. Dittmeyer R, Keim W, Kreysa G, Oberholz A, eds. Wiley-VCH Verlag, Weinheim, 2004, 273.

[34] Hempelmann U, in Industrial Inorganic Pigments, 3rd edn. Buxbaum G, Pfaff G, eds. Wiley-VCH Verlag, Weinheim, 2005, 11.

[35] Völz HG. Angew. Chem. 1975,87,721.

[36] Mie G. Ann Physik (Folge 4) 1908,25,377.

[37] Hergert W, Wriedt T, eds. The Mie Theory – Basics and Applications, Springer-Verlag, Heidelberg, New York, Dordrecht, London, 2012.

[38] Billmeyer JrFW, Abrams RL, Phillips DG. J. Paint Technol. 1973,45,23.

[39]   Kubelka P, Munk FZ. Tech. Phys. 1931,12,539.

[40]   Foldy LL. Phys. Rev. 1945,67,107.

[41]   Van de Hulst RC Light Scattering by Small Particles, John Wiley & Sons, New York, 1957.

[42]   Orchard SE J. Oil Colour Chem. Assoc 1968,54,44.

[43]   Auger J-C, Stout B. J. Coat. Technol. Res. 2012,9,287.

[44]   Völz HG. Ber. Bunsen-Ges.Phys. Chem. 1967,71,326.

[45]   Kubelka P. J. Opt. Soc. Am. 1948,38,448.

[46]   Alcaraz de la Osa R, Iparragirre I, Ortiz D, Saiz JM. Chem. Texts. 2020,6,2.

[47]   Hoffmann K. Farbe + Lack. 1974,80,118.

[48]   Gall L. Farbe + Lack. 1974,80,297.

[49]   Hempelmann U, in Colour Technology of Coatings, Kettler W et al., eds. Vincentz Network, Hannover, 2016, 137.

[50]   Völz HG. Angew. Chem. Int. Ed. Eng. 1975,14,688.

[51]   Gilchrist A, Nobbs J, in Encyclopedia of Spectroscopy and Spectrometry, 3rd edn. Lindon J, Tranter GE, Koppenaal D, eds. Elsevier, Amsterdam, 2017, 328.

[52]   Völz HG. Angew. Chem. Int. Ed. Eng. 1975,14,655.

[53]   Cramer WR, in Encyclopedia of Color, Dyes, Pigments, vol. 2. Pfaff G, ed. Walter de Gruyter GmbH, Berlin, Boston, 2022, 407.

[54]   Gauss S, in Colour Technology of Coatings, Kettler W et al., eds. Vincentz Network, Hannover, 2016, 15.

[55]   Kettler W, in Colour Technology of Coatings, Kettler W et al., eds. Vincentz Network, Hannover, 2016, 85.

[56]   Nassau K. The Physics and Chemistry of Color, Wiley-Interscience, New York, 1983.

[57]   Tilley R. Colour and the Optical Properties of Materials, John Wiley & Sons, Chichester, New York, Weinheim, Brisbane, Singapore, Toronto, 2000.

[58]   Reinen D. Angew. Chem. 1971,83,991.

[59]   Kober F. Grundlagen der Komplexchemie, Otto Salle Verlag, Frankfurt am Main, 1979.

[60]   Marfunin AS. Physics of Minerals and Inorganic Materials, Springer-Verlag, Berlin, Heidelberg, New York, 1979.

[61]   Figgis BN, Hitchman MA. Ligand Field Theory and Its Applications, Wiley-VCH Verlag, Weinheim, 1999.

[62]   Bersuker IB. Electronic Structure and Properties of Transition Metal Compounds, 2nd edn. John Wiley & Sons, Hoboken, 2010.

[63]   Klöckl I. Chemie der Farbmittel, Walter de Gruyter GmbH, Berlin, München, Boston, 2015.

[64]   Nassau K Spektrum der Wissenschaft 1980,12,64.

[65]   Briegleb G, Czekalla J. Angew. Chem. 1960,72,401.

[66]   Kaim W, Ernst S, Kohlmann S. Chem. Unserer Zeit 1987,21,50.

[67]   Buser HJ, Schwarzenbach D, Petter W, Ludi A. Inorg. Chem. 1977,16,2704.

[68]   Ludi A. Chem. Unserer. Zeit. 1988,22,123.

[69]   Seel F, Schäfer H, Güttler H-J, Simon G. Chem. Unserer Zeit. 1974,8,65.

[70]   Clark RJH, Franks MI. Chem. Phys. Lett. 1975,34,69.

[71]   Clark RJH, Cobbold DG. Inorg. Chem. 1978,17,3169.

[72]   Clark RJH, Dines TJ, Kurmoo M. Inorg. Chem. 1983,22,2766.

[73]   Tarling SE, Barnes P, Klinowski J. Acta Cryst. B 1988,44,128.

[74]   Reinen D, Lindner G-G. Chem. Soc. Rev. 1999,28,75.

[75]   Hund F. Angew. Chem. 1962,74,23.

[76]   Hund F. Farbe + Lack 1967,73,111.

[77]   Dondi M, Cruciani G, Guarini G, Matteucci F, Raimondo M. Ceram Int. 2006,32,393.

[78]   Kobayashi A, Sankey OF. Phys. Rev. B. 1983,28,946.

[79] Cohen ML, Chelikowsky JR. Electronic Structure and Optical Properties of Semiconductors, 2nd edn. Springer-Verlag, Berlin, Heidelberg, New York, 1989.

[80] Rohlfing M, Krüger P, Pollmann J. Phys. Rev. Lett. 1995,75,3489.

[81] Berger LI. Semiconductor Materials, CRC Press Inc, New York, 1996.

[82] Völz HG. Industrial Color Testing, 2nd edn. Wiley-VCH Verlag, Weinheim, 2001.

[83] Keifer S. Farbe + Lack. 1976,82,811.

[84] Völz HG. Progr. Org. Coat. 1987,15,99.

[85] Joppien GR. Farbe + Lack. 1975,81,1102.

[86] Marwedel G. Farbe + Lack. 1976,82,789.

[87] Keifer S. Farbe + Lack. 1973,79,1161.

[88] Kresse P. DEFAZET – Dtsch. Farben Z. 1974,28,459.

[89] Herrmann E. DEFAZET – Dtsch Farben Z. 1975,29,116.

[90] Miele M. Farbe + Lack. 1975,81,495.

[91] Kaluza U. DEFAZET – Dtsch Farben Z. 1973,27,427.

[92] Völz HG. Farbe + Lack. 1990,96,19.

[93] Kämpf G, Papenroth W. Farbe + Lack. 1977,83,18.

[94] Grassmann W. Farbe + Lack. 1960,66,67.

[95] Winkler J., Titanium Dioxide, Vincentz Network, Hannover, 2003.

[96] Auer G, in In Industrial Inorganic Pigments, 3rd edn. Buxbaum G, Pfaff G, eds. Wiley-VCH Verlag, Weinheim, 2005, 51.

2 General methods of manufacture

Various mechanical, chemical, and physical-chemical processes are used for the manufacture of inorganic pigments. The goal of the processes is the production of pigments with a suitable composition, crystal structure, and optimal particle size distribution, and [1]. Another objective is to achieve the required properties of the pigments in the different application systems. For this purpose, various pigments have to be provided with an additional surface treatment. The manufacturing processes begin with the provision of the raw materials and end with the packaging of the pigments for delivery to the customers. Today, all process steps in the pigment industry are accompanied by strict project management. An optimal project management is a complex task that requires close coordination between the various parties involved. Errors in planning and technology selection can lead to costly delays and increased risks. Project teams can be spread across the globe. Choosing suitable project management tools ensures that project timelines and quality goals are met.

Introductions of new pigments to the market usually start with a product development phase in which a decision must be made about the manufacturing technology to be used. Three cases are to be distinguished here:
- new product developments in which existing technologies can be used;
- new product developments in which existing technologies can be used in modified form;
- new product developments in which a new technology must be developed.

Applying a technology in a form that has already been used or modified is often advantageous. Various factors must be taken into account when developing, introducing, and using technologies. Requirements on the quality of the product as well as on the manufacturing costs can be summarized as follows:
- provision of raw materials, consumables, and supplies as well as their consumption, energy consumption, CO_2 footprint, and transportation expenses;
- environmental impacts (environmentally relevant extraction of resources, disposal and reuse of by-products, emissions, toxicity issues);
- further use of the technology for other products, recycling, environmentally sound disposal, dismantling, after-use of the equipment and the site.

All the factors mentioned here play an important role in the development of new inorganic pigments. This also applies to organic pigments and dyes. In each case, it is first checked whether existing technologies can be used or, if necessary, modified before the development of a new technology is initiated [2].

https://doi.org/10.1515/9783110743920-002

2.1 Mechanical preparation processes

Mechanical processes play a role in the manufacture of both natural and synthetic inorganic pigments. Ideal pigments should have mean particle sizes that exhibit an optimal diffuse reflection behavior, according to the theory of scattering. Other characteristics are a narrow particle size distribution and the smallest possible amount of agglomerates and aggregates. These are, however, typically included in natural and synthetic pigments and require mechanical processing. Figure 2.1 shows a scheme of mechanical processing as it is used, e.g., for the mechanical processing of natural iron oxide pigments.

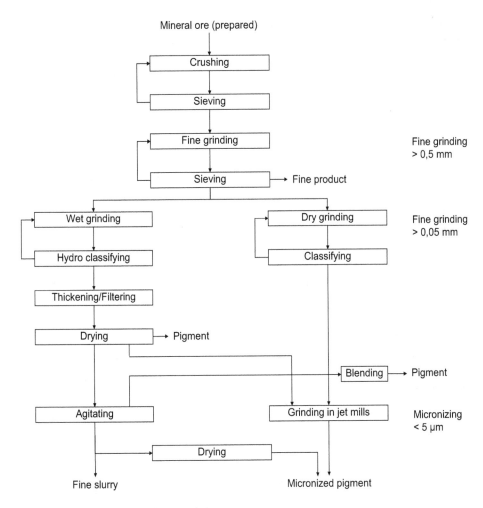

Fig. 2.1: Mechanical processes for natural pigments.

Natural inorganic pigments are obtained from ores by crushing, grinding, agitating, and further processing steps. The ore released from the gangue is crushed to the desired fineness in several steps. The oversized grains are separated from the fine particles in separation stages and are recycled into the process. According to hardness of the material, impurities, and indented application, grinding and classification are carried out, wet or dry.

The mineral raw materials used for the production of inorganic pigments are classified as industrial minerals and metalliferous minerals (ores). There is a fluid technological interface with the pigment industry in terms of raw material extraction and processing. Industrial minerals are extracted in open-pit mines or by underground mining. The interface with the processing is usually carried out after the extraction process, i.e., the transfer into a bulk material that can be extracted. In open pit mining, the maximum particle size can be 1000 mm and above. Primary raw material sources with good accessibility and medium-to-high valuable material content and easy processability are preferred. However, these are often already exploited.

2.1.1 Crushing and grinding

Two size ranges are distinguished for crushing of coarse material:
- coarse crushing, also referred to as breaking, proceeds from the size of meters down to particles of 1 to 2 cm. Breaking is at all times carried out in the dry state.
- fine crushing, also referred to as grinding, proceeds from 1 to 2 cm size down to sizes of about 50 μm. Grinding can be carried out wet or dry.

Pulverization and micronization are further grinding processes that lead to even smaller particle sizes. Table 2.1 contains a summary of the crushing and grinding devices used for the manufacture of inorganic pigments. The selection of a certain

Tab. 2.1: Crushing and grinding devices for the manufacture of pigments.

Crushing fineness	Devices used
Coarse crushing (breaking)	Jaw crusher, cone crusher, roller crusher, impact crusher, percussion crusher, hammer crusher
Fine crushing (grinding)	Edge mill, ball mill, crushing mill, disintegrating mill, cross-beater mill
Pulverization	
Dry	Pin mill, ball mill, pendulum roller mill, vibratory mill
Wet	Ball mill, vibratory mil, cane mill
Micronization	
Dry	Spiral mill, countercurrent jet mill
Wet	Corundum disc mill, stirrer mill, agitator ball mill, roller mill

crushing and grinding device depends mainly on the hardness and abrasiveness of the pigment to be produced. Possible wear and abrasion as well as impurities from the material of the device and the damage to sensitive pigments should be taken into account. The fracture of the crushed material is caused by tensile and shear stresses. The stress fields in the grains are caused by different types of stresses: stress between solid surfaces (pressure, friction, shear, cutting) and impact stress on solid surfaces (grinding media, e.g., grinding beads impact at high speed onto the surface of the pigment material to be shredded) [3–5].

Pulverization leads to particle sizes in the range from 5 to 50 μm. This range is already suitable for some inorganic pigments and their applications. Pulverization can be carried out dry or wet. Wet pulverization is used when soluble components must be separated or when a pigment is to be equipped with surface treatment directly in the suspension. The following necessary drying, however, is often unfavorable because agglomeration can take place. This is the reason why this process is often combined with disintegrating grinding.

Sophisticated applications require a very energy-intensive micronization of the pigments, which is also referred to as colloid grinding. Dry micronization is done in spiral mills and countercurrent jet mills with air or vapor as energy carriers. Wet micronization is carried out with grinding beads of 0.1 to 10 mm diameter in a suspension of medium viscosity.

Continuous or batchwise operating bead mills and sand mills have become established in the manufacture of pigment pastes and masterbatches (pigment concentrates).

2.1.2 Classifying

The term "classifying" covers mechanical separation processes such as sieving, flow classification, and air classification. Separation and recycling of the coarse parts (lump fraction) is required after each crushing step during the pigment manufacture. During dry processes, the fine material is separated with classifiers and sieves, which are directly integrated in the mills, but separate classifiers are also used.

In process engineering, classification is a separation process. The separated components are called fractions. In mechanical process engineering, classification refers to the separation of a disperse mixture of solids into fractions, preferably based on the criteria of particle size or particle density [6]. If the geometric particle size is the separation criterion, classification is carried out by means of sieving; if density or other equivalent diameters are decisive, sifting is used as the classification method. The result is at least two fractions, which differ in that the minimum limit of one fraction is also the maximum limit of the other fraction.

The aim of classification is:

- The production of at least two subsets of the original solid mixture, where each subset corresponds as completely as possible to the specified size criteria. In industrial practice, however, such ideal separation accuracy is not achievable.
- The separation of oversized and undersized particles, respectively, in order to avoid disturbances or overloads in subsequent processes.

Basically, a distinction is made between the following classification methods:
- screen classification by means of sieves;
- stream classification with the separating medium air or other gases, e.g., by means of air classifiers;
- stream classification with the separating medium water or other liquids, e.g., by means of the float/sink method.

Air classifiers that are used as spiral classifiers with stationary installations or as basket and centripetal classifiers with a blade fitted rotor are common for finer particle fractions as they are typical for pigments. The separation of the oversized material during wet crushing takes place in hydroclassifiers. Hydrocyclones have a simple design and are easy to handle. They are widespread in the pigment producing industry. The separation limit is in the particle size range of 5 to 100 μm. Gravity processes are only appropriate for the separation and thickening of flocculated pigment slurries.

Pigments must be thickened after wet micronization steps. They have to be dried carefully to avoid the formation of aggregates. The use of spray drying or even freeze drying processes is advantageous for very fine pigments, e.g., for transparent pigments (Chapter 7). Separation of the pigments after dry crushing takes place in electrostatic precipitators (ESP), cyclones, and tube filters [1].

2.1.3 Blending and conveying

Blending processes play a significant role in the manufacture of pigments. The main objective of blending steps is thorough mixing, so that a uniform composition (homogeneity) is achieved in the whole powder and the target color is met. Blending can take place in the dry, moist, or wet states. In principle, static and dynamic blenders can be used. There are various blender types in use such as turbine agitators (turbomixer), double-motion agitators, screw mixers, mixing mills, and drum mixers. Many manufacturing processes require blending steps for the adjustment of the desired pigment properties. A simple continuous mixing in silo containers is rarely sufficient. A better mixing efficiency can be achieved using a fluidized bed or stirred, staggering, and rotating mixing vessels.

Pigment powders are frequently transported with pneumatic conveyor systems, although this is associated with dust extraction problems. The effort of dust separation can be avoided by the use of cup conveyers or screws. The delivery to users takes place in drums or in multilayer valve bags. These can contain additional polyethylene foil to provide protection against soaking. More and more bulk packs and silo facilities are being used to reduce dust pollution related to the use of bags. Various inorganic pigments have been offered for a long time, not only as pure, free flowing powders but also as pigment preparations (granulates, pellets, pastes, color concentrates). Such pigment preparations contain the highest possible amount of pigment. Other components contained in the preparations are binders or mixtures of binder agents treated with wetting agents and special additives. The binder system can be solventborne or waterborne and should later be compatible with the application system. In addition to positive application characteristics, such as better dispersibility and wetting behavior, pigment preparations have the advantage of dust-free operation [1].

2.2 Precipitation reactions

A narrower particle size distribution than by grinding is achieved by the direct manufacture of pigments using chemical methods, especially by precipitation processes. An example of a pigment synthesized directly by precipitation is iron oxide hydroxide (iron oxide yellow), which is formed by the reaction of an iron(II) salt with sodium hydroxide and air in an aqueous solution. Without any further treatment, the precipitation reaction leads to a defined solid with the desired color, crystal structure, morphology, and other pigment properties. The characteristics of the particles formed can be influenced by the choice of the precipitation conditions.

The formation of nuclei and crystallization during precipitation reactions are complex processes [7, 8]. The rate of nuclei formation depends on the relative oversaturation in the solution. The stability of a crystal nucleus of critical size, i.e. in the equilibrium between resolution and growth, is also dependent on the interfacial tension solid/liquid, the lattice energy, and the activation energy [9]. The critical nucleus size decreases with increasing oversaturation. The theories of homogeneous nucleation are only valid for ideal, clean solutions. It has not been proven so far that homogeneous nucleation takes place during the synthesis of pigments by precipitation. The nuclei are formed instead on solid contaminations [10].

Due to the temporal randomness of nucleation associated with a reduction of the oversaturation in finite time, a size distribution of the crystals is generated. Controlling this distribution by temperature, pressure, concentration, type of solvent, and by other process parameters are part of the substantial know-how for the production of synthetic inorganic pigments. The formed crystal nuclei grow in the course of the reaction to the desired pigment size by maintaining a small oversaturation. The size distribution is broadened regularly. It often obeys a logarithmic standard distribution.

Nucleation can happen in a fraction of a second ($BaSO_4$), in minutes, or in hours (α-FeOOH). The processes have to be designed accordingly. Precipitations with rapid nucleation are carried out continuously in stirred vessels or tube mixing units. Reactions with slow nucleation are discontinuously operated. Thus, a broadening of the particle size distribution is avoided.

During the phase of crystal growth, care must be taken that no new, "wild" nuclei are formed [9]. Already, a wrong stirrer arrangement or very intensive stirring can lead to the formation of new nuclei as a result of collisions between particles. Thus, the size distribution of the crystallites can be broadened (secondary formation of nuclei by the so-called collision breeding). For hydrolytic precipitation, the exact control and regulation of temperature, pH value, and concentration of the reaction partners is essential. Particularly critical is the pH regulation near the neutral point because of the steepness of the titration curve.

After the precipitation reaction, the pigment suspension is filtered by using pressure and suction filters or continuously working rotating filters [11]. A large part of the soluble salts formed during the precipitation can be separated using an upstream gravity thickener. In larger production units, washing facilities with several successively arranged rotating filters are used, whereby the filter cake is mashed after each filtering stage and then transferred to the next filter.

Drying cabinets are sufficient for drying aggregates if the throughput is small. Larger production volumes are dried by using continuously working drying aggregates, such as belt dryers, disk dryers, cylinder dryers, pneumatic dryers, contact dryers, fluidized bed dryers, or spray dryers. Contact and spray dryers have an especially high factor of energy efficiency. Mechanical processes for an additional dehydration of filter cakes have grown in significance, e.g., screen belt pressing.

An elegant drying process is the displacement of water by organic solvents. The pigment is made hydrophobic and transferred in kneaders from the aqueous into the organic phase. This so-called flushing is used especially in the production of very fine powders, which tend to agglomerate (transparent and iron blue pigments). Another possibility is the manufacture of highly concentrated pigment pastes.

Inorganic pigments produced by precipitation reactions are typically characterized by low dispersing hardness, a lack of hard agglomerates, and good wettability in almost all binder systems. This behavior makes them suitable for high-gloss coatings.

Figure 2.2 shows the entire process of pigment manufacture via precipitation. Depending on the pigment, further steps follow after the precipitation. In some cases, these include aging of the fresh precipitate in the suspension. Next steps are usually filtering, washing, and drying the precipitate. In many cases, this is followed by a calcination step in order to achieve the final composition or a certain grain growth. After calcination, a grinding process is often necessary before the pigment powder is packaged for sale. Examples of pigments in whose synthesis precipitation plays a major role are zinc sulfide, iron oxide hydroxide, cadmium sulfide, lead chromate, or iron blue.

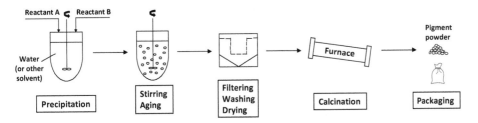

Fig. 2.2: Entire process of pigment production via precipitation including further steps up to the final product.

2.3 Calcination processes

Calcination refers to the pressureless heating of materials to temperatures above 200 °C with one of the following objectives:

- increase of the crystallite size (ZnS, TiO_2, $\alpha\text{-}Fe_2O_3$)
- refinement of the crystal lattice (Fe_3O_4)
- adjustment of a specific crystal phase (TiO_2 as anatase, rutile, or anatase/rutile mixture)
- solid-state transformation ($\alpha\text{-}FeOOH \rightarrow \alpha\text{-}Fe_2O_3$)
- gas/solid reaction ($Fe_3O_4 \rightarrow \gamma\text{-}Fe_2O_3$)
- solid-state reaction ($CoO + Al_2O_3 \rightarrow CoAl_2O_4$)

The most important process parameters are temperature, time, and heating rate. The temperatures must be sufficient to achieve a certain ion mobility. They should, however, not be too high, so that too excessive particle growth and sintering are avoided.

The optimal conditions in the furnace used have to be determined for each calcination process. Impurities or targeted doping can lower the annealing temperature and reduce the time. A small percent of a dopant is often enough to reduce the annealing temperature by several 100 °C. Dopants and impurities have a specific effect on each pigment. Their influence must be determined experimentally. They can be decisive for the preferred formation of a specific crystal modification, e.g., anatase or rutile, in the case of TiO_2. The atmosphere in the furnace is another parameter, besides temperature and time, which influences the high temperature reactions. Water vapor, for example, acts as mineralizer and leads to a faster growth of the particles.

Depending on the throughput, the calcination is performed continuously or discontinuously. Discontinuously operating muffle and drum-type furnaces enable fast product changes (batch production). Rotary kilns are only suitable for continuous production. They can be fired directly or indirectly. Indirect firing with gas or radiant heaters is used when a defined gas atmosphere has to be adjusted during calcination. One possibility of continuous operation for a small throughput is the use of a rotary ring hearth furnace. The material to be heated is thereby conveyed continuously on an indirectly heated rotating ring. There, it is dehydrated and annealed during one circulation [12].

2.4 Other methods

Inorganic pigments are also produced using special processes such as flux crystallization, hydrothermal methods, and gas-phase reactions.

2.4.1 Flux crystallization processes

Flux crystallization means growth of crystals from high temperature solutions, where the flux or solvent is typically a molten salt or oxide [13]. Salt melts are useful process media for the formation of crystals with suitable morphology. Flux crystallization processes are, therefore, used for monocrystal growth and, in rare cases, also for the manufacture of pigments. The crystallization of aluminum oxide platelets (alumina platelets) in a melt of metal salts is one example of the use of a flux process for the pigment manufacture. Such alumina platelets act as a substrate material for special effect pigments based on the substrate-layer principle. For this purpose, they are coated with thin layers of titanium dioxide (anatase or rutile), α-iron(III) oxide, or other oxides [14, 15].

2.4.2 Hydrothermal reactions

Hydrothermal conditions are also suitable for the formation of useful crystals. The presence of water, increased pressure, and increased temperature are necessary for crystal formation [16, 17]. The technological effort for the execution of a hydrothermal process is often high. This is why only a few industrial hydrothermal processes for pigments are in use. The high production costs of hydrothermally manufactured pigments are mainly the result of the necessary batch operation associated with long dwell times. However, the use of continuous hydrothermal processes for the production of fine powders, i.e., also of pigments, is possible in principle, as studies have shown [18].

Examples for the use of hydrothermal processes for the pigment synthesis are chromium(IV) oxide and chromium(III) oxide hydroxide. CrO_2 pigments for magnetic data carriers crystallize out of chromic acid and chromium(III) salts at temperatures above 200 °C and pressures above 100 bar. CrOOH pigments can be produced by hydrothermal recrystallization starting from a $Cr(OH)_3$ precipitate.

Platelet-like α-Fe_2O_3 (micaceous iron oxide), mainly used as effect pigment, is also produced by hydrothermal synthesis. The formation of the iron oxide platelets typically starts from α-FeOOH and takes place in an autoclave in alkaline media at temperatures above 170 °C. The incorporation of substantial amounts of dopants, e.g., Al_2O_3, SiO_2, and Mn_2O_3, leads to increased aspect ratios of up to 100 for the thin platelets formed. High aspect ratios of the platelets lead to stronger luster, while smaller ratios result in higher hiding power. The color can be influenced by the composition, and various red to brown tones are achieved [19, 20].

2.4.3 Gas-phase reactions

High-purity pigments can be produced by reactions in the gaseous phase [17]. An easily evaporable starting compound is converted in air, oxygen, or humid atmosphere at elevated temperature. Halides or carbonyls, which are available in high purity by distillation, are mostly used as starting compounds. The reaction chambers used are squish jet combustion chambers, plasma combustion chambers, or high-speed turbulent burners. TiO_2, ZnO, and transparent iron oxide pigments as well as SiO_2, Al_2O_3, and ZrO_2 powders are produced using one of these reactors. In the case of iron oxides, $Fe(CO)_5$ is used as starting compound. Depending on the reaction regime, fine products with corresponding high surface areas are obtained, which are suitable for use as pigments or fillers.

In a special process, aluminum flakes can be coated in a fluidized bed reactor with thin iron oxide layers to form golden, orange, and red effect pigments. The coating takes place by fluidizing the aluminum flakes at 450 °C in an atmosphere of nitrogen and inflowing of $Fe(CO)_5$ and oxygen. α-Fe_2O_3 is formed under these conditions and deposited on the surface of the aluminum flakes in the form of a thin film [20, 21].

2.5 Conclusions

Inorganic pigments can be produced using various mechanical and physical-chemical processes. The selection of the processes used depends on the nature of the raw materials and on the properties required for the pigments to be produced. The chemical composition, crystal structure, mean particle size and particle size distribution, the content of impurities, and surface characteristics are some of the most important properties of a pigment.

Mechanical preparation processes include crushing and grinding, classifying, blending, and conveying. Processes for pigment formation may be subdivided into precipitation reactions and calcination methods. Some inorganic pigments are manufactured using both precipitation and calcination. Further preparation methods, which are used only for a small number of pigments, are flux crystallization processes, hydrothermal reactions, and gas-phase reactions.

The mechanical and physical-chemical parameters of the single processes have to be adjusted carefully for each pigment to achieve the desired properties and to ensure reproducibility. Even small deviations in single pigment properties, e.g., with respect to the particle size or the crystal structure, can lead to an unacceptable color deviation visible to the naked eye.

? Study questions

2.1 What are the mechanical preparation processes used for the manufacture of inorganic pigments?

2.2 What are the objectives of calcination processes for the production of inorganic pigments and what are the most important process parameters?

2.3 What special preparation methods are used for the synthesis of inorganic pigments besides mechanical processes, precipitation, and calcination?

Bibliography

[1] Pfaff G, in Winnacker–Küchler: Chemische Technik, Prozesse und Produkte, vol. 7, 5th edn. Industrieprodukte, Dittmeyer R, Keim W, Kreysa G, Oberholz A, eds. Wiley-VCH Verlag, Weinheim, 2004, 290.

[2] Pfaff G. Sitzungsberichte Leibniz-Sozietät Wiss. 2021,146,67.

[3] Ring TA. Fundamentals of Ceramic Powder Processing and Synthesis, Academic Press, Cambridge, 1996.

[4] Schatt W, Wieters K-P, Kieback P, eds. Pulvermetallurgie: Technologie und Werkstoffe, Springer-Verlag, Berlin, Heidelberg, 2007.

[5] Salmang H, Scholze H. Keramik, 7th edn. Springer-Verlag, Berlin, Heidelberg, New York, 2007.

[6] Stieß M. Mechanische Verfahrenstechnik 1, 2nd edn. Springer-Verlag, Berlin, Heidelberg, 1995.

[7] Gösele W, Egel-Hess W, Wintermantel K, Faulhaber FR, Mersmann A. Chem. Ing. Tech. 1990,62,544.

[8] Hulliger J. Angew. Chem. 1994,106,151.

[9] Kallweit M. Ber. Bunsenges. Phys. Chem. 1974,78,997.

[10] Adamski T. DECHEMA-Monogr. 1974,73,229.

[11] Bender W, Redeker D. Chem.-Ing.-Tech. 1981,53,227.

[12] Patent DE 23 20 806 (Bayer AG) 1973.

[13] Elwell D. Fundamentals of flux growth, in Crystal Growth in Science and Technology, Arend H, Hulliger J, eds. Plenum Press, New York, 1989, 133.

[14] Teaney S, Pfaff G, Nitta K. Eur. Coat. J. 1999,4,90.

[15] Pfaff G. Effect pigments based on alumina flakes, in Special Effect Pigments, 2nd edn. Pfaff G, ed. Vincentz Network, Hannover, 2008, 72.

[16] Rabenau A. Angew. Chem. 1997,97,1017.

[17] Kriechbaum GW, Kleinschmidt P. Angew. Chem. Adv. Mater. 1989,101,1446.

[18] Darr JA, Zhang J, Makwana NM, Weng X. Chem. Rev. 2017,117,11125.

[19] Ostertag W. Nachr. Chem. Tech. Lab. 1994,42,849.

[20] Ostertag W, Mronga N. Macromol. Chem. Macromol. Symp. 1995,100,163.

[21] Ostertag W, Mronga N, Hauser P. Farbe + Lack. 1987,93,973.

3 White pigments

3.1 Introduction

White pigments are achromatic inorganic pigments whose optical action is mainly based on nonselective light scattering. White pigments do not show absorption in the range of visible light, but have a high scattering power, which leads to a high hiding power in applications. The larger the difference between the refractive indexes of the white pigment and that of the surrounding medium, the higher the scattering power.

White fillers, which are also used as pigments in emulsion paints and paper, are distinguished from white pigments by the refractive index and by the purpose of their use. Organic white pigments are unknown.

The first synthetically produced white pigment was white lead (basic lead carbonate). It was already known in Roman times. Around 1800, zinc white (zinc oxide) was developed, followed by antimony white (antimony oxide) and zinc sulfide. Lithopones (zinc sulfide, barium sulfate mixtures) were manufactured for the first time in the second half of the nineteenth century. During the twentieth century, the main processes for the production of titanium dioxide pigments were developed (sulfate and chloride process). Titanium dioxide in anatase and rutile modification are today the most important white pigments. Lithopones, zinc sulfide, and zinc oxide also have a significant importance, besides titanium dioxide pigments, whereas white lead and antimony white are not largely applied for toxicological reasons. White pigments are used not only for coloration of coatings, emulsion paints, printing inks, and plastics but also for brightening of

Tab. 3.1: Overview on white pigments and fillers.

Name	Formula	Refractive index n_D	C.I.
White pigments			
Titanium dioxide	TiO_2		PW 6
Anatase		2.55	
Rutile		2.75	
Zinc sulfide	ZnS	2.37	PW 7
Zinc oxide (zinc white)	ZnO	2.01	PW 4
Lithopones	$ZnS + BaSO_4$	1.84–2.08	PW 5
Basic lead carbonate (white lead)	$2\ PbCO_3 \cdot Pb(OH)_2$	1.94–2.09	PW 1
Antimony(III) oxide (antimony white)	Sb_2O_3	2.00–2.09	PW 11
Fillers			
Calcium carbonate	$CaCO_3$	1.48–1.65	PW 18
Barium sulfate	$BaSO_4$	1.64	PW 21
Talc	$3\ MgO \cdot 4\ SiO_2 \cdot H_2O$	1.54–1.59	PW 26

https://doi.org/10.1515/9783110743920-003

different colored systems. They show good stability against chemical substances and are highly weather-resistant.

Table 3.1 contains a summary of industrially relevant white pigments and fillers.

3.2 Titanium dioxide pigments

3.2.1 Fundamentals and properties

Titanium dioxide (TiO_2) occurs naturally in three modifications rutile, anatase, and brookite. Only rutile and anatase are manufactured industrially in large quantities. Brookite is difficult to produce and has no technical value. A large part of the titanium dioxide produced is used for pigment purposes (C.I. Pigment White 6). Catalysts and ceramic materials are further applications for industrially manufactured TiO_2 [1–3].

The history of titanium dioxide pigments began comparatively late, even though the element titanium was discovered as early as 1790 [4]. More than one hundred years later, the first industrial process for the production of titanium dioxide was developed [5–7]. In the years from 1916 to 1920, the first plants for the mass production of TiO_2 went into operation [4]. Production was carried out according to the sulfate process, which is used till today in industrial scale. For the next 40 years, all titanium dioxide pigments were produced using this process. The further development of processes for the production of TiO_2 powders led to the chloride process [8]. Both processes, the sulfate and the chloride process, are the basis for the production of millions of tons TiO_2 pigments every year.

The existing titanium reserves are relatively abundant. Titanium is the tenth most common element on the Earth's crust [9, 10]. As a structural metal, titanium is the fourth largest consumer product after iron, aluminum, and magnesium. There are hundreds of titanium minerals and compounds on the Earth, but only a few of them have industrial value [11]. Since 2020, the reserves of rutile and ilmenite are estimated to be about, respectively, 46 and 770 million metric tons in the world [12]. Therefore, ilmenite can be estimated to be the most important titanium resource, and it continues to dominate the future titanium industry. The titanium processing metallurgical industry produces large quantities of titania-containing slags. Both ilmenite and titania-containing slags play an extremely important role in the production of TiO_2 pigments [13].

Rutile is the thermodynamically most stable modification of titanium dioxide. The lattice energies of anatase and brookite are, however, similar to that of rutile. Both modifications are, therefore, also stable under normal conditions over long periods. A monotropic conversion to rutile takes place for anatase at about 915 °C and for brookite at about 750 °C. Rutile is stable up to its melting range of about 1830 to 1850 °C.

In all three TiO_2 modifications, titanium atoms are surrounded octahedrally by six oxygen atoms. TiO_6 octahedrons are, therefore, dominant structural units in the three crystal lattices. The single octahedrons, however, differ in their structural arrangement in the modifications. They are linked to one another in different ways via corners and edges. There are two common edges each between the TiO_6 octahedrons in rutile. The octahedrons in brookite are linked via three and in anatase via four common edges. The oxygen atoms in the three crystal lattices are surrounded by three titanium atoms in a trigonal arrangement. Rutile crystallizes in the tetragonal system, and anatase does so as well. Brookite, on the other hand, has a rhombic crystal structure. Crystal lattice sections of anatase, rutile, and brookite are shown in Figure 3.1.

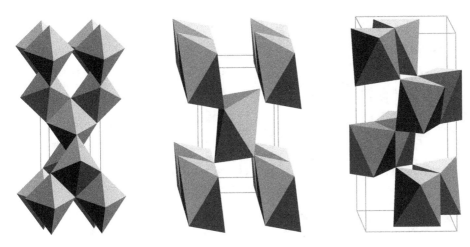

Fig. 3.1: Crystal lattice sections of the TiO_2 modifications, from left to right anatase, rutile and brookite (source: H. Ißbrücker).

Thermal treatment of titanium dioxide to temperatures above 1000 °C leads to an increase of the oxygen partial pressure. Oxygen is liberated continuously under these conditions and lower oxides of titanium are formed (TiO_{2-x}). Changes in color and electrical conductivity are the consequence of the oxygen release. Above 400 °C, a significant yellowing of the TiO_2 is observed, which can be attributed to the thermal expansion of the lattice. The effect is reversible when the temperature is lowered again.

Rutile has the most compact crystal structure of the three TiO_2 modifications and, therefore, the highest density. With a Mohs hardness of 6.0 to 6.5, it is also the hardest form of titanium dioxide (anatase and brookite 5.5 to 6.0). The lower hardness is the reason why anatase is used in applications where abrasiveness is of relevance, e.g., in synthetic fiber, plastics, and in paper industry.

The high refractive index of titanium dioxide, together with the absent absorption in the visible spectral range between 380 and 700 nm, is the basis for the use of this compound as white pigment. The normally used refractive index values for pigment applications are 2.7 for rutile and 2.55 for anatase.

The UV-VIS spectra of rutile and anatase exhibit characteristic differences (Figure 3.2) [2]. There is especially a noticeable difference concerning the absorption edge in the near-UV region. The significant absorption of rutile in the range of about

385 nm (absorption of anatase starts at lower wavelengths and not in the visible range) leads to the situation where anatase pigments exhibit a slight blue tint, and to the human eye, they appear really white, whereas rutile pigments have an extremely small yellow tinge in the white. Anatase can, therefore, be considered to be the whiter white.

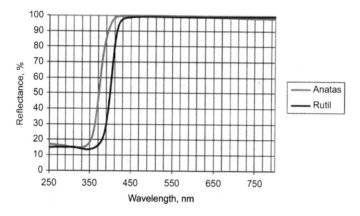

Fig. 3.2: UV/VIS spectra of rutile and anatase pigment pellets [2].

Titanium dioxide is photoactive and is allocated to the group of light-sensitive semiconductors. It absorbs electromagnetic radiation in the near-UV region. During light absorption, electrons from the valence band are raised to the conduction band. The energy difference between the valence and the conduction bands in the solid state is determined to be 3.05 eV for rutile and 3.29 eV for anatase. These values correspond to an absorption band at <415 nm for rutile and <385 nm for anatase. Raising of electrons to the conduction band leads to the formation of separated electron/hole pairs, which are called "excitons". The formation of excitons is the reason for the photosemiconductor properties of TiO_2. The photoactivity of titanium dioxide pigments is usually not required because the excitons can cause undesired redox reactions. The binder system can be damaged during these reactions. Great efforts are, therefore, being made to reduce the photoactivity of titanium dioxide pigments. On the other hand, the photoactivity of TiO_2 is important when it is used as material for photocatalytic applications.

Figure 3.3 shows the reflectance spectra of rutile and anatase from the UV range at about 250 nm up to the near infrared at about 2500 nm. Compared with rutile there is a slightly enhanced absorption of anatase starting at about 600 nm and continued in the near infrared. Rutile starts to absorb from 1300 nm on. Both spectra have no characteristic absorption bands in the near infrared.

Titanium dioxide belongs to oxides with amphoteric character. It is chemically very stable and insoluble in water and in organic reagents. Dissolution takes place in

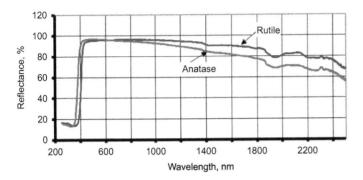

Fig. 3.3: UV/VIS/NIR spectra of rutile and anatase pigment pellets [2].

hot concentrated sulfuric acid, hydrofluoric acid, and hot alkaline solutions. Titanium dioxide is decomposed by potassium hydrogen sulfate (acid digestion). This is the usual way to obtain TiO_2 in dissolved form. Solved alkali metal titanates and free titanic acids are, however, unstable in aqueous systems. They tend to form amorphous titanium dioxide hydrates when reacting with water.

TiO_2 reacts at high temperatures with reducing agents, e.g., with hydrogen or ammonia, to form titanium suboxides (TiO, Ti_2O_3, Ti_3O_5, Ti_4O_7). Titanium dioxide reacts with chlorine above 700 °C in the presence of coke to form titanium tetrachloride ($TiCl_4$). The reduction of TiO_2 to metallic titanium does not succeed even under strongly reducing conditions. In the presence of carbon, nitrogen, or sulfur, titanium carbides, titanium nitrides, and titanium sulfides are formed. Even with hydrogen, the reduction is only successful up to divalent titanium compounds, e.g., TiO. Only when TiO_2 reacts with silicon, calcium, or magnesium at high temperatures, elemental titanium is formed together with the oxides of the metals involved.

When titanium dioxide is irradiated with UV light in the presence of reducing agents such as glycerol or $SnCl_2$ and in the absence of atmospheric oxygen, it turns blue-gray. Any differences that occur in the tendency to gray can serve as an indicator of the photostability of the different TiO_2 samples. Even without the presence of a reducing agent, anatase pigments of lower purity, in particular, show slight graying when exposed to intense light. This behavior is referred to as phototropy.

The specific surface area of commercially available TiO_2 powders can vary from 0.5 to more than 300 m^2/g, depending on the manufacturing conditions. Coordinatively bonded water in the form of hydroxyl groups are located at the surface of the TiO_2 particles. Depending on the nature of the bonding of the hydroxyl species to the titanium atoms, these groups possess acidic or basic properties [14, 15]. The surface of titanium dioxide particles, therefore, always has a polar character. Covering of the surface with hydroxyl groups has a decisive influence on the properties of TiO_2 pigments, e.g., on dispersibility and weather resistance. Photochemically induced reactions of titanium dioxide are strongly related with the presence of hydroxyl groups at the surface of the particles, examples include the decomposition of water into hydrogen and oxygen and the reduction of nitrogen to ammonia and hydrazine [16].

Titanium dioxide as a white pigment is of outstanding importance. Its scattering behavior is unsurpassed compared to all other white pigments. This extremely relevant property, together with the lack of absorption in the visible range, chemical stability, and nontoxicity, is the reason why TiO_2 is the most commonly used pigment.

3.2.2 Production of titanium dioxide pigments

Two main processes are industrially used for the production of high-quality titanium dioxide pigments with controlled mean particle size, particle size distribution, particle shape, crystal structure (anatase, rutile), and composition:
1. sulfate process
2. chloride process

Inorganic and organic surface treatment of the titanium dioxide pigments produced plays an important role in achieving the required final application properties.

The sulfate process, as the older one of the two production processes, starts with a digestion reaction of the titanium-containing raw material (ilmenite or titanium slag) with concentrated sulfuric acid at 170 to 220 °C, resulting in "black liquor". Relatively pure titanium dioxide hydrate (hydrated titanium dioxide) is precipitated by hydrolysis of the black sulfate solution, which, besides titanyl sulfate and iron sulfate, contains colored heavy metal sulfates, sometimes in high concentrations. Purification stages are installed to remove the iron and the other coloring metal impurities. The titanium dioxide hydrate is calcined to form TiO_2, which is ground and coated with a suitable surface treatment (aftertreatment).

The chloride process uses predominantly natural rutile or synthetic rutile (from ilmenite or leucoxene) as raw materials. These are chlorinated in the presence of coke in the temperature range from 700 to 1200 °C. The resulting $TiCl_4$ is purified by distillation and oxidized at temperatures of 900 to 1400 °C to form TiO_2. The so-obtained raw white pigment is ground and equipped with an appropriate surface treatment.

3.2.2.1 Raw materials

The raw materials for the production of titanium dioxide pigments include not only natural products such as ilmenite, leucoxene, and rutile, but also synthetic materials such as titanium slag and synthetic rutile.

3.2.2.1.1 Natural raw materials

Titanium, as the ninth most abundant element on the Earth's crust, occurs naturally in oxide-containing minerals. Important titanium minerals are shown in Table 3.2. Only ilmenite, rutile and leucoxene, a weathering product of ilmenite, are of economic importance for the production of titanium dioxide pigments. Anatase and titanomagnetite reserves are the largest in the world, but they are usually difficult to process. By far, the largest part of the mined ilmenite and rutile is used for the manufacture of TiO_2 pigments. The remaining part of these mined minerals is used for the production of titanium metal and for welding electrodes.

Tab. 3.2: Titanium minerals [1].

Mineral	Formula	TiO_2 content (wt%)
Rutile	TiO_2	92–98
Anatase	TiO_2	90–95
Brookite	TiO_2	90–100
Ilmenite	$FeTiO_3$	35–60
Leucoxenes	$Fe_2O_3TiO_2$	60–90
Perovskite	$CaTiO_3$	40–60
Sphene (titanite)	$CaTiSiO_5$	30–42
Titanomagnetite	$Fe(Ti)Fe_2O_4$	2–20

Ilmenite ($FeTiO_3$) occurs worldwide in primary massive ore deposits or as secondary alluvial deposits (sands). Easily accessible ilmenite ore deposits exist in Canada, USA, Brazil, Russia, Norway, Australia, South Africa, and India. All ilmenite sources contain significant amounts of other metals besides titanium and iron. Ilmenite from massive ores is often associated with intermediary intrusions.

The use of ilmenite from beach sands in existing or fossil coastlines for TiO_2 production is possible because the action of surf, currents, and wind has led to the enrichment of ilmenite and other heavy minerals (e.g., rutile, zircon, monazite, and other silicates) on dunes or beaches. A consequence of this enrichment process is the layering of the minerals. Interaction of seawater and air with minerals leads to corrosion of the ilmenite over geological periods of time. Iron is partly removed from the $FeTiO_3$ lattice with the result that TiO_2 is enriched in the remaining material. The ilmenite lattice is stable with TiO_2 contents of up to 65%. Further removal of iron results in the formation of a submicroscopic mixture of mineral phases, which may contain anatase, rutile, and amorphous components.

Mixtures with TiO_2 contents of about 90% are known as leucoxenes. These are typically formed in corroded ilmenite and are separately mined and treated in some deposits. The removed leucoxene quantities are small in comparison to those of ilmenite.

Concentrates from ilmenite sands with reduced iron contents are generally richer in TiO_2 than those from the massive deposits. These concentrates contain larger quantities of magnesium, manganese, vanadium, aluminum, calcium, chromium, and silicon that originate from mineral intrusions.

Rutile ores were primarily formed by the crystallization of magma with high titanium and low iron contents. Another way was the metamorphosis of titanium-bearing sediments or magmatites. The rutile content in primary rocks is too low to be used commercially. As a consequence, only sands in which rutile is accompanied by zircon, ilmenite, and other heavy minerals can be regarded as deposits. The largest producers of natural rutile are located in Australia, South Africa, and Sierra Leone. The quantities of mineable rutile are limited and too low to fulfill all volume demands. Synthetic rutile is, therefore, used in part as a substitute for the natural variety [1].

The preparation of the titanium-containing ores for the production of TiO_2 pigments starts mostly from heavy mineral sands. Due to the association of ilmenite with rutile and zircon in the raw material, ilmenite production is linked to the recovery of these minerals. Usually, raw sands containing 3% to 10% heavy minerals are obtained, wherever possible, by wet dredging. After sieving, the raw sand is subjected to gravity concentration in several stages with Reichert cones and/or spirals to give a product containing 90% to 98% heavy minerals. The heavy minerals are separated by this procedure from the lighter ones [17].

The magnetic components, particularly ilmenite, are then separated from the nonmagnetic parts, especially from rutile, zircon, and leucoxenes, by dry or wet magnetic separation. The separation of harmful, electrically nonconducting mineral impurities (granite, silicates, and phosphates) from the conducting ilmenite is achieved in an electrostatic separation step. The nonmagnetic components are subjected to further hydromechanical processing to remove the remaining low-density minerals, mostly quartz. In the next process step, the weakly magnetic weathered ilmenites and leucoxenes are recovered by high-intensity magnetic separation. Rutile is then separated from zircon electrostatically based on the different conductivities of the two materials. Finally, the residual quartz is removed by an air blast [1].

3.2.2.1.2 Synthetic raw materials

Synthetic raw materials with a high TiO_2 content have gained a great significance for the production of titanium dioxide pigments. Titanium slag for the sulfate process and synthetic rutile for the chloride process are the two materials produced, starting from ilmenite and leucoxene ores. Processes used for the manufacture of these synthetic materials are focused on the removal of iron.

The production for titanium slag is based on a metallurgical process in which the iron from ilmenite is reduced by anthracite or coke to iron metal at 1200 to 1600 °C in an electric arc furnace, followed by separation (Sorel process). Titanium slag containing 70% to 85% TiO_2 (depending on the ore used) is formed together with titanium-free pig iron. The slag can be digested with sulfuric acid because it has high Ti^{3+} and low carbon content. Titanium slags of this type are primarily produced in Canada, South Africa, and Norway.

Only a minor number of natural rutile deposits exist that can be economically mined and subsequently used for the production of TiO_2 pigments. The price of natural rutile is correspondingly high. Consequently, processes were developed for the manufacture of synthetic rutile. The most important producers of synthetic rutile are located in Australia, India, and Malaysia.

The core objective here is to remove iron from ilmenite (Benilite process) or leucoxene (Becher process), without changing the grain size of the minerals, because this is highly suitable for the subsequent fluidized bed chlorination process. The processes involve the reduction of Fe^{3+} with carbon or hydrogen. In some executions of the Benilite process (often misspelt as Benelite process), ilmenite is preliminarily activated by oxidation before the reduction process takes place. The reducing conditions decide whether Fe^{2+} in an activated ilmenite lattice is formed or metallic iron is generated. Although, there were always attempts to improve and to optimize the processes, the best known are still the Becher and the Benilite processes [18, 19].

Activated Fe^{2+}-containing ilmenite can be treated with hydrochloric or diluted sulfuric acid (preferably under pressure) to form "synthetic rutile" with a TiO_2 content of 85% to 96% [20]. The resulting iron(II) salt-containing solutions are concentrated and then thermally decomposed to form iron oxide and hydrochloric acid. The latter can be used again in the digestion process [21]. Metallic iron, when, formed can be removed in various physical and chemical processes such as magnetic separation and flotation, dissolution in acids, oxidation, chlorination, or reaction with carbon monoxide (formation of iron carbonyl) [1].

If leucoxenes are used as raw material (Becher process), these are also initially oxidized, followed by a reduction process. Thus, pores and cracks on the surface of the ores are generated. Simultaneously, the iron is transferred into the elemental form, from where it is oxidized (utilizing an ammonium chloride catalyst) and brought into solution in the following step. This results in a titanium raw material with a TiO_2 content of 90% to 92%. The advantage of this process is that the iron is formed as iron oxide and not as chloride, which is difficult to utilize. Recent developments for the concentration of titanium ores are aimed for the optimization of the processes with regard to energy saving and environment protection.

It has to be noted that the two main production processes for TiO_2 pigments, the sulfate and the chloride processes, do not compete with one another for the same raw materials, because they have different requirements. In general, the chloride process is reliant on raw materials with higher titanium contents because all metal atoms are transferred from the oxide into the chloride. This leads to higher chlorine consumption and to higher generation of waste materials. The waste products from the chloride process, mainly $FeCl_2$, are more problematic than those from the sulfate process because of missing the recycling possibilities. Overall, the chloride process offers tighter product control, is less labor intensive, and is environmentally safer. It is estimated that currently about 60% of the worldwide TiO_2 pigment production is generated by the chloride process [19].

3.2.2.2 The sulfate process

The sulfate process for the production of titanium dioxide pigments is industrially used since 1916. The process (Figure 3.4) consists of the following steps:
- grinding of the raw materials
- digestion
- re-digestion (dissolution and reduction)
- clarification
- crystallization
- hydrolysis
- purification of the hydrolyzate
- doping of the hydrate
- calcination
- grinding

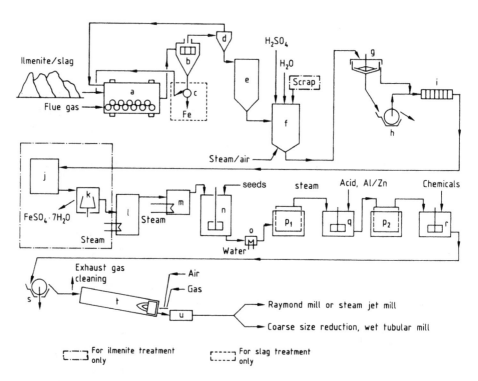

Fig. 3.4: Flow diagram of TiO_2 production by the sulfate process. (a) ball mill/dryer; (b) screen; (c) magnetic separator (optional); (d) cyclone; (e) silo; (f) digestion vessel; (g) thickener; (h) rotary filter; (i) filter press; (j) crystallizer; (k) centrifuge; (l) vacuum evaporator; (m) preheater; (n) stirred tank for hydrolysis; (o) cooler; (p_1) and (p_2) Moore filters; (q) stirred tank for bleaching; (r) stirred tank for doping; (s) rotary filter for dewatering; (t) rotary kiln; (u) cooler [1].

The raw materials, ilmenite or titanium slag from the Sorel process, are dried at the beginning of the process to a moisture content of less than 0.1%. Drying is done in order to prevent heating and premature reaction during the subsequent mixing of ilmenite with sulfuric acid. Before this, however, grinding of the raw materials in ball mills [Figure 3.4(a)] takes place to give a mean particle size of less than 40 µm. The ground material passes through a screen [Figure 3.4(b)], optionally a magnetic separator [Figure 3.4(c)], and cyclones [Figure 3.4(d)] and enters a silo for intermediate storage [Figure 3.4(e)]. Figure 3.5 shows ground ilmenite before it is introduced into the process for the production of TiO_2 pigments.

Fig. 3.5: Ground ilmenite as raw material for the sulfate process.

Digestion is usually employed in batches [Figure 3.4(f)]. The ground raw materials are mixed with 80% to 98% H_2SO_4. It is also possible to use 80% sulfuric acid and start the reaction by a cautious addition of oleum. The reaction can also be started by adding water to a suspension of raw materials in concentrated sulfuric acid. The mixing enthalpy initiates the process in all cases and a vigorous digestion reaction starts reaching temperatures of about 170 to 220 °C. During the digestion reaction, the oxysulfates and the sulfates of titanium, iron, and other metals are formed.

$$FeTiO_3 + 2\ H_2SO_4 \rightarrow TiOSO_4 + FeSO_4 + 2\ H_2O \tag{3.1}$$

The ratio of sulfuric acid to raw materials is chosen in such a way that the weight ratio of free H_2SO_4 to TiO_2 in the suspension produced by the hydrolysis, the so-called "acid number" is between 1.8 and 2.2. If diluted acid or sparingly soluble raw materials are used, external heating is required. The reaction mixture must be maintained to mature for up to 12 h, after the maximum temperature has been reached. The maturing period depends on the raw material so that the titanium-containing components become as soluble as possible. Acceleration of the digestion can be achieved by blowing air through the reaction mixture during the heating process and also during the maturing period.

The re-digestion step of the sulfate process consists of dissolution and reduction. The cake obtained after cooling of the digested reaction mixture is dissolved in cold water or in diluted sulfuric acid, recycled from the process. The temperature is kept below 85 °C to avoid premature hydrolysis. Air is blown into the suspension to keep it moving during the dissolution. In the case of the ilmenite derived product, the TiO_2 concentration of the resulting solution is 8 to 12 wt%. The slag-derived product contains 13 to 18 wt% TiO_2.

Because solved Fe^{3+} hydrolyzes together with Ti^{4+} and accumulates on the surface of the formed titanium dioxide hydrate, it has to be removed prior to the later following hydrolysis step. Therefore, all Fe^{3+} ions derived from ilmenite have to be reduced to Fe^{2+} by the addition of scrap iron during the dissolution or immediately afterwards. Reoxidation of Fe^{2+} ions during the subsequent processing is prevented by the addition of small amounts of Ti^{3+}. The deeply violet colored Ti^{3+} ions give the digestion solution a black appearance. Therefore, and also because of the high viscosity of the solution, it is called "black liquor". Another way to avoid reoxidation of Fe^{2+} is the reduction of Ti^{4+} to Ti^{3+}. This may be performed in a part of the solution under optimized conditions. The thus obtained concentrated Ti^{3+} solution is then added to the reaction solution [22]. Solutions derived from titanium slag need a decrease of the Ti^{3+} content prior to the next steps of the sulfate process to prevent lower yields during hydrolysis. This is achieved by oxidation with atmospheric oxygen.

Digestion with ilmenite – titanium slag mixtures can be carried out as well. The Ti^{3+} content of the slag in this case reduces all the Fe^{3+} ions to Fe^{2+}. It is likewise possible to mix the dissolved products obtained from the separate digestion of ilmenite and titanium slag [23, 24].

The aim of the clarification step is the removal of all undissolved solid material from the solution, as completely as possible. A preliminary settling in a thickener [Figure 3.4(g)] is carried out, followed by filtration of the sediment with a rotary vacuum drum filter [Figure 3.4(h)] or a filter press. The filtrate and the supernatant from the thickener are put through a filter press [Figure 3.4(i)] in order to remove fines. Because of the poor filtering properties of the solution, the rotary filter operates as a precoat filter. The preliminary settling in the thickener is supported by suitable sedimentation-promoting chemicals.

Solutions from the slag digestion contain 5 to 6 wt% $FeSO_4$ and those from the ilmenite digestion, 16 to 20 wt% $FeSO_4$ (after reduction of the Fe^{3+}). For the crystallization step, the ilmenite-derived solution is cooled under vacuum in a crystallizer [Figure 3.4(j)]. $FeSO_4 \cdot 7\,H_2O$ crystallizes under these conditions and can be separated. The amount of $FeSO_4$ discharged with the waste acid is strongly reduced at this stage. The converted concentration for TiO_2 in the solution now amounts to approximately 25 wt%. The crystallized iron(II) sulfate heptahydrate (copperas salt) is separated by filtration or centrifugation [Figure 3.4(k)]. It is used for water purification and as raw material for the production of iron oxide pigments (Section 4.2.4.1). There is also a growing demand for $FeSO_4$ as chromate-substituting substance in cement. Iron(II) sulfate, formerly treated as a waste material, is therefore attaining increasingly more importance as a commercial product, with corresponding effects on the titanium dioxide industry.

Before hydrolysis, the titanyl sulfate solution passes through a vacuum evaporator [Figure 3.4(l)] and a preheating unit [Figure 3.4(m)]. The hydrolysis step is the central part of the sulfate process. Titanium dioxide hydrate ($TiO_2 \cdot x\ H_2O$) is precipitated by diluting the previously received solution with water in a temperature range of 94 to 110 °C.

$$TiOSO_4 + (x+1)H_2O \rightarrow TiO_2 \cdot x\ H_2O + H_2SO_4 \qquad (3.2)$$

Other components of the raw material, which are soluble in sulfuric acid and not yet crystallized together with the iron(II) sulfate heptahydrate, are precipitated simultaneously with titanium dioxide hydrate, mainly niobium oxide hydrate (hydrated niobium oxide). The hydrolysis reaction is performed in brick-lined, stirred tanks [Figure 3.4(n)] into which steam is injected. The generated hydrolyzate, which has the character of a precipitate, does not show any pigment properties. Physical and chemical properties of the hydrolyzate strongly depend on the particle size and the degree of flocculation. The mean primary particle size of the hydrolyzate is about 5 nm, whereas the typical particle sizes of TiO_2 pigments are in the range of 200 to 300 nm.

The hydrolysis of concentrated solutions of titanium sulfate proceeds sluggishly and incompletely unless suitable nuclei are added or formed to accelerate the reaction. The nuclei are usually produced by the Mecklenburg method or by the Blumenfeld method. The Mecklenburg method generates colloidal titanium dioxide hydrate, which is precipitated with sodium hydroxide at 100 °C from titanium oxysulfate solutions. Only 1% of this precipitate is sufficient for nucleation. The Blumenfeld method takes a small part of the sulfate solution for a separate hydrolysis reaction in boiling water. The thus obtained precipitate is added to the bulk solution [25]. The particle size of the hydrolyzate depends on the number of nuclei added.

The following parameters influence the hydrolysis process and, with it, the properties of the formed hydrolysate:

- the agitation (in particular, during the initial stage of the hydrolysis);
- the titanium concentration (this is adjusted, if necessary, by vacuum evaporation, to give a converted TiO_2 content of 170 to 230 g/l during hydrolysis; lower concentrations result in coarser particle size);
- the "acid number" (this should be between 1.8 and 2.2; has a considerable effect on the TiO_2 yield and on the particle size of the precipitate; typical hydrolysis periods are of 3 to 6 h and have a TiO_2 yield of 93% to 96%);
- the concentration of other salts present, especially $FeSO_4$ (high concentrations lead to finely divided hydrolyzates);
- the temperature regime (this mainly affects the volume-time yield and the purity of the hydrolyzate).

After hydrolysis, the titanium dioxide hydrate suspension is cooled down [Figure 3.4(o)]. The liquid phase of the suspension contains 20% to 28% H_2SO_4 and various amounts of dissolved sulfates, depending on the raw material used. The hydrate is filtered off with Moore filters [Figure 3.4(p_1)] and washed with water and diluted acid to remove the undesired heavy metal ions. Washing, however, is still not sufficient to remove all these ions, which are partly adsorbed on the surface of the hydrate particles. The hydrate is, therefore, not directly usable for the production of a white pigment. Most of the remaining impurities can be removed in the next step by reduction (bleaching). The filter cake is slurried for that purpose with 3% to 10% diluted acid at 50 to 90 °C and mixed with zinc or aluminum powder in a stirred tank [Figure 3.4(q)]. Bleaching may also be performed by adding Ti^{3+} ions or strong nonmetallic reducing agents. The bleaching step finally leads to the chemical reduction of the impurities to lower valence states associated with better solubility of the same. Consistently, a second filtration and washing process follows [Figure 3.4(p_2)], after which the hydrate has only very low quantities of colored impurities. However, the titanium dioxide hydrate still contains chemisorbed 5% to 10% H_2SO_4. This acid cannot be removed by washing; it is only driven off by the subsequent calcination step [1].

Titanium dioxide of maximum purity is obtained when the hydrate is calcined without any further additions (dopants). A relatively coarse TiO_2, with a rutile content that depends on the heating temperature, is formed, which is mostly not usable for pigment purposes. For the production of specific pigment grades, however, the hydrate has to be treated in a tank [Figure 3.4(r)] with alkali metal compounds and phosphoric acid as mineralizers (<1%), prior to calcination. Rutile pigments are produced with less phosphoric acid than anatase pigments. Furthermore, production of rutile pigments needs the addition of rutile nuclei (<10%). Zinc oxide, aluminum oxide, and/or antimony(III) oxide (<3%) are also added for certain rutile pigments to stabilize the crystal structure. The latter takes place in such a way that the ions of the metals mentioned above are incorporated in the rutile lattice (preferably on Ti^{4+} sites; in higher concentration, only on interstitial sites) during the subsequent calcination process and from there, contribute to the improvement of the rutile pigments with respect to weather resistance and UV stability.

Rutile nuclei are formed by the reaction of the purified titanium dioxide hydrate with sodium hydroxide. Sodium titanate is formed, which is washed until it is free from sulfates. It is then treated with hydrochloric acid to form rutile nuclei. Such nuclei can also be prepared by precipitation from titanium oxychloride solutions, with sodium hydroxide as precipitant.

The aim of the subsequent calcination step is the conversion of the titanium dioxide hydrate into the final titanium dioxide pigment. The doped hydrate is filtered with rotary vacuum filters [Figure 3.4(s)] to remove water, up to a TiO_2 content of 30% to 40%. The use of pressure rotary filters or automatic filter presses is also suitable. TiO_2 contents of about 50% are possible in these cases. Some of the dopants are soluble in water, and are lost during filtering in the filtrate. They can be complemented by adding new dopant amounts to the filter cake before it is charged into the kiln.

Calcination is carried out in rotary kilns [Figure 3.4(t)], which are directly heated with gas or oil using the countercurrent principle. In the first phase (approximately two-thirds of the residence time, which is typically 7 to 20 h in total), the material is dried. Above 500 °C, sulfur trioxide is released from the sulfate, which partially decomposes to sulfur dioxide and oxygen at higher temperatures. The temperature is increased to a maximum of 800 to 1100 °C, depending on the pigment type required, the desired throughput, and the temperature profile of the kiln. Water is removed from the hydrate during the main reaction, for the formation of titanium dioxide.

$$TiO_2 \cdot x\,H_2O \rightarrow TiO_2 + x\,H_2O \tag{3.3}$$

The set operating program for the kiln controls the rutile content and the rutile/anatase ratio, the mean particle size, the particle size distribution, and the formation of aggregates. The product coming out of the kiln is a TiO_2 clinker, which can be indirectly cooled or directly air-cooled in drum coolers [Figure 3.4(u)]. The temperature of the exhaust gas must be above 300 °C at the exit of the kiln to prevent condensation of sulfuric acid in the ducting. Energy saving during the calcination process is possible by recirculating some of the hot gas to the combustion chamber of the kiln and mixing it with the fuel gases as a partial replacement for air. Figure 3.6 shows TiO_2 clinker as it comes out of the kiln after calcination.

Fig. 3.6: TiO_2 clinker (rutile type) after the calcination step of the sulfate process.

Grinding is the final step of the sulfate process for the production of TiO_2 pigments. Agglomerates and aggregates in the clinker are destroyed to a high degree by wet or dry grinding, and the required pigment fineness is adjusted effectively. Clinker lumps are treated in hammer mills, prior to wet grinding in tube mills. Dispersion agents are added to optimize the grinding process. Centrifugation is the method of choice to remove the coarse fraction from the suspension and return it to the mills. Roller milling is another established technology, which is used in combination with the subsequent deagglomeration by wet milling. Suitable aggregates for dry grinding are hammer mills, cross beater mills, roller mills, and, particularly, pendular and steam jet mills. Additives are often used to achieve wetting of the particles during the grinding process and to improve the dispersibility of the final untreated pigments. An impression of a ground and ready-to-use TiO_2 pigment is given in Figure 3.7.

Fig. 3.7: TiO_2 pigment (rutile type) at the end of the sulfate process.

3.2.2.3 The chloride process

The chloride process for the industrial production of titanium dioxide pigments was developed in the 1950s. The process (Figure 3.8) consists of the following steps:
- chlorination
- gas cooling
- purification of $TiCl_4$
- oxidation of $TiCl_4$ and recovery of TiO_2

The process starts with the chlorination of the raw materials (natural and synthetic rutile). These are converted to titanium tetrachloride in a reducing atmosphere. Calcined petroleum coke acts as the reducing agent. It has a very low ash and a volatile content. The latter is the reason why only very little hydrogen chloride is formed. The reaction of the rutile with chlorine gas and coke, also referred to as carbo-chlorination, proceeds exothermically.

$$TiO_2 + 2\,Cl_2 + C \longrightarrow TiCl_4 + CO_2 \tag{3.4}$$

An endothermic reaction of carbon with the generated carbon dioxide proceeds with the formation of carbon monoxide at rising temperatures to an increasing extent ($C + CO_2 \rightarrow 2\,CO$). The more complex situation during the chlorination in the reactor can, therefore, be described by the following equation:

$$TiO_2 + 2\,Cl_2 + 3\,C + CO_2 \longrightarrow TiCl_4 + 4\,CO \tag{3.5}$$

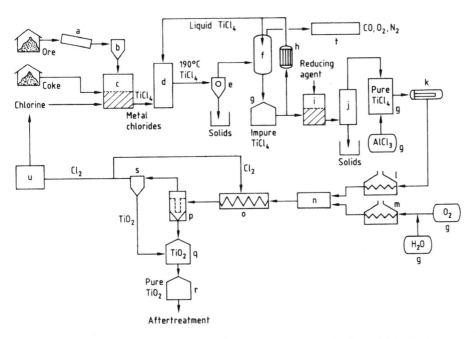

Fig. 3.8: Flow diagram of TiO$_2$ production by the chloride process. (a) mill; (b) silo; (c) fluidized bed reactor; (d) cooling tower; (e) unit for the separation of metal chlorides; (f) TiCl$_4$ condensation unit; (g) tank; (h) cooler; (i) vanadium reduction unit; (j) distillation unit; (k) evaporator; (l) TiCl$_4$ superheater; (m) O$_2$ superheater; (n) combustion reactor; (o) cooling coil; (p) filter; (q) TiO$_2$ purification unit; (r) silo; (s) gas purification unit; (t) waste gas cleaning unit; (u) Cl$_2$ liquefaction unit [1].

The carbon monoxide formed contributes to the withdrawal of oxygen from the rutile.

$$TiO_2 + 2\,Cl_2 + 2\,CO \rightarrow TiCl_4 + 2\,CO_2 \qquad (3.6)$$

Oxygen is blown into the reactor, together with chlorine, to maintain a reaction temperature in the range between 800 and 1200 °C. The coke consumption per ton of TiO$_2$ is 250 to 300 kg when pure chlorine is used. It is also possible to use CO$_2$-containing chlorine gas from the combustion of TiCl$_4$. In this case, 350 to 450 kg of coke per ton of TiO$_2$ are necessary [1].

Carbochlorination is carried out using a fluidized bed process. The ground [Figure 3.8(a)] and stored [Figure 3.8(b)] titanium–containing raw material used has particle sizes comparable with the grains of sand. The particle sizes of the petroleum coke are five times larger. Raw material and coke react with chlorine and oxygen in a brick-lined fluidized bed reactor [Figure 3.8(c)] at 800 to 1200 °C. The materials used must be as dry as possible to avoid the formation of hydrogen chloride; 98% to 100% of the chlorine are converted to TiCl$_4$. The titanium in the raw material reacts to nearly 100%, depending on the reactor design and the gas velocity. Other metals available in the raw material, besides titanium, are also converted into volatile chlorides under the process conditions used, and are found in the reaction gases. Magnesium chloride and calcium chloride, however, tend to accumulate in the reactor due to their low volatility. Zirconium silicate also accumulates because its chlorination occurs only very slowly at the temperatures used during the process. The fluidized-bed reactor is equipped with a ceramic cladding, which is usually worn down rather quickly by abrasion and corrosion. Interruption of the chlorination before the end of the process is problematic because of the possible raw material sintering associated with difficulties in the further fluidization.

Gas cooling of the reaction gases is done with liquid $TiCl_4$, either indirectly or directly [Figure 3.8(d)]. Crystallization of metal chlorides during cooling causes problems because the solid chlorides tend to deposit on the cool surfaces of the equipment. Particularly, larger quantities of iron(II) and iron(III) chlorides formed during chlorination of iron-containing raw materials are separated. The reaction gases are only cooled down in the first phase to a temperature slightly below 300 °C. Under these conditions, the accompanying chlorides can be separated from the volatile $TiCl_4$ by condensation or sublimation [Figure 3.8(e)]. The gas now consists mainly of $TiCl_4$. It is cooled down to a temperature below 0 °C to allow the $TiCl_4$ to condense [Figure 3.8(f)]. Small amounts of titanium tetrachloride and chlorine remaining in the exhaust gas (consisting mainly of carbon dioxide, carbon monoxide and nitrogen) are removed by wet scrubbing with alkaline compounds [Figure 3.8(t)].

Purification of the $TiCl_4$ produced takes place by means of distillation. The remaining metal chlorides and, possibly, entrained dust can easily be removed from $TiCl_4$ in this way. Only the volatile $TiCl_4$ evaporates and can be collected after condensation as a clear and colorless distillate. Dissolved chlorine is removed by reduction with iron, copper, or tin powder or simply by heating [1]. A challenge is the removal of vanadium tetrachloride (VCl_4) and vanadium oxychloride ($VOCl_3$) from $TiCl_4$ by distillation, due to the closeness of their boiling points. The two vanadium compounds are, therefore, reduced to form solid, low valence vanadium chlorides, mainly VCl_3 [Figure 3.8(i)]. Copper, titanium(III) chloride, hydrogen sulfide, hydrocarbons, and amines can be used as convenient reducing agents. After the subsequent distillation [Figure 3.8(j)], titanium tetrachloride should contain only less than 5 ppm vanadium. Larger vanadium quantities lead to a yellow tinge of the TiO_2 produced. Possibly, existing phosgene and $SiCl_4$ can be removed by fractional distillation.

In the following process step, the purified titanium tetrachloride reacts with oxygen at 900 to 1400 °C in a combustion reactor to form TiO_2 and chlorine [Figure 3.8(n)]. The purified $TiCl_4$ is first vaporized [Figure 3.8(k)] and the vapor is indirectly heated in a superheater [Figure 3.8(l)] to temperatures of 500 to 1000 °C. Oxygen is heated in a separate superheater to more than 1000 °C [Figure 3.8(m)]. The reaction can be described according to the following equation:

$$TiCl_4 + O_2 \rightarrow TiO_2 + 2\,Cl_2 \tag{3.7}$$

The resulting TiO_2 already has pigment character. The reaction is moderately exothermic and requires a high reaction temperature. This is why titanium tetrachloride and oxygen must be heated up prior to the reaction. In the case of oxygen, this can be done by using an electric plasma flame (a part of the oxygen reacts there with carbon monoxide) or by indirect heating. Oxygen is introduced with an excess of 110% to 150% with respect to the stoichiometric amount of the reaction. Hot $TiCl_4$ and oxygen are blown separately into the reaction chamber and are mixed there as fast and complete as possible to achieve a high transformation rate. Reactor designs take these requirements into account. It must also be considered that the TiO_2 formed has a strong tendency to cake onto the walls [26–28]. In order to avoid such caking, the material is prevented from sticking to the walls by the addition of abrasive particles, e.g., sand or coarse TiO_2 grains, which can be easily separated from the TiO_2 formed [29, 30]. The cooling unit [Figure 3.8(o)] is constructed in such a way that the pigment is cooled down very quickly to temperatures below 600 °C.

The titanium dioxide formed during the process and the remaining gas mixture consisting of chlorine, oxygen, and carbon dioxide are cooled further during the dry separation of TiO_2, either indirectly or directly by solid particles, mostly by sand. Filtration [Figure 3.8(p)] is used to separate the TiO_2 from the gas mixture and from the sand, which is much coarser than the pigment particles. The gas is returned to the cooling unit [Figure 3.8(o)] of the combustion reactor and also directly to the carbo-chlorination process as oxygen-containing chlorine via the liquefaction unit [Figure 3.8(u)]. Chlorine adsorbed on the surface of the TiO_2 particles can be removed by heating or by flushing with nitrogen or air.

Various process parameters have an influence on the properties of the TiO_2 pigment formed, in particular, on the mean particle size and the particle size distribution. The most important of these parameters are the reaction temperature, the excess of oxygen, and the rheological conditions in the combustion reactor. Each reactor must be designed individually to allow the adjustment of optimal conditions for selected pigment qualities. Caking of TiO_2 on the reactor walls almost always leads to a loss in quality.

Specific nuclei are formed during the combustion of $TiCl_4$ when water vapor and alkali compounds are present in the reaction mixture. These nuclei promote the controlled formation of finely divided TiO_2 particles with high scattering power [31]. The presence of $AlCl_3$ promotes the formation of TiO_2 in the rutile modification. At the same time, a finer pigment is formed. Therefore, $AlCl_3$ is added in quantities of a few percent when rutile pigments are to be produced. Efforts have been made for the fast generation of such rutile nuclei via the direct introduction of $AlCl_3$ vapor into the vaporized $TiCl_4$. The formation of anatase modification in the combustion reactor is promoted by the addition of PCl_3 and $SiCl_4$. Rutile formation is suppressed in this case and anatase pigments are obtained [32].

The combustion of vaporized $TiCl_4$ with oxygen using the conditions described normally results in rutile pigments. The calcination of TiO_2 hydrates, on the other hand, leads either to anatase or to rutile pigments, depending on doping and lattice stabilization. However, many of the TiO_2 pigments produced contain anatase, together with rutile. The ratio of the two TiO_2 modifications in a specific pigment can be adjusted by the reaction conditions. Especially, the calcination temperature can be used to adjust this ratio (as a rule, the higher the calcination temperature, the more rutile is formed).

Most of the TiO_2 pigments are additionally coated with selected metal oxides or hydroxides. Pigments of all grades are also available with an organic treatment. Whether such surface treatments are necessary depends on the specific application. Overall, the variety of TiO_2 pigment grades that are produced and marketed ranges from anatase to rutile types and from untreated to surface treated pigments. Moreover, the diversity of TiO_2 pigments can be attributed to many different requirements for the granulometric properties (mean particle size, particle size distribution, particle shape, and specific surface area).

3.2.2.4 Inorganic and organic surface treatment

Surface treatment (also referred to as aftertreatment or posttreatment) of the pigment particles improves weather resistance and lightfastness of the pigment itself and of the pigmented organic matrix (application medium, application system). It also improves the dispersibility of the pigments in the system chosen for the application [33–36]. Surface treatment is the coating of the individual pigment particles with colorless inorganic and organic compounds of low solubility by depositing them onto the surface. The optical performance of surface-treated pigments, compared to untreated pigments, is thereby only little affected. Suitable coatings on the pigment surface prevent direct contact between the application medium and the reactive surface of the TiO_2 particles. The effect of these coatings depends on their composition and quality. However, the surface treatment can affect the dispersibility of the pigment in a somewhat unfavorable manner, which means that a compromise for the coating often has to be found. High weather resistance as well as good dispersibility of the pigment in the application system are usually desired. These pigment properties can be controlled by the adjustment of suitable coating densities and porosities.

Several preparation routes for inorganic surface treatments are in use:
- Precipitation of the components of surface treatment from aqueous solutions onto the suspended TiO_2 particles. Typically, metal oxides and oxide hydrates are deposited using specific precipitation conditions in a batch process in stirred tanks. Deposition of the treatment components can be proceeded either successively or jointly. A variety of production sequences is used for the generation of sophisticated treatments. The most common compounds used for the coating, besides oxides and oxide hydrates, are silicates and phosphates of titanium, zirconium, and aluminum. Typical reagents used as starting materials for the surface treatment are Na_2SiO_3, $NaAlO_2$, $Al_2(SO_4)_3$, $ZrOSO_4$, and $TiOSO_4$. In some special cases, cerium, tin, vanadium, boron, zinc, manganese, or antimony are used as components of the treatment [37, 38]. In practice, the treatment procedure starts

with the adjustment of the pH value of the TiO_2 suspension. The pH level is chosen in such a way that no precipitation occurs during the addition of the reagents. The pH value is now changed (typically to the neutral point), and the inorganic components of the surface treatment are precipitated in the form of very fine particles, which are deposited on the surface of the suspended TiO_2 [39].

- Adsorption of metal oxides, hydroxides, or other suitable compounds on the surface of the TiO_2 particles during grinding. This process leads mostly to a partial coating of the pigment surface.
- Deposition of the components of the surface treatment from the gas phase by hydrolysis with water vapor or decomposition of volatile substances, e.g., of chlorides or suitable organometallic compounds. This route is used especially for TiO_2 pigments produced under dry conditions – by the chloride process.

TiO_2 pigments with an inorganic surface treatment can be divided into the following groups:

- Pigments with dense surface coatings for paints and plastic applications, prepared by homogeneous precipitation of SiO_2 in aqueous suspension (with precise control of temperature, pH value, and precipitation rate), followed by a drying step [40] by successive precipitation of the treatment components in aqueous suspension, combined with calcination at temperatures of 500 to 800 °C in between the single deposition steps or at the end of the completed surface treatment [41] by aftertreatment with zirconium, titanium, aluminum, and silicon compounds [42, 43], or by aftertreatment with merely 1% to 3% of alumina.
- Pigments with porous surface coatings for the use in emulsion paints prepared by treatment with titanium, aluminum, and silicon compounds (silica content of 10%, TiO_2 content of 80% to 85%).
- Pigments with dense surface coatings and excellent lightfastness for the paper laminate industry, prepared by surface coating of highly lattice-stabilized TiO_2 types with silicates and phosphates of zirconium, titanium, and aluminum (TiO_2 content of about 90%).

Normally, TiO_2 pigments used in plastics have only less than 3% of inorganic surface treatment (TiO_2 content of more than 95%).

The wet chemical coating route for the surface treatment proceeds in such a way that the pigments are washed after precipitation on a rotary vacuum filter or a filter press until they are salt-free. They are then dried in belt dryers, spray dryers, or fluidized bed dryers. For certain TiO_2 types, an additional organic surface treatment is carried out on the already inorganically treated pigment surface. The aim of this treatment is the improvement of pigment dispersibility and the facilitation of the subsequent process steps. The choice of the organic chemistry used here depends on the intended application of the pigment. The pigment surfaces may be equipped either hydrophilically (alcohols, esters, ethers and their polymers, amines, or organic acids are used) or hydrophobically (silicones, organophosphates, and alkyl phthalates are used). In special cases, a combination of hydrophilic and hydrophobic compounds is used on the pigment surface [44].

After all treatments have been applied, the pigments are micronized using air jet or steam jet mills. Problems may occur from the micronization step when the temperature of the pigment powders rises to more than 100 °C. This can cause sticking problems during filling and packaging (especially when polyethylene and polypropylene are used) and also decomposition problems of the organic components of the surface treatment. Direct and indirect cooling of the pigments during micronization can overcome such difficulties.

3.2.2.5 Waste management

Waste management with regard to waste water and waste gas is of great importance for TiO_2 production using the sulfate and the chloride processes.

About 2.4 to 3.5 tons of concentrated sulfuric acid is generated per ton of TiO_2 during the sulfate process. A part of this acid is converted into iron(II) sulfate; the rest is free sulfuric acid [1]. It was common practice in earlier times to dilute this sulfuric acid to so-called "dilute acid" and discharge it directly into the open sea or into coastal waters (dumping of dilute acid). Many controversial discussions concerning this procedure finally led to the decision of the responsible committees of the European Community to stop dumping of dilute acid into open waters by the year 1993.

Producers of TiO_2 have for years been developing various treatment processes for effluents to fulfill the environmental requirements [45]. An established procedure today is the precipitation of gypsum from dilute acid [46]. Another way to use this acid is its concentration and reutilization. The gypsum process starts with the treatment of the acid effluent with finely divided $CaCO_3$. The carbonate is dissolved under the action of the acid, and white gypsum is precipitated. Further process steps are filtering, washing, and drying. The so obtained white gypsum can be used for the manufacture of plasterboard. Residual metal sulfates in the filtrate are precipitated in a second step as metal hydroxides and further gypsum by the addition of calcium hydroxide. The mixture formed is referred to as red gypsum, and can be used to a limited extent for the recultivation of dumps.

The recycling process includes recovering of sulfuric acid and of metal sulfates from the diluted acid. This can be achieved by acid evaporation and by metal sulfate calcination in a suitable furnace. The process consists of two steps:
- concentration and recovery of the free acid by evaporation
- thermal decomposition of the metal sulfates, followed by the production of sulfuric acid from the sulfur dioxide formed

Only acid containing more than 20% H_2SO_4 can be economically recovered by evaporation. The dilute acid is concentrated to about 28% with as little energy effort as possible. This is achieved by using waste heat from sulfuric acid produced by the contact process or from the waste gases of the calcination kilns used for the production of TiO_2 [1]. After the preliminary evaporation, further concentration is achieved in multieffect vacuum evaporators. The water vapor pressure decreases strongly with the increase of the H_2SO_4 concentration. Only a two-stage evaporation procedure can, therefore, effectively exploit the water vapor as a heating medium. Evaporation leads to a suspension of metal sulfates in a 60% to 70% sulfuric acid. Cooling the suspension down to 40 to 60 °C in a series of stirred tanks gives a product with good filtering properties and an acid of suitable quality for recycling to the digestion process [47]. Pressure filters are normally used in the next filtering step to obtain a filter cake with an extremely low residual liquid content [48].

The quality and concentration of the sulfuric acid recycled for the digestion process depend on the titanium-containing raw material. High titanium contents in the raw material lead to acids of 65% to 70%. These are separated from the metal sulfates and further concentrated to give acids of 80% to 87%.

The thermal decomposition of the metal sulfates in the gypsum process leads to the formation of metal oxides, sulfur dioxide, water, and oxygen. The energy consumption is high but the process is reasonable because of the advantageous ecological aspects. Decomposition takes place at temperatures of 850 to 1100 °C in a fluidized bed furnace. Sulfur dioxide produced under these conditions is purified, dried, and converted into sulfuric acid. The acid obtained is pure and can be mixed with the recycled sulfuric acid and then used in the digestion step of the sulfate process of the TiO_2 manufacture.

An environmentally friendly process has also been developed for the recycling of the remaining 5% to 30% sulfate in the acidic wash water [49]. All in all, up to 99% of the sulfuric acid used can be recovered and reused in the sulfate process for the production of TiO_2 [1].

The chloride process also sets requirements for environmental friendliness. In this regard, it is important to solve potential waste water problems, particularly when raw materials with TiO_2 contents of 90% or less are used. Metal chlorides formed during the process as byproducts can be disposed off as metal chloride solutions. Such solutions are pumped through deep boreholes into porous geological layers. There are high demands on suitable geological formations in order to avoid contamination of the groundwater by impurities. Iron(III) chloride – one of the most relevant byproducts – is used for industrial water purification and as flocculation agent.

Problems with waste gas from the chloride process also have to be solved. Gases formed in the calcination kiln are cooled in a heat exchanger. Entrained pigment is removed from the gas, washed, and fed back into process. The SO_2 and SO_3 generated during the calcination step are scrubbed from the other gases and form diluted sulfuric acid, which is recycled [1].

3.2.3 Pigment properties and uses

The TiO$_2$ pigments are characterized by various optical, chemical, and application technical properties. The most important of these properties are scattering power, hiding power (tinting strength), whiteness, dispersibility, lightfastness, and weather resistance. The properties strongly depend on the chemical purity, lattice stabilization, particle size and size distribution, and surface treatment. There is also a significant dependency on the nature of the application system.

The optical behavior of rutile and anatase pigments is determined by their exceptionally high refractive indices, which with 2.70 and 2.55 are even higher than the refractive index of diamond (2.42). The incorporation of a TiO$_2$ pigment in a binder (with a typical refractive index of 1.5 to 1.6) or another application medium leads to a relative refraction coefficient (refractive index of the pigment/refractive index of the binder) of 1.5 to 2.0. The scattering power, which is strongly related to the visible impression of hiding (hiding power), depends on the refractive index of the TiO$_2$ powder and on its particle size. According to Mie theory, the scattering maximum for rutile is at a particle size of 0.19 µm and for anatase, at 0.24 µm (calculated for a wavelength of 550 nm) [50]. The optimal particle sizes for rutile and anatase pigments consistently correspond with these values. The scattering power is also dependent on the wavelength. TiO$_2$ pigments with smaller particle sizes scatter light of shorter wavelengths more strongly. They, therefore, exhibit a slightly bluish tone, while larger particles show a more yellowish shade. A scanning electron micrograph of a rutile pigment is shown in Figure 3.9. The size and a certain sintering of the TiO$_2$ particles caused by the manufacturing process are clearly visible.

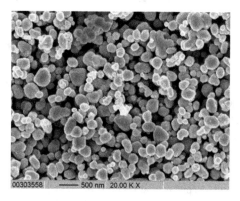

00303558 ——— 500 nm 20.00 K X

Fig. 3.9: Scanning electron micrograph of a rutile pigment.

The whiteness, brightness, and also the mass tone of titanium dioxide pigments mainly depend on the purity, crystal modification, and particle size of TiO$_2$. Because of the mentioned shift of the absorption band of anatase pigments into the UV region (Section 3.2.1), these have a less yellow undertone than rutile pigments. Transition elements in the TiO$_2$ crystal structure have an adverse effect on the whiteness and must, therefore, be avoided in the final products by suitable manufacturing conditions. Pigments produced by the chloride process typically have a higher color purity (neutral hue) and higher brightness values than pigments from the sulfate process. The main reason for this is the effective distillation step in the chloride process, which leads to very pure $TiCl_4$ and, after oxidation, to very white and bright TiO_2 pigments. Purification steps during the sulfate process, on the other hand, are effective enough to produce high-quality pigments with sufficiently good whiteness and brightness properties as well.

The TiO_2 pigments produced, on the one hand, by the sulfate process and, on the other hand, by the chloride process are exchangeable for a wide range of uses. For a variety of applications, sulfate derived pigments offer advantages, e.g., in printing inks, because they are less abrasive than chloride-derived products.

Good dispersion of the TiO_2 particles in the application medium is necessary to obtain high gloss and low haze on the pigmented surface or product. These requirements are normally fulfilled by intensive grinding and by the use of pigments with suitable organic surface treatment. The composition of this treatment depends on the specific application.

Lightfastness and weather resistance of TiO_2 pigments in the application system play an exceptionally important role. Weathering of TiO_2-containing paints and coatings often leads to pigment chalking [51]. No chalking is observed in binder systems with low permeability to oxygen, e.g., in melamine–formaldehyde resins, and when weathering occurs in the absence of oxygen. On the other hand, graying occurs under these conditions, which decreases during exposure to air. Water is another factor that influences graying. This is greatly reduced in the absence of water. Both chalking and graying are more severe with anatase pigments. The higher activity of anatase in the binder, compared to rutile, is a result of the different crystal structures. A suitable stabilizing inorganic surface treatment improves the situation for both pigment types significantly. Doping and lattice stabilization are additional methods to stabilize TiO_2. In practice, these are used especially for rutile pigments.

Problems with regard to lightfastness and weather resistance observed for TiO_2 pigments (anatase as well as rutile) in various binder systems can be explained with the following reaction cycle (chalking cycle) [2, 52]:

1. Water molecules interact with TiO_2 particles. Hydroxyl groups are thereby formed and are bound to the TiO_2 surface.
2. Absorption of UV light by the TiO_2 leads to a charge separation in the crystal lattice. Using the band model, bond electrons are raised to a higher energy level (from valence band to the conduction band) where they move freely. Electrons in the conduction band and positively charged "holes" in the valence band are thus formed.

$$TiO_2 + h\nu \rightarrow TiO_2(e^- + p^+) \tag{3.8}$$

The holes also move freely. They use "hole conduction" for their motion, which means charge exchange with neighboring lattice sites. The separated electron/ hole pairs are called "excitons". They can be understood as chemically very active states of the titanium dioxide particles. Excitons that have migrated to the surface of the pigment interact with the components of the organic binder system. This leads to a degradation of the organic material, which is initially accompanied by a loss of gloss and finally by chalking. The name chalking derives from the fact that in such weathered surfaces, the pigment particles are no longer firmly embedded. They can be removed from such surfaces by simply wiping.

3. The OH^- ions are now oxidized by the positively charged holes to $OH^.$ radicals on the TiO_2 surface.

$$p^+ + OH^- \rightarrow OH^. \tag{3.9}$$

The $OH^.$ radicals are extremely reactive and attack polymer molecules of the binder easily. Usually, there is a separation of hydrogen from these radicals in the first step under the formation of water and polymer radicals $R^.$. These radicals react further, for example, with oxygen to peroxy radicals $R-O-O^.$. They can also disintegrate to other polymer radicals of the general type $R_1^.$ or $R_2^.$. Erasing of holes by the reaction with OH^- ions leads to an excess of electrons in the conduction band of TiO_2.

4. These electrons have the ability to reduce Ti^{4+} in the TiO_2 lattice to Ti^{3+}.

$$Ti^{4+} + e^- \rightarrow Ti^{3+} \tag{3.10}$$

5. In the next step, Ti^{3+} ions can be oxidized again by the existing or newly adsorbed oxygen. Ti^{4+} ions, and oxygen radical anions are formed.

$$Ti^{3+} + O_2 \rightarrow Ti^{4+} + {}^.O_2^- \tag{3.11}$$

The thus generated oxygen radical anions are adsorbed on the TiO_2 surface.

6. They react with moisture under the formation of hydroperoxyl radicals.

$$Ti^{4+} + O_2- + H_2O \rightarrow Ti^{4+} + {}^.O_2H \tag{3.12}$$

Hydroperoxyl radicals are able to desorb from the TiO_2 surface and attack the polymer molecules of the binder. The formed OH^- anions react with TiO_2 to OH groups bound to the surface.

The chalking cycle is thus finished. It can be summarily regarded as the reaction of water and oxygen to form $OH^.$ and $HO_2^.$ radicals under the influence of UV radiation and the catalytic activity of the TiO_2 surface.

$$H_2O + O_2 + hv \rightarrow OH^. + HO_2^. \tag{3.13}$$

The exclusion of air or water interrupts the described cycle after the reduction of Ti^{4+} to Ti^{3+} (3.10). The concentration of Ti^{3+} in the TiO_2 lattice is built up when oxygen is excluded or a binder that has a very low diffusion rate for oxygen is chosen. This leads to graying of the polymer due to the inherent color of the Ti^{3+} ions. Graying is reversible and decreases with gradual exposure to oxygen. The reaction according to (3.12) under formation of surface hydroxyl groups and hydroperoxyl radicals does not take place when water is excluded. The degradation of the binder can, therefore, be stopped by moisture exclusion. The photochemical degradation of the binder is decelerated or nearly prevented with the use of suitable surface treated, doped, and lattice-stabilized rutile pigments. Nonpigmented application media such as coatings are degraded by exposure to light and weathering. The addition of TiO_2 pigments prevents light penetration into deeper areas of the coating film and thus inhibits the degradation, and finally the breakdown of the binder. Highly sophisticated TiO_2 pigments must fulfill very high requirements with respect to weather resistance. Several tests under severe climatic conditions must be passed from coated panels or polymer sheets containing practice-related amounts of TiO_2 pigments. The Florida test, for example, includes a two-year outdoor exposure of pigmented panels in Florida. The panels must not show appreciable chalking or gloss deterioration in this period of time. In order to reduce times and weathering efforts, laboratory tests with respect to photoactivity and weathering behavior of pigments in application systems are used. The influence of UV radiation is tested, for example, with a high-pressure xenon lighting arc lamp or a xenon high-pressure steam lamp, in combination with suitable optical filters. Weathering tests are done in special chambers by a combination of UV radiation, moisture, and temperature treatment. Practice conditions are thereby simulated to be as realistic as possible. After a specified period of time, e.g., after 1000 h, short term weathering results are obtained, which already give a realistic picture of the pigment stability in the chosen application medium. A complete statement on the behavior of a pigmented system in the final application is, however, only possible after suitable outdoor tests.

Titanium dioxide pigments are used universally in all relevant application systems such as paints and coatings, printing inks, plastics, fibers, papers, cosmetics, and ceramics. Paints and coatings constitute the largest part of the use of TiO_2 pigments. Films of paints and coatings pigmented with TiO_2 vary in thickness depending on the application and the quality of the pigment. High-quality TiO_2 grades, when used in a suitable pigment concentration in a coating, allow film thicknesses of only a few micrometers to fully cover the underground. Commercially available pigments enable the paint manufacturer to work with simple dispersion equipment, such as disk dissolvers. The pigment volume concentration of TiO_2 is usually 10% to 35% in gloss paints and coatings, but may exceed 80% in matt emulsion paints.

Most of the printing processes operate with layer thicknesses of less than 10 µm. They, therefore, need TiO_2 pigments of a very small particle size. Such film thicknesses are only possible with TiO_2 pigments that have a high lightening power. Other white pigments, such as lithopone powders, do not fulfill the requirements. The neutral mass tone of TiO_2

pigments is the basis for their application for the brightening of colored pigments. The TiO_2 pigment concentrations in printing inks are normally in the range from 8% to 15%.

Titanium dioxide pigments are widely used in plastics. White coloration of durable and nondurable goods such as toys, appliances, automobiles, furniture, and plastic packaging films are important application fields for TiO_2 pigments. Brightening of colored plastics by the addition of TiO_2 to the color-giving pigment used is another application. The ability of TiO_2 pigments to absorb UV radiation in the wavelength range of less than 415 nm protects pigmented goods against these harmful rays. Pigment concentrations of TiO_2 in plastics and comparable products vary strongly, depending on the polymer and the desired effect.

The use of TiO_2 pigments in synthetic fibers gives a solid appearance. At the same time, it eliminates the greasy appearance caused by the translucent properties of the fibrous material. Anatase pigments are preferred for this application. Their abrasive character is only one quarter of that of rutile pigments, which means that in this case, there is no negative effect on the spinning operation. Lightfastness deficiencies of anatase pigments in polyamide fibers can be overcome by an appropriate surface coating.

TiO_2 pigments are also widely used in paper applications. However, fillers such as kaolin, chalk, or talc are preferred as brightening materials and opacifiers in the paper industry. Titanium dioxide pigments are used for very white paper that has to be opaque even when it is very thin, e.g., for air mail or thin printing paper. TiO_2 can be incorporated directly into the paper during paper manufacturing or applied as a coating to give a superior quality ("art" paper). Laminated papers are usually pigmented with extremely lightfast rutile grades before they are impregnated with a melamine-urea resin for use as decorative layers or films.

Titanium dioxide pigments also play a significant role in cosmetic formulations. Especially, anatase pigments are used in cosmetics. TiO_2 in cosmetic products is responsible for white effects and for covering the skin. It shows the best coverage ability amongst all white pigments due to the very high refractive index. The pigment concentration regulates the coverage ability of a cosmetic foundation. Such foundations generally achieve their hue due to inorganic pigments such as titanium dioxide and iron oxides. TiO_2 pigments can be used in all types of cosmetic emulsions, pastes, pencils, and suspensions and are also applied in soaps and toothpastes. Typical decorative applications are day creams and eye shadows. Titanium dioxide pigments are also used as UV absorbers in sunscreen products. The basis for this application is the spectral properties of anatase and rutile. Nanosized rutile pigments with narrow particle size distribution and sufficient dispersibility are mainly used for sun protection purposes.

Due to their nontoxicity, TiO_2 pigments are also applied as food additives. They are approved for food without any quantity restriction (quantum satis) and are particularly used for food coatings, chewing gum, and coloring of medicinal products, e.g., coating of tablets.

Other application areas for TiO_2 pigments include the enamel and ceramic industries, manufacture of white cement, and coloring of rubber and linoleum. The pigments survive the high temperatures used in some of these industries, without color changes. The pigment concentrations vary strongly depending on the specific application.

TiO_2 pigments can be equipped with electric conductivity by an additional coating with conductive oxidic compositions, such as indium tin oxide (ITO) or antimony tin oxide (ATO) [53]. Such pigments have a slight or more intensive gray color, depending on the specific composition. The use of these pigments is common in photosensitive papers for electrophotography as also in coatings and plastics where an antistatic behavior is required.

There are many industrial products requiring titanium dioxide powders with well-defined properties as starting materials. These requirements concern, for example, highly specific surface areas, small particle sizes, a very high reactivity, or a specific purity of TiO_2. Major applications for nonpigmentary TiO_2 are vitreous enamels, glass, glass ceramics, structural ceramics, electroceramics, catalysts and catalyst carriers, welding fluxes, and chemical intermediates, e.g., fluorotitanates.

3.2.4 Toxicology and occupational health

Titanium dioxide pigments are nontoxic unless they are contaminated with heavy metal compounds. They do not exhibit acute toxicity (LD_{50} value rat oral: > 5000 mg/kg, inhalational LC_{50} value rat > 6.82 mg/l, 24 h); TiO_2 pigments are not irritating to the skin or mucous membranes. They are, therefore, not classified as hazardous materials and are not subject to international transport regulations.

Mild irritation of the eyes and the respiratory tract due to mechanical abrasion is possible. Animals that had been fed TiO_2 over a long period of time did not show any indication of titanium uptake [2]. Absorption of finely divided titanium dioxide in the lungs of animals does not lead to TiO_2-related specific carcinogenic effects [54]. The general principle is that inhalation of smaller particles into the lung is more detrimental, compared to inhalation of larger particles. In any case, it is important to avoid the inhalation of pigment dust, no matter whether it is TiO_2 or any other pigment. No specific effects of TiO_2, resulting in inflammation of living beings, are known.

Chronic effects observed during the manufacture and application of titanium dioxide over many years have not been reported. Physical hazards caused by the use of TiO_2 have not been identified.

While the production processes for TiO_2 pigments are submitted to critical discussions from time to time, no adverse effects caused by the use of these pigments on the environment have been found to date; TiO_2 pigments do not show toxicity toward aquatic organisms. They are insoluble and practically inert in the environment.

The question concerning special risks of nanosized TiO_2 pigments (also referred to as transparent pigments, Section 7.5) in manufacture and application must be considered in conjunction with the general evaluation of nanopowders (powders with a substantial proportion of particles with size smaller than 100 nm). It must be borne in mind in this context that TiO_2 nanopigments as well as other nanosized pigments are typically agglomerated and aggregated into larger units, which are not nanomaterials per se. In use, the pigments are dispersed in a liquid matrix, which is mostly transferred into cured products, such as paints, plastics, or printing inks. Once dispersed, there is no potential for inhalation of nanoparticles, which can be regarded as the most critical risk. Consequently, prevention of pigment dust inhalation has a very high priority.

The chemical classification of titanium dioxide, which for many years was considered unproblematic from a regulatory point of view, has been the subject of intense discussion in the last few years. In 2021, the responsible EU authority changed the classification in such a way that several titanium dioxide powder qualities (depending on the particle size) are to be classified as carcinogenic. Mixtures containing titanium dioxide in powder form with 1% or more of particles with aerodynamic diameters ≤10 μm are consequently classified as dangerous [55].

3.3 Zinc sulfide pigments

3.3.1 Fundamentals and properties

Zinc sulfide pigments were developed and first manufactured around 1850 in France. In a patent from that year, pigments, later known as lithpone, were also described for the first time [56]. Zinc sulfide pigments still belong to the most important white pigments, although they have been steadily losing market share since TiO_2 was introduced in the twentieth century. Besides pure ZnS pigments (C.I. Pigment White 7), pigment mixtures consisting of zinc sulfide and barium sulfate (C.I. Pigment White 5) are also available. The latter are on the market under the name lithopone. Lithopone pigments have a much larger sales volume than pure zinc sulfide pigments.

Zinc sulfide pigments are used not only where good light scattering behavior is required, but also where properties such as low abrasion, low oil number, or low Mohs hardness are needed [57–61]. In some cases, industrial zinc-containing effluents can be used for the production of zinc sulfide pigments. This is a suitable way to reuse zinc from wastes and to take care of the environment at the same time.

Selected physical and chemical properties of ZnS and $BaSO_4$ are summarized in Table 3.3.

Tab. 3.3: Properties of the components of zinc sulfide pigments [60].

Property	Zinc sulfide	Barium sulfate*
Physical properties		
Refractive index n	2.37	1.64
Density, g/cm^3	4.08	4.48
Mohs hardness	3	3.5
Solubility in water (18 °C), wt%	6.5×10^{-5}	2.5×10^{-4}
Chemical properties		
Resistance to acids/bases	Soluble in strong acids	Insoluble
Resistance to organic solvents	Insoluble	Insoluble

*Component of lithopone

A necessary requirement for white pigments is that they do not absorb light in the visible spectral range. In addition, they should scatter incident light of this range as completely as possible. The spectral properties of zinc sulfide (Co-doped) and barium sulfate fulfill these demands to a large extent (Figure 3.10). The absorption band for ZnS at around 700 nm is caused by the intended lattice stabilization with cobalt ions. The absorption edge in the near-UV range is responsible for the bluish-white tinge of zinc sulfide. ZnS crystallizes in the sphalerite (zinc blende) or wurtzite lattice, depending on the specific production parameters.

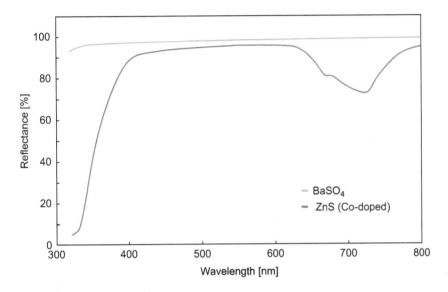

Fig. 3.10: Reflectance spectra of barium sulfate and zinc sulfide (Co-doped) [60].

The refractive index of zinc sulfide is 2.37, which is decisive for the scattering properties; it is much higher than that of binders or other application media (1.5 to 1.6). The maximum scattering power of spherical ZnS particles is achieved when their diameters are in the range of 290 to 300 nm. The contribution of barium sulfate to the light scattering phenomenon is only small due to its relatively low refractive index of 1.64, which is comparable with that of the binder. $BaSO_4$ acts as a filler and increases the scattering efficiency of zinc sulfide.

Pure zinc sulfide and lithopone pigments are thermally stable in air up to temperatures of about 550 °C. Their hardness is low and their texture is soft, compared with other white pigments, which explains that they are less abrasive. Zinc sulfide is stable in aqueous systems in the pH range of 4 to 10 and largely inert in organic solvents. Oxidative decomposition of ZnS, under irradiation of UV light, takes place when oxygen and water are present. Barium sulfate is chemically almost inert. It is insoluble in acids, bases, and organic media.

3.3.2 Production of zinc sulfide pigments

3.3.2.1 Raw materials

Zinc sulfide pigments are synthesized starting from the following raw materials: zinc oxide and zinc slags from smelters, ammonium chloride slags from hot dip galvanizing, and liquid zinc-containing waste such as pickle liquors from galvanizing plants.

The production of barium sulfate powders starts with the preparation of barium sulfide of sufficient purity. Water-soluble BaS is produced by the reaction of an intimate mixture of crushed natural barite, which has a low silica and strontium content, with petroleum coke at temperatures of 1000 to 1300 °C in a rotary kiln.

$$BaSO_4 + 2C \rightarrow BaS + 2CO_2 \tag{3.14}$$

The product formed is dissolved in warm water and the so obtained solution is adjusted to a concentration of ca. 200 g/l BaS. The solution is now filtered and immediately pumped into the lithopone precipitation vessel. Insoluble unreacted components are collected in the filter cake. The almost clear barium sulfide solution can be stored only for a short period of time. Longer storage leads to the undesired formation of polysulfides.

Barium sulfide of high purity can be easily synthesized in smaller quantities by annealing of barium carbonate in a stream of hydrogen sulphide at 1000 °C.

$$BaCO_3 + H_2S \rightarrow BaS + H_2O + CO_2 \tag{3.15}$$

3.3.2.2 Pure zinc sulfide pigments

At the beginning of the pigment synthesis, the raw materials are dissolved in acids, e.g., in sulfuric acid, to form zinc-containing solutions. Impurities in the solutions, which could affect the final white pigments coloristically in a negative manner have to be separated. Disturbing impurities include the salts of iron, nickel, chromium, manganese, silver, cadmium, and other heavy metals. The amounts of impurities depend on the composition of the raw materials. The necessary purification starts with the precipitation of iron and manganese as oxide hydroxides and of cobalt, nickel, and cadmium as hydroxides. The next step is the addition of zinc dust to the solution at 80 °C. All metals that are above zinc in the electrochemical voltage series, e.g., chromium, cadmium, indium, thallium, cobalt, nickel, lead, iron, copper, and silver, are almost reduced and completely precipitated under these conditions, while zinc dissolves. The resulting metal slime is filtered off and recycled. Small amounts of a soluble cobalt salt are now added to the so-obtained purified zinc salt solution in order to stabilize the ZnS that should be formed by the incorporation of cobalt ions (0.02% to 0.5%). The incorporation of cobalt ions into the ZnS crystal lattice leads to significant stabilization against light. Zinc sulfide pigments without such a treatment become gray when exposed to sunlight.

The central part of the pigment synthesis is the next reaction of the purified and cobalt-containing zinc salt solution with a highly purified sodium sulfide solution under suitable reaction conditions (pH value, temperature, concentration, etc.). Finely precipitated ZnS is formed during this reaction.

$$ZnSO_4 + Na_2S \rightarrow ZnS + Na_2SO_4 \qquad (3.16)$$

The white precipitate obtained is separated from the suspension by filtration. It is washed until it is free of sulfate, dried, and then annealed at 700 °C to obtain the optimal narrow particle size distribution of about 290 to 300 nm and, therefore, the best scattering power for the pigment. Cobalt ions are added to the zinc salt solution before the precipitation step and are incorporated during the annealing process. An impression of a pure zinc sulfide pigment is given in Figure 3.11.

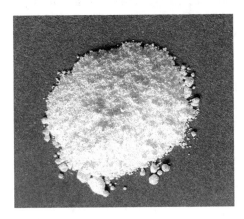

Fig. 3.11: Photograph of a pure zinc sulfide pigment.

Zinc sulfide pigments can also be produced using a hydrothermal process. In this case, the zinc sulfide is precipitated with a slight excess of sulfide at a pH value of 8.5. After pH adjustment to a value of 12 to 13 and addition of sodium carbonate, the suspension is autoclaved for 15 to 20 min at 250 to 300 °C. ZnS generated by the hydrothermal route crystallizes in the sphalerite structure, unlike the product formed by precipitation and calcination. The latter has a wurtzite structure with about 10% less scattering power. In general, ZnS pigments manufactured by the hydrothermal process have a higher quality. The pigment synthesis by using hydrothermal conditions, however, is less economical due to the higher production costs [60].

3.3.2.3 Lithopone

The process for the production of lithopone pigments is based on the reaction of a soluble zinc salt with barium sulfide in aqueous solution; $ZnSO_4$ is the preferred zinc salt, which is obtained by the reaction of the above described raw materials with sulfuric acid. The reaction of equimolar quantities of $ZnSO_4$ and BaS leads to a white water insoluble coprecipitate with the theoretical composition 29.4 wt% ZnS and 70.6 wt% $BaSO_4$.

$$ZnSO_4 + BaS \longrightarrow ZnS + BaSO_4 \qquad (3.17)$$

The 1:1 composition of the precipitate can be changed by the use of different molar ratios. As an example, a starting solution containing $ZnSO_4$, $ZnCl_2$, and BaS in the molar ratio 1:3:4 leads to a coprecipitate with the composition 62.5 wt% ZnS and 37.5 wt% $BaSO_4$.

$$ZnSO_4 + 3\,ZnCl_2 + 4\,BaS \longrightarrow 4\,ZnS + BaSO_4 + 3\,BaCl_2 \qquad (3.18)$$

The production process for lithopone pigments is shown in Figure 3.12. The purified solutions of the zinc salt and of barium sulfide are mixed thoroughly in a precipitation reactor [Figure 3.12(a)]. Parameters such as temperature, pH value, salt concentration, stirring speed, and reactor geometry have to be controlled carefully. Precipitated "raw lithopone", consisting of very fine ZnS and $BaSO_4$ crystallites, is formed. The precipitate has no pigment properties yet. It is filtered [Figure 3.12(b_1)], washed, and dried using a turbo dryer [Figure 3.12(c_1)]. The filter cake is crushed and the obtained lumps of ca. 2 cm size are calcined in a rotary kiln, which is directly heated up with natural gas to temperatures of 650 to 700 °C [Figure 3.12(d)]. Crystal growth during calcination is adjusted by traces of sodium, potassium, and magnesium salts. Temperature profile and residence time in the kiln are controlled to obtain the desired particle sizes for ZnS and $BaSO_4$. In particular, the ZnS particles should have sizes in the optimal range of 300 nm. The size of the $BaSO_4$ particles achieves values of up to 1 µm.

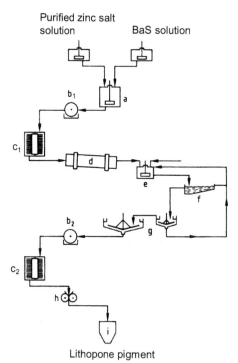

Fig. 3.12: Flow diagram for lithopone production: (a) precipitation vessel; (b_1) and (b_2) rotary filters; (c_1) and (c_2) turbo dryers; (d) rotary kiln; (e) chilling vessel; (f) rake classifier; (g) thickener; (h) grinder; (i) silo [60].

The hot $ZnS/BaSO_4$ mixture is quenched after calcination in water [Figure 3.12(e)]. The product must now pass classifiers and hydro-separators [Figure 3.12(f)], a thickener unit [Figure 3.12(g)], and rotary filters [Figure 3.12(b_2)]. It is washed on the filter until it is free of salts and then dried [Figure 3.12(c_2)]. The dry product is ground in high-intensity mills [Figure 3.12(h)]. The so obtained pigment can be equipped with an inorganic or organic surface treatment, depending on the intended use [60]. Figure 3.13 shows an example of a lithopone pigment at the end of the manufacturing process.

Fig. 3.13: Photograph of a lithopone pigment.

3.3.3 Pigment properties and uses

A scanning electron micrograph of a pure zinc sulfide pigment is shown in Figure 3.14. The size of the ZnS particles varies between 100 and 400 nm. A larger quantity of the particles has diameters close to the optimal value of 300 nm. Figure 3.15 shows a scanning electron micrograph of a lithopone pigment. The particle size distribution is comparable with that of the pure zinc sulfide pigment. The larger proportion of small particles is due to the $BaSO_4$ content in the pigment.

00302743 ———— 300 nm 50.00 K X

Fig. 3.14: Scanning electron micrograph of a pure zinc sulfide pigment.

Fig. 3.15: Scanning electron micrograph of a lithopone pigment.

Commercially available lithopone pigments contain different amounts of zinc sulfide. The most common types have a ZnS content of 30% (red seal) or 60% (silver seal). Other qualities are of minor importance. They contain 15% (yellow seal), 40% (green seal), or 50% (bronze seal) ZnS. A specialty is Sachtolith, which has ZnS content above 97%. Pure zinc sulfide pigments belong to this category. Various surface treatments, consisting of inorganic and organic compounds, and of hydrophilic or hydrophobic nature, are used to provide zinc sulfide pigments for common and special applications.

Lithopone pigments are mainly used in applications where relatively high pigment concentrations are needed. Primers, plastics, fillers, putties, artists' colors, and emulsion paints are examples of the use of lithopone. Lithopone pigments do require only low binder amounts. Therefore, especially paint products containing lithopone have favorable flow and application properties. Lithopone pigments are characterized by a good wetting and dispersion behavior. Good dispersion can be achieved by the action of a dissolver. The combined application of lithopone and TiO_2 pigments can be technologically and economically advantageous. The high hiding power contributed by the TiO_2 pigment makes such a combination particularly attractive.

Lithopone is very useful as a white pigment for UV-cured paint systems. The reason for this is the strong shift of the absorption band toward the blue caused by ZnS. Zinc compounds, and, therefore, also pure zinc sulfide and lithopone pigments, have fungicidal and algicidal effects. Their application in paint formulations for exterior use, therefore, prevents the growth of algae or fungi on the hardened paint surface.

Lithopone pigments are widely used in plastics, in particular because of their good lightfastness and the clear bluish-white shade. Their advantageous behavior in liquid plastic materials during the extrusion process results in high throughput rates and economical operation of the extruder equipment. Lithopone pigments are also used in flame protection. In fire resistant systems, they can replace up to 50% of the toxic flame retardant antimony trioxide without any adverse effect [60].

Pigments of the Sachtolith type with very high ZnS contents are mainly used in plastics. Favorable properties such as good lightening and high hiding power are criteria for their use. They are also very useful for the coloring of many thermoplastics and duroplastics. The pigments have lubricating properties in polymers and do not contribute to the abrasion of the metallic processing tools and molds. They do not affect the polymer in an unfavorable manner, even at relatively high operating temperatures. Coloring of ultra-high-molecular-weight thermoplastics is equally simple. The soft texture of Sachtolith pigments also prevent the damage of glass fibers during the manufacturing process of reinforced plastics containing these.

The lifetime of stamping tools applied in the production of rubber articles is extended by the use of Sachtolith. The reason for this is once again the low abrasiveness of the pigments. The lightfastness and long term stability of many elastomers are improved when Sachtolith is used in a suitable concentration as a component of the formulation. The pigments are also used for coloring of greases and oils and as dry lubricants for roller and plain bearings.

The unsurpassed properties of titanium dioxide have led to a decrease in the quantity of zinc sulfide pigments in the last decades. A replacement of ZnS by TiO_2 pigments may be beneficial, especially in coating materials. The high-quality ZnS types, however, will most probably maintain their importance for the market. They will retain their importance among the white pigments where special technical properties, in addition to light scattering, are required. The changing regulations for titanium dioxide pigments [55] could lead to an increase in the quantities of zinc oxide pigments used in the future.

3.3.4 Toxicology and occupational health

Zinc sulfide pigments are nontoxic unless they are contaminated with heavy metal compounds. They do not exhibit acute toxicity (LD_{50} value rat oral: > 5000 mg/kg). Studies on rats, which had zinc sulfide pigments administered, only showed extremely low toxicity. Inhalation of pigment dust may cause mechanical respiratory tract irritation. Contact with skin or eyes can lead to irritation by mechanical friction [62–66].

The effects of zinc ions on the organism are very complex. Soluble zinc in large quantities is toxic. However, the human body contains 2 g of zinc and requires daily quantities of 10 to 20 mg for metabolic processes. The low solubility of zinc sulfide is the reason why it is harmless in the human body. The concentration of stomach acid and the rate of zinc sulfide dissolution are not sufficient to generate physiologically significant amounts of soluble zinc. Toxic effects attributed to zinc in earlier times were caused by contamination with other metals. No cases of poisoning or chronic health damages have been reported from the manufacture and processing of zinc sulfide pigments. The application of both zinc sulfide and barium sulfate in contact with foods is permitted by the FDA (United States) and by most relevant European authorities.

Thermal treatment of zinc sulfide pigments at temperatures above 570 °C in oxidizing atmosphere leads to the formation of ZnO and SO_2. BaS is formed when lithopone pigments are heated up to temperatures above 875 °C in a reducing flame. Measureable quantities of H_2S are developed at pH values below 2.5. Larger H_2S amounts are liberated at pH values lower than 1.5.

An environmental aspect that should be considered for the production of zinc sulfide pigments is the liberation of sulfur dioxide during the reduction of barite and during the calcination of zinc sulfide lithopone, respectively; SO_2 is removed from the remaining waste gas by utilizing the reversible, temperature-dependent solubility of sulfur dioxide in polyglycol. The absorbed SO_2 can be regained in a liquefied form or as gaseous raw material for the production of sulfuric acid. Another aspect is the removal of soluble barium amounts in the residue from the dissolution of the fused BaS by treatment with Na_2SO_4. The resulting $BaSO_4$ precipitate can easily be recycled after filtration.

3.4 Zinc oxide pigments

3.4.1 Fundamentals and properties

Zinc oxide, known in medicine as lana philosophorum since the Middle Ages, is generally believed to have been used as a pigment only around 1780. It can be found in paintings from 1800 onward [67]. Synthetic zinc oxide (ZnO) was developed and first manufactured around 1840 in France by Leclaire (indirect or French process) [68]. One of the early applications was the use as white pigment, known as zinc white (C.I. Pigment White 4). Other terms for zinc oxide pigments are Chinese white, snow white, eternally white, and flowers of zinc. The term zinc white now only applies to zinc oxide pigments produced by the combustion of zinc metal according to the indirect or French process. In nature, zinc oxide occurs in form of the mineral zincite. ZnO crystallizes in the hexagonal wurtzite structure.

Pure zinc oxide is a white powder that turns yellow when heated to temperatures above 300 °C. The color change is reversible, this means the ZnO is white again after cooling (thermochromism). Weak luminescence is observable in the dark after heating up and cooling. The color change from white to yellow during heating is caused by a small excess of zinc (ca. 0.03%) generated by the liberation of oxygen during the thermal treatment. The amount of the zinc excess and the concentration of oxygen defects in the ZnO lattice depends on the synthesis route chosen.

Zinc oxide is a direct semiconductor characterized by a band gap of 3.2 to 3.4 eV. Consequently, it absorbs UV radiation (at wavelengths below 366 nm) and is transparent in visible light. The conductivity of ZnO can be increased by doping with suitable elements. Aluminum and boron are preferred for n-doping of zinc oxide; gallium and indium can also be used. The preparation of p-doped ZnO is much more difficult.

Zinc oxide powders produced by thermal processes are crystalline, have granular or nodular particle shapes, and have particle sizes in the range from 0.1 to 5 µm. Particles with an acicular shape with sizes from 0.5 to 10 µm are possible as well. Wet chemically generated ZnO particles are amorphous, have a sponge form, and have particle sizes of up to 50 µm.

Selected physical and chemical properties of ZnO are summarized in Table 3.4.

Zinc oxide is soluble in organic and inorganic acids. The addition of alkalis, e.g., of sodium hydroxide leads to precipitates of hydrated zinc oxide. These are dissolved

Tab. 3.4: Properties of zinc oxide pigments.

Property	
Physical properties	
Refractive index n	1.95–2.10
Density, g/cm^3	5.61–5.68
Mohs hardness	4.0–4.5
Solubility in water (29 °C), mg/l	1.6
Chemical properties	
Resistance to acids/bases	Soluble in diluted acids and concentrated bases
Resistance to organic solvents	Insoluble

again when an excess of hydroxide is used (formation of soluble zincates). ZnO reacts at high temperatures in a solid-state reaction with other oxides, e.g., with iron or manganese oxides to form ternary oxides such as zinc ferrites.

3.4.2 Production of zinc oxide pigments

Several processes are common for the production of zinc oxide pigments [68–74]. Among these, the direct process (American process), the indirect process (French process), and the precipitation process (wet process) have the greatest importance.

3.4.2.1 Raw materials

Whereas, earlier, mainly zinc ores and concentrates (direct process) or metal from zinc producers (indirect process) were used as raw materials for the industrial synthesis of ZnO pigments, today, zinc oxide manufacturers prefer to use zinc residues and secondary zinc (zinc ashes from hot dip galvanization, zinc concentrates, metallization residues, skimming from casting furnaces, and oxide residues from the indirect process). The production processes had to be modified over time in order to meet the increasing demands regarding chemically pure and user friendly white pigments. Improved purification steps and a sophisticated surface treatment are decisive for the final pigment quality.

3.4.2.2 Direct process (American process)

The direct process or American process is considered as the most simple, cost-effective, and energy-efficient route for the manufacture of zinc oxide pigments. It consists of two steps, the high temperature reduction of a zinc-containing raw material to zinc metal and the oxidation of the zinc formed to zinc oxide by oxygen. The reduction of the raw material, typically containing a zinc oxide amount of 60% to 75%, proceeds with coal as the reducing agent in rotary kilns and at temperatures of 1000 to 1200 °C. Zinc vapor and carbon monoxide are formed.

$$ZnO + C \rightarrow Zn + CO \tag{3.19}$$

The carbon monoxide acts as reducing agent as well, and carbon dioxide is generated as a further reaction gas besides zinc vapor.

$$ZnO + CO \rightarrow Zn + CO_2 \tag{3.20}$$

A part of the zinc vapor and the carbon monoxide is already oxidized to zinc oxide and carbon dioxide above the reaction bed or at the exit of the furnace by oxygen present in the kiln.

Two types of rotary kilns are in use. The first one is heated by gas or oil and has a length of ca. 30 m (a diameter of 2.5 m). The mixture of zinc-containing raw material and coal is continuously charged to the combustion gases using either the counter-current or the cocurrent principle. Unconverted residues containing zinc compounds and unburnt coal leave the kiln continuously. The residual coal is sieved out and re-cycled. The second type of rotary kiln has a length of only 5 m (a diameter of ca. 3 m). Charging with the starting mixtures is continuous as well, but the reaction residues are removed batchwise.

The combustion gases from both types of rotary kiln contain zinc vapor, ZnO, CO, and CO_2. They are passed into a reactor where the zinc and the carbon monoxide are oxidized.

$$Zn + CO + O_2 \rightarrow ZnO + CO_2 \tag{3.21}$$

Impurities are separated at the same time because they do not react in a similar way. The conditions during zinc oxidation are controlled carefully to obtain high yields and to adjust for particle shape and size as required. The reaction gases are cooled after the oxidation step in a heat exchanger or by dilution with air. The zinc oxide is extracted and collected in filters. It can be directly used as a pigment without further processing. However, thermal treatment at temperatures of up to 1000 °C improves the pigment properties of the ZnO. An additional improvement of the pigment proper-ties can be achieved by suitable surface treatment.

3.4.2.3 Indirect process (French process)

The indirect process or French process essentially consists of the vaporization of zinc by boiling, followed by the oxidative combustion to ZnO in air. The combustion condi-tions have to be adjusted carefully to achieve the desired physical and chemical prop-erties of the resulting zinc oxide. The chemical purity of the formed ZnO is primarily a function of the composition of the zinc vapor.

Several furnace types are in use for the production of zinc vapor of the required quality. They are selected according to the raw materials and the highest possible yield. Pure metallic zinc as well as zinc-containing metal residues or secondary zinc are used as raw materials. Various interfering metals, such as aluminum, cadmium, iron, or lead have to be separated by liquid or vapor phase techniques before the zinc is oxidized.

In a first process variant, muffle furnaces or retorts of graphite or silicon carbide are used to vaporize the zinc and/or the zinc-containing metal. Loading of the furnace with the metal takes place either batchwise as a solid or continuously as a liquid. Vaporization in a retort is achieved by heating the retort outside with a burner. If a muffle furnace is used, the vaporizing section is separated from the heating chamber by a silicon carbide arch. The heat from burning gas or oil is transferred to the zinc bath from the arch by radiation. Solid aluminum, iron, lead, and other nonvolatile residues are collected in the muffle furnace or in the retort. They must be removed at given intervals. The vapor contains significant amounts of aluminum, cadmium, copper, iron and lead, which have to be removed before the subsequent zinc oxidation. Fractional distillation in columns with silicon carbide plates can be used to purify the vapor. The purified zinc vapor leaves the columns at an exit and is oxidized with atmospheric oxygen in a combustion chamber connected with the distillation unit.

$$2\,Zn + O_2 \rightarrow 2\,ZnO \tag{3.22}$$

The conditions for the zinc oxide formation have to be controlled carefully. High yields and the required pigment properties can thus be achieved. The zinc oxide formed is extracted after cooling and collected in filters. A direct use of the obtained ZnO powder as a pigment is possible. However, suitable surface treatment improves the pigment properties similar to the zinc oxide derived from the direct process.

In a second variant, a furnace with two separate chambers is used. In the form of smaller or larger pieces, the metallic raw material is fed into the first chamber of the furnace. The metal is heated and melted in this chamber. A second chamber is connected with the first one. Distillation of the molten metal in the absence of air takes place in this electrically heated chamber. Nonmetallic residues are removed from the surface of the melt beforehand. Metallic impurities in the melt, such as aluminum, iron, and lead are collected in the residue from the distillation and are removed from there at intervals. Last traces of lead are separated by fractional distillation. The molten zinc is converted into zinc vapor and then combusted in an oxygen-containing atmosphere to form zinc oxide.

Finally, in a third variant, zinc oxide is produced in a rotary kiln, starting again from zinc or zinc-containing metal. In this case, melting, distillation, and also a part of the oxidation take place in the same zone of the kiln. A large part of the heat resulting from the zinc combustion can be used here for the other process steps. The impurity content as well as the shape and the size of the resulting ZnO particles can be controlled and adjusted by the temperature and the gas atmosphere in the kiln [68].

3.4.2.4 Precipitation process

Zinc oxide pigments can also be produced using a wet chemical route. In the first process step, zinc carbonate is precipitated from purified solutions of zinc sulfate or zinc chloride by addition of Na_2CO_3 solution.

$$ZnSO_4 + Na_2CO_3 \rightarrow 2\,ZnCO_3 + Na_2SO_4 \tag{3.23}$$

It is also formed by the reaction of zinc salts with $KHCO_3$.

$$ZnSO_4 + 2\,KHCO_3 \rightarrow ZnCO_3 + K_2SO_4 + CO_2 + H_2O \tag{3.24}$$

Depending on the reaction conditions, basic zinc carbonate can be precipitated.

$$5\,ZnSO_4 + 5\,Na_2CO_3 + 3\,H_2O \rightarrow 2\,ZnCO_3\ 3\,Zn(OH)_2 + 5\,Na_2SO_4 + 3\,CO_2 \tag{3.25}$$

The choice and concentration of the reactants, pH value, temperature, and other parameters are decisive for the chemical composition and the physical properties of the precipitate generated. The latter is filtered, washed, and finally calcined to form zinc oxide.

$$ZnCO_3 \rightarrow ZnO + CO_2 \tag{3.26}$$

Zinc oxide pigments with a very fine particle size and a high specific surface area are obtained. An impression of a ZnO pigment is given in Figure 3.16.

Fig. 3.16: Photograph of a zinc oxide pigment.

3.4.2.5 Surface treatment

Additional thermal treatment of the zinc oxide powders produced at temperatures of up to 1000 °C can improve the pigment properties of ZnO. Such treatment is mainly applied to zinc oxide produced by the direct method. The photoconductive properties of high purity zinc oxide used in photocopying machines can be improved by controlled calcination in a suitable atmosphere.

Surface treatment of the particles improves the weather resistance, lightfastness, and dispersibility of zinc oxide pigments in most application systems. The surface of the ZnO particles is, therefore, mostly equipped with an organophilic coating generated from oil and propionic acid. Zinc oxide pigments are offered not only as powders but also in the form of pellets and granules. These dosage forms facilitate the handling of the pigments, e.g., by easier dispersing or by dust reduction.

3.4.3 Pigment properties and uses

A scanning electron micrograph of a zinc oxide pigment is shown in Figure 3.17. It is interesting to note that a larger proportion of particles is acicular or cubical corresponding to the crystallographic structure.

00302253 ──── 300 nm 50.00 K X

Fig. 3.17: Scanning electron micrograph of a zinc oxide pigment. A major percentage of the particles is acicular or cubical.

Zinc oxide pigments are mainly used in paints, coatings, and artists' colors. Clear benefits of ZnO, compared with TiO_2 or ZnS pigments, exist only in special applications. Zinc oxide pigments are used as additives in exterior paints for wood preservation. They are also applied in antifouling and anticorrosion paints [75]. ZnO improves film formation and durability of cured films. The application of the pigments in paints and coatings contributes to the avoidance of mildew. The efficacy of ZnO pigments is based on the reaction with acidic oxidation products and on the absorption of UV radiation.

Zinc oxide is not only used for pigment purposes. The most important use is in the rubber industry, where it acts as an activator for vulcanization accelerators in natural and synthetic rubber materials. The specific surface area of a zinc oxide powder correlates with its reactivity. The higher the surface area, the more reactive the ZnO. The reactivity is also influenced by impurities in the zinc oxide, e.g., by lead and sulfate. ZnO also contributes to a good durability of the vulcanized rubber and increases its thermal conductivity. The ZnO content in rubbers is in the range of 2% to 5%.

ZnO powders and medical ointments containing zinc oxide are used in the pharmaceutical and cosmetic industries because of their antibacterial properties. The reaction of ZnO with eugenol is important for the formation of dental cements (zinc oxide eugenol cement) [76, 77]. Zinc oxide is also used in glasses, ceramics, and enamels. These applications are based on the ability of ZnO to reduce thermal expansion, to lower the melting point, and to increase resistance against chemicals. An additional aspect is its ability to modify gloss and improve opacity.

Zinc oxide acts as a raw material for various chemical compounds. Among these are chromates, phosphates, stearates, bromates, ferrites, and organic dithiophosphates. ZnO is also used as an additive in animal food to ensure supply with the essential amounts of zinc. Electrogalvanization and desulfurizing of gases are further applications of zinc oxide. It is also a component in several adhesive compositions. The catalytic potential of ZnO, in combination with other metal oxides, is used for organic syntheses, e.g., for the production of methanol [78–80].

Varistor materials based on zinc oxide are produced by calcination of highly purified ZnO with additives such as Bi_2O_3 [81, 82]. The photoconductive behavior of ZnO is used for photoreproduction purposes. The electrical conductivity of zinc oxide can be increased by doping with aluminum (aluminum doped zinc oxide, AZO) [83].

3.4.4 Toxicology and occupational health

Zinc oxide pigments are nontoxic unless they are contaminated with heavy metal compounds. They do not exhibit acute toxicity (LD_{50} value rat oral: > 5000 mg/kg). ZnO is soluble in acids and can, therefore, be resorbed from living organisms. As an essential element, zinc plays a role in the growth of cells, skin, and hair and in the functioning of the liver. In mammals, zinc deficiency is more problematic than excess zinc. Exposure to zinc oxide mainly takes place by inhalation of dust during manufacture or application. Such inhalation can be avoided by suitable means of protection. Some zinc oxide qualities that are not used for pigment purposes can contain up to 5% lead. They, therefore, require diligent handling to avoid poisoning by inhalation of dust or oral intake. There are national occupational exposure limits for ZnO dust, which should be supervised while handling the pigment. There is no indication that zinc oxide has skin or eye irritating properties. The pigment is used in skin ointments in concentrations of up to 40%. It is obviously resorbed through intact skin only to a very limited extent. There is also no evidence of carcinogenicity, genotoxicity, or reproduction toxicity in human beings [63, 84].

Zinc oxide pigments are almost insoluble in water and can, therefore, be separated in water installations by mechanical means. Zinc ions, however, have to be removed from the waste water by precipitation or flocculation. Whereas zinc is essential for mammals, even very small quantities can be fatal for aquatic organisms. The presence of too much zinc in the water inhibits growth and photosynthesis, and finally leads to the death of organisms. The EC_{50} value for ZnO in water (Selenastrum capricornicum 72 h) has been defined to be 170 µg/l. Zinc oxide is, therefore, consequently classified as dangerous for the environment (EU Classification 67/548/EEC, 29ATP 2004) with the Risk Phrases R50/53 (very toxic to aquatic organisms, may cause long term adverse effects in the aquatic environment) [68].

3.5 Other white pigments

Two additional representatives of white pigments are basic lead carbonate and anti-mony(III) oxide. Both are nowadays only of historical significance as white pigments. They were replaced by titanium dioxide, zinc sulfide, and zinc oxide pigments because of their less favorable properties, in particular also because of their toxicity.

3.5.1 Basic lead carbonate (white lead)

Basic lead carbonate, also referred to as white lead, lead white, Kremser white, Venetian white, or cerusa (C.I. Pigment White 1), has the composition $2\,PbCO_3 \cdot Pb(OH)_2$ or $Pb(OH)_2 \cdot 2\,PbCO_3$. It occurs naturally as hydrocerussite, a hydrated form of cerussite. It was formerly used in various pigment applications, especially in paints. Another use was as a skin whitener (Venetian Ceruse) in cosmetic formulations of former centuries. This application caused lead poisoning, skin damage, and hair loss, and could even lead to death. Basic lead carbonate was also used as lubricant for bearings. White lead was the most important white pigment used in the classical period of European oil painting. It was replaced in artists' paints for a long time by titanium dioxide (titanium white), which has the advantages of much higher tinting strength and nontoxicity.

Earlier methods for the preparation of white lead used metallic lead, which was corroded with vinegar (acetic acid) in the presence of carbon dioxide [85]. The modern route to produce basic lead carbonate pigments is based on the reaction of a soluble lead compound; preferred are lead acetate and lead propionate, with carbon dioxide.

$$3\,Pb(OAc)_2 + 2\,CO_2 + 4\,H_2O \rightarrow Pb(OH)_2 \cdot 2\,PbCO_3 + 6\,HOAc \qquad (3.27)$$

A white precipitate is obtained without the need of sophisticated reaction conditions. The precipitate is filtered, washed, and dried and is ready to use without further processing. Precisely controlled reaction parameters (pH value, temperature, concentration of the lead acetate solution, and CO_2 dosage) allow the synthesis of thin basic lead carbonate platelets, which can be used with restrictions as an effect pigment [86].

Basic lead carbonate pigments have to be handled carefully, considering the dangerous potential derived from their toxicity. Pigment dust can lead to irritation of the eyes, the respiratory tract, and the skin. Poisoning incidents are rare because larger lead amounts are resorbed only very slowly and poorly through the gastrointestinal mucous membrane. Basic lead carbonate is classified as teratogenic and belongs to the water endangering substances.

3.5.2 Antimony(III) oxide (antimony white)

Antimony(III) oxide (Sb_2O_3) is found in nature as the minerals, senarmontite and valentinite. It is the antimony compound with the highest industrial importance. One of the earlier applications was the use as white pigment, known as white antimony, antimony white, or flowers of antimony (C.I. Pigment White 11).

Sb_2O_3 belongs to the amphoteric oxides; it dissolves in aqueous hydroxide solutions under formation of antimonites, such as $NaSbO_2$. It also dissolves in concentrated acids to form the corresponding salts, which hydrolyze upon dilution with water. The reaction of Sb_2O_3 with nitric acid leads to the formation of antimony(V) oxide (Sb_2O_5).

Antimony(III) oxide shows thermochromic properties. Thermal treatment to temperatures above 600 °C leads to a yellow discoloration, which is reversible if cooled down again. The cause of the color change is the polymorphic, enantiotropic transformation of the white cubic crystal form (senarmontite) to the yellow rhombic crystal form (valentinite) at 606 °C.

The main application of Sb_2O_3 today is as flame retardant synergist in combination with halogenated materials. This combined use is the key to the flame retardant action for polymers because it supports the formation of less flammable components. Applications of Sb_2O_3-containing flame retardants are found in electrical devices, textiles, leather, and coatings.

Antimony(III) oxide is also used as an opacifying agent for glasses, ceramics, and enamels. It is also a useful catalyst in the production of polyethylene terephthalate (PET) and the vulcanization of rubber. Antimony-doped tin oxide (ATO) is slightly grayish colored and exhibits usable electrical conductivity. It is, therefore, used for transparent conductive coatings on glasses, e.g., in display technology. It is applied as functional pigment for bright and transparent antistatic coatings [87, 88].

The Sb_2O_3 pigments can be produced starting from elementary antimony or antimony(III) sulfide (stibnite). In the first process, antimony reacts with atmospheric oxygen in a furnace to form antimony(III) oxide.

$$4\,Sb + 3\,O_2 \longrightarrow 2\,Sb_2O_3 \qquad\qquad (3.28)$$

The reaction is exothermic and Sb_2O_3 is sublimated after its formation. Filters are used to collect the freshly formed powder. The particle size of the Sb_2O_3 is controlled by the process conditions, mainly by the gas flow and the temperature.

The second process uses crude stibnite, which reacts in a furnace at temperatures of 850 to 1000 °C with oxygen under formation of antimony(III) oxide.

$$2\,Sb_2S_3 + 9\,O_2 \longrightarrow 2\,Sb_2O_3 + 6\,SO_2 \qquad\qquad (3.29)$$

The crude Sb_2O_3 formed is vaporized and condensed in a suitable reactor. Impurities are separated during this procedure and a pure white pigment is obtained.

Antimony(III) oxide has suspected carcinogenic potential for human beings. Its threshold limit value is defined to be 0.5 mg/m^3. Other human health hazards have not been identified. The usual precautions must be taken for the handling of dangerous substances during its manufacture and application [89].

3.6 Conclusions

Inorganic white pigments are manufactured today by modern chemical processes in industrial scale production. The most important representative among the white pigments is titanium dioxide. Both rutile and anatase pigments are produced and used in a broad variety of applications. Zinc sulfide pigments – this means pure zinc sulfide and lithopone, as well as zinc oxide pigments – are other important whites. They often find their use in special pigment applications. The advantages of TiO$_2$, compared with other white pigments, include broad applicability, chemical and thermal stability, and value in use. Equipped with suitable inorganic and organic surface treatment, they also exhibit an excellent lightfastness. Organic white pigments are not available.

White pigments are applied in all pigment relevant systems, such as coatings, paints, plastics, artists' paints, cosmetics, printing inks, leather, construction materials, paper, glass, and ceramics. As pigments without an absorption color, they are applied not only for whitening of surfaces and materials but also for brightening of all colored pigments. They are produced by different processes. Wet chemical, solid-state, and vapor-phase processes are used for their manufacture. The largest pigment volumes are produced with the sulfate and the chloride processes for TiO$_2$.

Chemical purity, particle size distribution, and surface quality determine the coloristic and application technical properties of the individual pigments. Surface treatments are used in many cases to improve the quality of pigments with respect to stability and compatibility with the application system. Major efforts have been made in regard to particle management and sophisticated surface treatment.

Study questions

3.1 How do white pigments interact with visible light?
3.2 How can the structure of the three TiO$_2$ modifications be described?
3.3 What raw materials are used for the production of TiO$_2$ pigments?
3.4 What are the main steps of the sulfate process for the production of TiO$_2$ pigments?
3.5 What are the main steps of the chloride process for the production of TiO$_2$ pigments?
3.6 Why do TiO$_2$ pigments often need surface treatment and what are the components of such treatment?
3.7 What are other white pigments besides titanium dioxide?
3.8 What are the application fields for white pigments?

Bibliography

[1]   Auer G, in Industrial Inorganic Pigments, 3rd edn. Buxbaum G, Pfaff G, eds. Wiley-VCH Verlag, Weinheim, 2005, 51.

[2]   Winkler J. Titanium Dioxide, Vincentz Network, Hannover, 2003.

[3]   Elfenthal L, Schmelzer J, in Kittel – Lehrbuch der Lacke und Beschichtungen, vol. 5, Pigmente, Füllstoffe und Farbmetrik, 2nd edn. Spille J, ed. S. Hirzel Verlag, Stuttgart, Leipzig, 2003, 24.

[4]   de Keijzer M. The history of modern synthetic inorganic and organic artists' pigments, in Contributions to Conservation: Research in Conservation at the Netherlands Institute for Cultural Heritage (ICN Instituut Collectie Nederland), James & James, London, 2002, 42.

[5]   Patent US 1,106,409 (Titanium Alloy Manufacturing Company of New York) 1914.

[6]   Patent US 1,206,798 (Titanium Alloy Manufacturing Company of New York) 1916.

[7]   Patent US 1,223,357 (Titanium Alloy Manufacturing Company of New York) 1917.

[8]   Patent US 2,559,638 (E. I. du Pont de Nemours & Company) 1947.

[9]   Knittel D, James BC. Kirk-Othmer Encycl. Chem. Technol. 1983,23,98.

[10] Rudnick RL, Gao S. Treatise Geochem. 2003,3,659.

[11] Zhang W, Zhu Z, Cheng CY. Hydrometallurgy 2011,108,177.

[12] US Geological Survey, Mineral Commodity Summaries. 2021. https://pubs.er.usgs.gov/publication/mcs2021.

[13] Yan Z, Lv X, Li ZJ. Iron Steel Res. Int. 2021, V, https://doi.org/10.1007/s42243-021-00678-z(0123456789().

[14] Boehm HP. Chem. Ing. Tech. 1974,46,716.

[15] van Veen AR. Z. Phys. Chem. Suppl. 1989,162,215.

[16] Schrauzer GN, Guth TD. Am. Chem. Soc. Div. Org. 1977,99,7189.

[17] Leutz R. Erzmetall 1989,42,383.

[18] Filippou D, Hudon G. Miner. Metal Mater. Soc. 2009,61,36. http://dx.doi.org/10.1007/s11837-009-0150-3.

[19] Gázquez MJ, Bolivar JP, Garcia-Tenorio R, Vaca F. Mater. Sci. Appl. 2014,5,441. http://www.scirp.org/journal/msa; http://dx.doi.org/10.4236/msa.2014.57048.

[20] Patents DE 20 38 244 – 20 38 248 (British Titanium Products) 1970.

[21] Patent DE 15 33 123 (Ruthner Industrieplanungs AG) 1966.

[22] Patent DE 20 15 155 (Bayer AG) 1970.

[23] Patent JP 30–5166 (Ishihara Sangyo Kaisha, Ltd.) 1955.

[24] Patent DE 29 51 799 (Bayer AG) 1979.

[25] Barksdale J. Titanium: Its Occurrence, Chemistry and Technology, 2nd edn. Ronald Press, New York, 1966, 264.

[26] Patent EP 0 265 551 (Kronos Titan) 1990.

[27] Patent DE 15 92 960 (PPG Industries Inc.) 1967.

[28] Patent DE 17 67 798 (DuPont) 1968.

[29] Patent DE 19 08 747 (Cabot Corp.) 1969.

[30] Patent US 5,266,108 (DuPont) 1992.

[31] Patent US 5,201,949 (DuPont) 1992.

[32] Patent DE 23 42 889 (DuPont) 1973.

[33] Tyler F. Paint Coat. Ind. 2000,16,32.

[34] Nanetti P. Coatings from A to Z, Vincentz Network, Hannover, 2006, 179.

[35] Brock T, Groteklaes M, Mischke P. European Coatings Handbook, 2nd edn. Vincentz Network, Hannover, 2010, 127.

[36] Veronovski N, Lesnik M, Verhovsek D. J. Coat. Technol. Res. 2014,11,255.

[37] Patent US 4,781,761 (DuPont) 1987.

[38] Patent EP 0 406 194 (Kemira Oy) 1990.

[39] Gesenhues U. J. Colloid Interface Sci. 1994,168,428.

[40] Patent US 2,885,366 (DuPont) 1956.

[41] Patent GB 1008652 (Tioxide Group Ltd.) 1961.

[42] Patent DE 12 08 438 (Titangesellschaft mbH) 1960.

[43] Patent GB 1073338 (British Titan Products Co. Ltd.) 1965.

[44] Patent DE 29 46 549 (Bayer AG) 1977.

[45] Rothe U, Velleman K, Wagner H. Polym. Paint Colour J. 1992,182,10.

[46] Ainley D. Chem. Rev. 1993,93,18.

[47] Patent EP 0 133 505 (Bayer AG) 1985.

[48] Patent EP 0 194 544 (Bayer AG) 1986.

[49] Patent EP 0 393 430 (Bayer AG) 1990.

[50] van de Hulst HC. Light Scattering by Small Particles, John Wiley & Sons, New York, 1957.

[51] Kämpf G, Papenroth W, Holm R. Farbe + Lack 1973,79,9.

[52] Völz HG, Kämpf G, Klärn A. Farbe + Lack 1976,82,805.

[53] Patent JP 58025-363 (K. K. Ricoh) 1981.

[54] Mühle H. Am. J. Ind. Med. 1989,15,343.

[55] European Chemicals Agency. Guide on the Classification and Labeling of Titanium Dioxide, ECHA, Helsinki, 2021.

[56] Patent GB 13,092 (Guillaume Ferdinand de Douhet) 1850.

[57] Clausen H. Zinc-based pigments, in Pigment Handbook, vol. 1, 2nd edn. Lewis PA, ed. John Wiley & Sons, New York, 1988.

[58] Issel M. Mod. Paint Coat. 1991,9,35.

[59] Cremer M. Non TiO$_2$ White Pigments with Special References to ZnS Pigments, Industrial Minerals, Pigment & Extenders Supplement, 1985.

[60] Griebler WD, in Industrial Inorganic Pigments, 3rd edn. Buxbaum G, Pfaff G, eds. Wiley-VCH Verlag, Weinheim, 2005, 81.

[61] Küppers HJ, in Kittel – Lehrbuch der Lacke und Beschichtungen, vol. 5, Pigmente, Füllstoffe Und Farbmetrik, 2nd edn. Spille J, ed. S. Hirzel Verlag, Stuttgart, Leipzig, 2003, 105.

[62] Fosmire GJ. Am. J. Clin. Nutr. 1990,51,225.

[63] ATSDR. Toxicological Profile for Zinc. US Department of Health & Human Services, Agency for Toxic Substances and Disease Registry, Atlanta, Georgia, 1993.

[64] Frederickson CJ, Koh JY, Bush AI. Nat. Neurosci. 2005,6,449.

[65] Maret W, Sandstead HH. J. Trace Elem. Med. Biol. 2006,20,3.

[66] Nriagu J. Zinc Toxicity in Humans, Elsevier B.V, Amsterdam, 2007.

[67] Klöckl I. Chemie der Farbmittel, Walter de Gruyter GmbH, Berlin, München, Boston, 2015, 225.

[68] Jahn B, in Industrial Inorganic Pigments, 3rd edn. Buxbaum G, Pfaff G, eds. Wiley-VCH Verlag, Weinheim, 2005, 88.

[69] Brown H. Zinc Oxide Rediscovered, New Jersey Zinc Co., New York, 1957.

[70] Brown H. Zinc Oxide Properties and Applications, International Lead Zinc Research Organ, New York, 1976.

[71] Ulbrich KH, Backhaus W. Kautsch. Gummi. Kunstst. 1974,27,269.

[72] Hänig G, Ulbrich K. Erzmetall 1979,32,140.

[73] Heiland G, Mollwo E, Stockmann F. Electronic Processes in Zinc Oxide, vol. 8, Solid State Physics, Academic Press, New York, 1959.

[74] Jahn B, in Kittel – Lehrbuch der Lacke und Beschichtungen, vol. 5, Pigmente, Füllstoffe Und Farbmetrik, 2nd edn. Spille J, ed. S. Hirzel Verlag, Stuttgart, Leipzig, 2003, 179.

[75] Meyer G. New Application for Zinc, Zinc Institute Inc, New York, 1986.

[76] Brauer GM. Deutsche zahnärztliche Zeitschrift 1976,31,824.

[77] Hume WR. J. Am. Dent. Assoc. 1986,113,789.

[78] Wöll C. Prog. Surf. Sci. 2007,82,55.

[79] Moezzi A, McDonagh AM, Cortie MB. Chem. Eng. J. 2012,185/186.1.

[80] Kolodziejczak-Radzimska A, Jesionowski T. Materials 2014,7,2833.

[81] Ziegler E, Helbig R. Phys. Unserer Zeit 1986,17,171.

[82] Levinson L, Philipp HR. Am. Ceram. Soc. Bull. 1986,65,639.

[83] Zhang P, Hong RY, Chen Q, Feng WG, Badami D. J. Mater. Sci.: Mater. Electron. 2014,25,678.

[84] Keerthana S, Kumar A. Crit. Rev. Toxicol. 2020,50,47.

[85] Holley CD. The Lead and Zinc Pigments, John Wiley & Sons, New York, 1909.

[86] Pfaff G, in Special Effect Pigments, 2nd edn. Pfaff G, ed. Vincentz Network, Hannover, 2008, 34.

[87] Hocken J, Griebler W-D, Winkler J. Farbe + Lack 1992,98,19.

[88] Pfaff G, in Special Effect Pigments, 2nd edn. Pfaff G, ed. Vincentz Network, Hannover, 2008, 67.

[89] Saerens A, Ghosh M, Verdonck J, Godderis L. Int. J. Environ. Res. Public Health 2019,16,4474.

4 Colored pigments

4.1 Introduction

Colored pigments are inorganic or organic colorants, which are insoluble in the application media where they are incorporated for the purpose of coloration. Their optical action is based on selective light absorption together with light scattering. There are also colored pigments that are not only used because of their colored character but also because of their functional properties such as corrosion protection and magnetism. Moreover, white pigments, black pigments, gray pigments, effect pigments, and luminescent pigments do not belong to colored pigments.

Inorganic pigments are, in some cases, also termed mineral colors or earth colors. There are historical names for natural inorganic pigments such as ocher, umbra, and ultramarine, which hint at the mineral origin. These names were also used for all other inorganic pigments; however today their use is becoming increasingly lost.

Each pigment is characterized by its Color Index (C.I.) Generic Name. This internationally accepted nomenclature system for pigments and dyes is taken from the Society of Dyers and Colorists, Bradford, England, and the American Association of Textile Chemists and Colorists [1, 2].

Inorganic colored pigments were already used in prehistoric times. Natural pigments were obtained from minerals or their fired products. Weathered products of iron oxide-containing minerals and rocks were yellow, orange, red, brown, and black powders, which could be used as pigments. In ancient times and in the Middle Ages, several mineral oxides, sulfides, and carbonates of the metals lead, mercury, copper, and arsenic were utilized to achieve red, blue, or green color shades. Table 4.1 gives an overview on natural and older synthetic inorganic pigments [1].

Tab. 4.1: Natural and older synthetic inorganic colored pigments [1].

Name	Chemical formula	C.I. Pigment	Historical significance
Natural products			
Egyptian blue	$CaCuSi_4O_{10}$	C.I. Pigment Blue 31	Most important blue pigment in ancient times
Orpiment	As_2S_3	C.I. Pigment Yellow 39	Used by Egyptians and Romans; book painting
Azurite	$2\,CuCO_3 \cdot Cu(OH)_2$	C.I. Pigment Blue 30	Used in Greece, Egypt; in Europe until the seventeenth century one of the most important blue pigments
Iron oxide red	Fe_2O_3	C.I. Pigment Red 101	Used in prehistorical caves (Altamira, Lascaux); Roman paintings
Yellow ocher	FeOOH	C.I. Pigment Yellow 42	Most important yellow pigment in Egypt; in Europe used in the Middle Ages

https://doi.org/10.1515/9783110743920-004

Tab. 4.1 (continued)

Name	Chemical formula	C.I. Pigment	Historical significance
Green earth (Veronese green)	silicates with FeO	C.I. Pigment Green 23	Belongs to the largest known mineral pigments
Malachite	$CuCO_3 \cdot Cu(OH)_2$	C.I. Pigment Blue 30	Used in Greece, Egypt; in Europe until seventeenth century one of the most important blue pigments
Realgar (ruby arsenic)	As_4S_4	C.I. Pigment Yellow 39	Yellow pigment in ancient times
Burnt sienna (Terra di Sienna)	Fe_2O_3	C.I. Pigment Brown 7	Used in the Middle Ages
Ultramarine (lazurite)	$Na_7(Al_6Si_6O_{24})S_n$	C.I. Pigment Blue 29	
Umber	with $n = 2–4$ $Fe_2O_3 \cdot MnO_2$	C.I. Pigment Brown 7	Important pigment in Egypt; in Europe used in the Middle Ages
Cinnabar	HgS	C.I. Pigment Red 106	Most important red pigment beside iron and lead oxides; used for mural painting, book painting; first synthetic manufacture in the eighth century

Synthetic products

Iron blue (Prussian blue)	$Fe_4^{III}[Fe^{II}(CN)_6]_3$	C.I. Pigment Blue 27	Since 1704 in Europe the most important blue pigment for artist paints and coating compounds
Burnt sienna	Fe_2O_3	C.I. Pigment Red 102	Used in the Middle Ages
Burnt umber	$Fe_2O_3 \cdot MnO_2$	C.I. Pigment Brown 7	Used in the Middle Ages
Massicot (lead ocher)	PbO	C.I. Pigment Yellow 46	Mural painting in Thracia in the fourth or third century B.C.
Minium (red lead)	Pb_3O_4	C.I. Pigment Red 105	Used in Egypt, Japan, China, India, and Persia

Natural inorganic colored pigments can be distinguished from each other with regard to their composition and therewith their properties in dependence on the finding location. Some of them have to be classified as hazardous substances in today's assessment. They contain environmentally relevant heavy metals. Some of them no longer fulfill the current high technical demands. They were, therefore, replaced by synthetic colored pigments in modern times. On the other hand, the historical designations for the characterization of color shades are still used, e.g., sienna, ocher, or Naples yellow.

With the exception of natural iron oxides, other natural pigment types are only utilized in artists' paints and for the restoration purposes. Only a few companies are specialized today on the recovery and supply of natural colored pigments.

The industrial manufacture of synthetic colored pigments started in the eighteenth and nineteenth centuries with the aim of enlarging color variety with new product classes. How ever, the production of some of these newly developed pigments was stopped for toxicological reasons. Examples of this are Schweinfurt green (cupric acetate arsenite), Scheele's green (copper(II) arsenite), and Naples yellow (lead antimonite).

Synthetic pigments, which are no longer industrially produced for application technological or toxicological reasons or which have only a limited importance today, are grouped in Table 4.2 [1].

Tab. 4.2: Inorganic colored pigments, which have lost their practical relevance [1].

Name	Chemical formula	C.I. Pigment	Color
Barium yellow (lemon yellow)	$BaCrO_4$	C.I. Pigment Yellow 31	Very bright yellow
Coeline blue	$CoSnO_3$	C.I. Pigment Blue 35	Cyanic blue
Yellow ultramarine (lemon yellow)	$BaCrO_4/SrCrO_4$		Bright yellow
Cobalt violet	$Co_3(PO_4)_2$	C.I. Pigment Violet 14	Violet
Manganese blue	$BaMnO_4 \cdot BaSO_4$	C.I. Pigment Blue 33	Greenish bright blue
Manganese violet	$NH_4MnP_2O_7$	C.I. Pigment Violet 16	Reddish violet
Naples yellow	$Pb(SbO_3)_2$	C.I. Pigment Yellow 41	Bright to dark yellow
Scheele's green	$Cu(AsO_3)_2$	C.I. Pigment Green 22	Brilliant green
Schweinfurt green	$Cu(CH_3COO)_2 \cdot 3\ Cu(AsO_2)_2$	C.I. Pigment Green 21	Bright brilliant green
Zinc green	Zinc chromate/iron blue		Bright to dark green
Cinnabar	HgS	C.I. Pigment Red 106	Bright red

With the exception of some natural iron oxides, the predominant amount of the presently used inorganic colored pigments is today produced on industrial scale under well controlled conditions. The pigments are optimized with respect to their properties for their respective application purposes. The inorganic colored pigments are summarized together with the black pigments in Table 4.3, classified according to their chemical composition. Table 4.4 contains the classification of organic pigments that are colored in all cases.

Tab. 4.3: Classification of inorganic colored and black pigments according to their chemical composition.

Pigment	Composition	Color index
Elements		
Carbon black	C	Pigment Black 7
Oxides/oxide hydrates		
Iron oxide yellow	$\alpha\text{-}FeOOH$	Pigment Yellow 42
Iron oxide red	$\alpha\text{-}Fe_2O_3$	Pigment Red 101
Chromium oxide green	Cr_2O_3	Pigment Green 17
Iron oxide black	Fe_3O_4	Pigment Black 11
Mixed metal oxides		
Bismuth molybdenum vanadium yellow	$(Bi,Mo,V)O_3$	Pigment Yellow 184

Tab. 4.3 (continued)

Pigment	Composition	Color index
Chromium titanium yellow	$(Ti,Cr,Sb)O_2$	Pigment Yellow 53
Spinel blue		
Reddish blue	$CoAl_2O_4$	Pigment Blue 28
Greenish blue	$Co(Al,Cr)_2O_4$	Pigment Blue 36
Iron manganese brown	$Mn_2O_3 \cdot Fe_2O_3, Fe(OH)_2$	Pigment Brown 7+8
Iron chromium brown	$(Fe,Cr)_2O_3$	Pigment Brown 29
Zinc iron brown	$ZnFe_2O_4$	Pigment Yellow 119
Iron manganese black	$(Fe,Mn)_3O_4$	Pigment Black 26
Spinel black	$CuCr_2O_4 \cdot Fe_2O_3$	Pigment Black 28
Cadmium sulfides/selenides		
Cadmium yellow	$(Cd,Zn)S$	Pigment Yellow 35
	Cds	Pigment Yellow 37
Cadmium red	$Cd(S,Se)$	Pigment Red 108
Chromates/molybdates		
Chromium yellow	$Pb(Cr,S)O_4$	Pigment Yellow 34
Molybdate orange/red	$Pb(Cr,Mo,S)O_4$	Pigment Red 104
Complex salts		
Iron blue	$Fe_4^{III}[Fe^{II}(CN)_6]_3$	Pigment Blue 27
Ultramarines		
Ultramarine	$Na_7(Al_6Si_6O_{24})S_n$ with $n = 2–4$	Pigment Blue 29
Blue		Pigment Violet 15
Violet/red		

Tab. 4.4: Classification of organic pigments according to their chemical composition.

Pigment	Color index (selected examples)
Monoazo pigments	
Acetoacetarylide	Pigment Yellow 1
Benzimidazolon	Pigment Orange 36
Naphth-2-ol	Pigment Red 3
Naphthol AS	Pigment Red 112
Pyrazolone	Pigment Yellow 10
β-Naphthol pigment lake	Pigment Red 57:1
β-Naphthol pigment lake	Pigment Red 53:1
Disazo pigments	
Azocondensation pigments	Pigment Red 144
Bisacetoacetarylide	Pigment Yellow 16

Tab. 4.4 (continued)

Pigment	Color index (selected examples)
Diarylide	Pigment Yellow 83
Dipyrazolone	Pigment Orange 13
Polycyclic pigments	
Anthanthrone	Pigment Red 168
Anthraquinone	Pigment Red 177
Anthrapyrimidine	Pigment Yellow 108
Azomethines (metal complex)	Pigment Yellow 129
Quinacridone	Pigment Violet 19
Quinophthalone	Pigment Yellow 138
Dioxazine	Pigment Violet 23
Flavanthrone	Pigment Yellow 24
Indanthrene	Pigment Blue 60
Isoindoline	Pigment Orange 66
Isoindolinone	Pigment Yellow 100
Perinone	Pigment Orange 43
Perylene	Pigment Red 149
Phthalocyanine	Pigment Blue 15
Pyranthrone	Pigment Orange 51
Thioindigo	Pigment Red 88
Triphenylmethan	Pigment Violet 1

New developments for industrial inorganic colored pigments are strongly limited by suitable chemistry. There have only been a few new developments with relevance for the market in the last years. The most important of these are bismuth vanadate pigments. Cerium sulfides, rare earths containing oxonitrides, or YInMn Blue have not yet come to a breakthrough.

There are, on the other hand, newly developed types of the existing pigments, which show special properties that have been achieved by the use of chemical and physical routes. The optimization of these pigments is driven from special purposes of application and from the target to improve industrial safety.

The world production of colored inorganic and organic pigments today is above 2.2 million tons. More than 1.7 million tons of these are iron oxide pigments [3].

The main application fields for colored inorganic and organic pigments are coatings, paints, plastics, artists' paints, cosmetics, printing inks, leather, construction materials, paper, glass, and ceramics. Most of the inorganic colored pigments have the following advantageous properties:

– strong hiding power
– highest lightfastness and weather fastness
– very high color shade and temperature stability
– solvent fastness and nonbleeding behavior

In many cases, inorganic pigments have the better value in use compared with organic pigments with comparable coloristical properties. A disadvantage compared with organic pigments is their weaker chroma, which is especially relevant for lightening. With the exception of lead chromate, cadmium sulfide/selenide, bismuth vanadate, and ultramarine pigments, inorganic pigments show a more cloudy color shade. A solution for many coloration problems is the combined use of inorganic pigments with their strong hiding power and organic colored pigments with their high chroma.

4.2 Iron oxide pigments

4.2.1 Fundamentals and properties

Iron oxide pigments have continuously achieved the highest importance amongst the colored pigments, which is based on their wide variety of colors ranging from yellow, orange, red, brown, to black, chemical stability, nontoxicity, and on their good performance/price ratio. An additional advantage is their relatively low production cost and, therefore, their low price. Iron oxide pigments are classified into natural and synthetic iron oxide types. Both consist of well-defined compounds with known crystal structures [4–8]:

-   α-FeOOH (goethite): diaspore structure, pigment iron oxide yellow C.I. Pigment PY 42, pigment particles with preferably acicular shape, color change with increasing particle size from green-yellow to brown-yellow.
-   γ-FeOOH (lepidocrocite): boehmite structure, use only as magnetic pigment, color change with increasing particle size from yellow to orange.
-   α-Fe$_2$O$_3$ (hematite): corundum structure, pigment iron oxide red C.I. Pigment PR 101, pigment particles with spherical, acicular, and cubical shape, color change with increasing particle size from light red to dark violet.
-   γ-Fe$_2$O$_3$ (maghemite): spinel superlattice, ferromagnetic, brown color.
-   Fe$_3$O$_4$ (magnetite): inverse spinel structure, ferromagnetic, pigment iron oxide black C.I. Pigment PB 11, black color.

From the end of the nineteenth century, increasing demands and quality requirements led to the replacement of natural iron oxide pigments, which had been used from prehistoric times on, with synthetic iron oxides. Processes for the technical synthesis of iron oxide pigments were developed at this time. The importance of the synthetic types grew strongly over the decades. Today, only less than 20% of all iron oxide pigments are natural ones [3].

4.2.2 Natural iron oxide pigments

Natural iron oxides and iron oxide hydroxides were used as colorants already in prehistoric times, e.g., for cave paintings (Altamira cave, Grotte de Lascaux, Grotte Chauvet).

Ancient cultures (Egypt, Greece, the Roman Empire), as well as developed countries in later times, used natural iron oxides as pigments for coloring purposes [9].

The iron oxide content of natural iron oxides varies in dependence on mineral deposits. Natural hematite is the source for the important class of natural red pigments, natural goethite for yellow pigments, and umber and sienna for brown pigments. Deposits with high iron oxide content are used as raw materials for natural iron oxide pigments. Naturally occurring magnetite has found only little use as a raw material for black pigments because of the poor tinting strength of the resulting black powders. Natural iron oxides and hydroxides are described in detail in [8].

Hematite is widely spread on Earth. Larger quantities with sufficient quality are found in the surroundings of Malaga in Spain (Spanish red) and near the Persian Gulf (Persian red). Other deposits are of only local relevance. The Fe_2O_3 content of natural hematite attains values from 45 to 95%. Especially the Spanish reds have a Fe_2O_3 content of more than 90%. Natural hematite is responsible for the red color of many ores and rocks. Micaceous iron oxide is a special variety of natural iron oxide which, due to its layered structure, can easily be cleaved into thin platelets, comparable to mica. It is found in larger quantities in Austria and is mainly used in corrosion protection coatings.

Goethite is the coloring component of yellow ocher. Ocher are light yellow to light brown weathering products of iron-containing ores with an iron oxide content of 20 to 80%. Important deposits for the extraction of natural iron oxide yellow are found in South Africa and France. The Fe_2O_3 content gives an indication of the iron oxide hydroxide content of the ocher and is ca. 20% in the French deposits and ca. 55% in the South African ones [4].

Umber is iron and manganese-containing clay, which is formed by the weathering of ores. The main deposits for umber are found in cyprus. Umbers contain iron oxide amounts of 45 to 70% and manganese dioxide amounts of 5 to 20%. They are deep brown to greenish brown in the raw state and become dark brown with a red undertone after calcination (burnt umbers).

Terra di Siena (sienna) is the name for a group of brown to orange-colored weathering products of iron-containing ores with an iron oxide content of 45 to 80% and a manganese content of <1%. Calcined Siena has a red-brown color.

Magnetite is a black mineral with an iron content of 80 to 95%. It is the most widespread iron ore. Grinding of magnetite ore is the only step used for the production of black magnetite pigments.

The manufacturing steps for natural iron oxide pigments depend on the composition of the raw materials. These are typically ground, washed, converted into a slurry, dried, and then ground in ball mills or more often in disintegrators or impact mills [10]. Sienna and umber ores are ground, followed by calcination, to remove the water. The hue of the products is dependent on the calcination period, the temperature, and the composition of the raw material [11].

Natural iron oxide pigments are mainly used in inexpensive marine coatings or in coatings with a glue, oil, or lime base. Other applications of these pigments are the coloring of cement, artificial stone, and wallpaper as well as the production of crayons, drawing pastels, and chalks.

4.2.3 Synthetic iron oxide pigments

The chemical composition of synthetic iron oxide pigments corresponds to that of the minerals hematite, goethite, lepidocrocite, and magnetite in their pure forms. In dependence on the raw materials used for the manufacture, synthetic pigments typically exhibit higher and more constant iron oxide contents. Fe_2O_3 amounts of 95% and more can be found for synthetic iron oxide pigments. Further components are Al_2O_3 and SiO_2 (together up to 5%) and water soluble salts, mostly sulfates and chlorides (up to 0.5%), in the final pigments. Typical particle sizes of synthetic iron oxide pigments are in the range of 0.1 to 1.0 μm. The mostly acicular crystallizing iron oxide yellow pigments consist of needles with a length of 0.3 to 0.8 μm and a diameter of 0.05 to 0.2 μm. The predominantly cubical iron oxide black particles, on the other hand, exhibit diameters of 0.1 to 0.6 μm.

The increasing importance of synthetic iron oxide pigments is mainly driven by their pure hue, consistent, reproducible quality, and their tinting strength. Chemically pure iron oxides and oxide hydroxides are mainly produced for red, yellow, orange, and black pigments. Brown pigments usually consist of mixtures of red and/or yellow and/or black pigments. Pigments based on homogeneous brown phases are also synthesized, in particular, $(Fe,Mn)_2O_3$ and $\gamma\text{-}Fe_2O_3$, but only in small quantities compared to the pigment mixtures. The relevance of $\gamma\text{-}Fe_2O_3$ consists in its use as magnetic pigment.

4.2.4 Production of iron oxide pigments

Several processes are currently used for the production of high-quality iron oxide pigments with controlled mean particle size, particle size distribution, particle shape, and composition:
1. solid-state processes for red, black, and brown pigments;
2. precipitation processes for yellow, red, orange, and black pigments;
3. Laux process for black, yellow, and red pigments.

Raw materials used are mainly byproducts from other industrial processes, especially iron(II) sulfate and iron(III) chloride from the steel industry and iron scrap from the metal working industry.

4.2.4.1 Solid-state processes

Iron oxide red pigments are produced in high quantities by thermal decomposition of iron compounds. The basis process for these pigments is the oxidative calcination of decomposable iron compounds, particularly of iron(II) sulfate and α-FeOOH. The oxidation of black iron oxides also leads to red pigments. The resulting pigments are typically ground to the desired particle size in pendular mills, pin mills, or jet mills, depending on their hardness and intended use.

Processes using iron(II) sulfate mostly start with $FeSO_4 \cdot 7\,H_2O$. After prior dehydration to the monohydrate, the iron(II) sulfate is oxidized and roasted at temperatures above 650 °C. Red-colored α-Fe_2O_3 pigments (copperas reds) are formed. Iron(III) sulfate is generated as an intermediate product. The entire process is shown Figure 4.1:

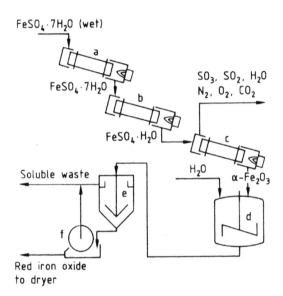

Fig. 4.1: Production of copper as red: (a) dryer; (b) rotary kiln (dewatering); (c) rotary kiln; (d) tank; (e) thickener; (f) filter [4].

$$6\,FeSO_4 \cdot H_2O + 1\tfrac{1}{2}O_2 \rightarrow \alpha\text{-}Fe_2O_3 + 2\,Fe_2(SO_4)_3 + 6\,H_2O \tag{4.1}$$

$$2\,Fe_2(SO_4)_3 \rightarrow 2\,\alpha\text{-}Fe_2O_3 + 6\,SO_3 \tag{4.2}$$

Sulfur trioxide can be processed to sulfuric acid. A certain disadvantage of this process is the contamination of the wastewater with the soluble unreacted iron sulfate. Lower qualities of Fe_2O_3 are obtained by single-stage calcination of $FeSO_4 \cdot 7\,H_2O$ in an oxidizing atmosphere. The resulting pigments have a relatively poor tinting strength and a blue tinge [4].

Iron(III) chloride, a salt, that is available in large quantities from pickling plants, can also be used for the thermal decomposition. The so-obtained iron oxide pigments do not, however, possess the usual pigment qualities. They are normally pelletized and reused in the manufacture of steel. $FeCl_2$ solution can also be used in a combined spray-drying calcination process for the production of α-Fe$_2$O$_3$ (Ruthner process). The resulting powders are utilized for the manufacture of hard and soft ferrites:

$$2\,FeCl_2 + 2\,H_2O + \tfrac{1}{2}\,O_2 \rightarrow \alpha\text{-Fe}_2O_3 + 4\,HCl \qquad (4.3)$$

The calcination of iron(III) oxide hydroxide produces pure red iron oxide pigments with a high tinting strength:

$$2\,\alpha\text{-FeOOH} \rightarrow \alpha\text{-Fe}_2O_3 + H_2O \qquad (4.4)$$

Fe$_3$O$_4$ obtained from the Laux process (see Section 4.2.4.3) or from other processes may be calcined in rotary kilns in an oxidizing atmosphere under countercurrent flow to synthesize a wide range of different red colors, depending on the starting material and the reaction conditions:

$$2\,Fe_3O_4 + \tfrac{1}{2}\,O_2 \rightarrow 3\,\alpha\text{-Fe}_2O_3 \qquad (4.5)$$

Controlled oxidation of Fe$_3$O$_4$ at 500 °C can be used for the production of single-phase brown γ-Fe$_2$O$_3$ with a neutral hue [12].

Micaceous iron oxide is obtained by the reaction of iron(III) chloride with iron at 500 to 1000 °C in an oxidizing atmosphere in a tubular reactor [4, 13].

Black Fe$_3$O$_4$ pigments with a high tinting strength are accessible by thermal treatment of iron salts under reducing conditions [14]. There is no industrial use of this process because of the furnace gases produced [4].

Solid-state reactions of iron oxides or iron oxide hydroxides with small quantities of manganese compounds can be used to produce homogeneous brown pigments with the composition (Fe, Mn)$_2$O$_3$ [15]. The calcination of iron and chromium compounds that decompose at elevated temperatures leads to corresponding pigments with the composition (Fe, Cr)$_2$O$_3$ [16].

4.2.4.2 Precipitation processes

α-FeOOH, α-Fe$_2$O$_3$, and Fe$_3$O$_4$ pigments can be obtained by combined precipitation-oxidation processes starting from iron(II) sulfate solutions and using air as the oxidizing agent. An appropriate choice of reaction conditions is important to achieve soft pigments with a pure bright hue. Nucleation seeds are used in some process variations. The production of the yellow α-FeOOH is carried out typically with FeSO$_4 \cdot$ 7 H$_2$O as the iron salt source. Alternatively, solutions from iron and steel pickling can be used [17]. The pickling solutions usually contain free acid, which can be neutralized by a reaction with scrap iron. The presence of other metal ions should be avoided in order to ensure that there are no negative effects on the color of the final iron oxide pigments.

The first step for the production of iron oxide yellow using a precipitation process is the addition of alkali, mostly sodium hydroxide, to an iron(II) sulfate solution in open reaction vessels (Figure 4.2, left). An air stream is blown into the suspension of freshly formed iron hydroxide oxidizing the iron in the precipitate and forming α-FeOOH:

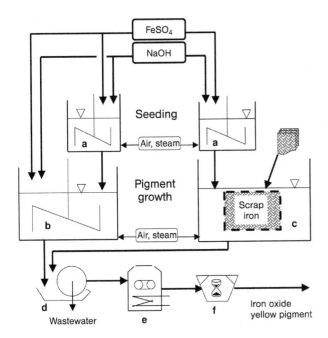

Fig. 4.2: Production of iron oxide yellow by the precipitation (left) and the Penniman process (right): (a) seeding tank; (b) precipitation pigment reactor; (c) Penniman pigment reactor with scrap basket; (d) filter; (e) dryer; (f) mill.

$$2\,FeSO_4 + 4\,NaOH + \tfrac{1}{2}\,O_2 \rightarrow 2\,\alpha\text{-FeOOH} + 2\,Na_2SO_4 + H_2O \tag{4.6}$$

The quantity of alkali used for this reaction is such that the pH value in the vessel remains acidic. The reaction time of up to 100 h depends on the temperature (10 to 90 °C) and on the desired particle size of the final yellow pigments. If α-FeOOH seeds are produced in a separate reaction (Figure 4.2, tank a), highly consistent yellow iron oxide pigments with a pure color can be obtained [18].

Black iron oxide pigments with magnetite structure and a good tinting strength are obtained by precipitation at 90 °C while air is passed into the suspension at a pH value of ≥7 (one-step precipitation). Such pigments are formed when the reaction is stopped at a FeO: Fe_2O_3 ratio of 1: 1:

$$3\,FeSO_4 + 6\,NaOH + \tfrac{1}{2}\,O_2 \rightarrow Fe_3O_4 + 3\,Na_2SO_4 + 3\,H_2O \tag{4.7}$$

Improved Fe_3O_4 pigments are possible to achieve by using this process when the reaction takes place under pressure at 150 °C [19]. Rapid heating of a suspension of iron oxide hydroxide with the necessary quantity of $FeSO_4$ and NaOH (intermediate formation of $Fe(OH)_2$) to 90 °C also produces black iron oxide of pigment quality (two-step precipitation) [20, 21]:

$$2\,\alpha\text{-FeOOH} + \text{FeSO}_4 + 2\,\text{NaOH} \rightarrow \text{Fe}_3\text{O}_4 + \text{Na}_2\text{SO}_4 + 2\,\text{H}_2\text{O} \tag{4.8}$$

Orange-colored iron oxide pigments with a lepidocrocite structure are obtained if the precipitation occurs in diluted iron(II) sulfate solutions with sodium hydroxide or other alkalis until almost neutral pH conditions. The suspension formed is heated for a short period, rapidly cooled, and oxidized [22, 23]:

$$2\,\text{FeSO}_4 + 4\,\text{NaOH} + \tfrac{1}{2}\,\text{O}_2 \rightarrow 2\,\gamma\text{-FeOOH} + 2\,\text{Na}_2\text{SO}_4 + \text{H}_2\text{O} \tag{4.9}$$

Very soft iron oxide pigments with a pure red color are accessible by preparing $\alpha\text{-Fe}_2\text{O}_3$ nuclei in a reactor followed by continuous addition of iron(II) salt solution and oxidation with air at 80 °C:

$$2\,\text{FeSO}_4 + 4\,\text{NaOH} + \tfrac{1}{2}\,\text{O}_2 \rightarrow \alpha\text{-Fe}_2\text{O}_3 + 2\,\text{Na}_2\text{SO}_4 + 2\,\text{H}_2\text{O} \tag{4.10}$$

Formed hydronium ions are neutralized by adding alkali and keeping the pH value constant [24]. Pigmentary $\alpha\text{-Fe}_2\text{O}_3$ is also obtained when solutions of an iron(II) salt, preferably in the presence of small amounts of other cations, are reacted at 60 to 95 °C with an excess of sodium hydroxide and oxidized with air [25].

The process that is probably the most used for the production of iron oxide yellow pigments is the Penniman process [26, 27]. An advantage of this route is the noticeable reduction of neutral salts formed as byproducts during the $\alpha\text{-FeOOH}$ formation. Raw materials are iron(II) sulfate, sodium hydroxide solution, and scrap iron. If the sulfate contains considerable quantities of impurities, these must be removed by special precipitation in a first step. The iron used must be free of disruptive alloying components. The process usually consists of two stages (Figure 4.2, right).

In the first stage of the Penniman process, nuclei of $\alpha\text{-FeOOH}$ are formed by the addition of caustic soda or other alkaline lye at 20 to 50 °C to an iron(II) sulfate solution while air is blown through the suspension with the precipitate (Figure 4.2, tank a). Yellow, orange, or red nuclei may be formed, depending on the reaction conditions. The nuclei suspension is pumped into a vessel charged with scrap iron (Figure 4.2, reactor c) and diluted with water. The main process step now takes place in this vessel, in which the completion of the $\alpha\text{-FeOOH}$ formation occurs by growing the precipitated iron oxide hydroxide onto the nuclei. This nuclei-driven process is responsible for the formation of sulfuric acid:

$$2\,\text{FeSO}_4 + \tfrac{1}{2}\,\text{O}_2 + 3\,\text{H}_2\text{O} \rightarrow 2\,\alpha\text{-FeOOH} + 2\,\text{H}_2\text{SO}_4 \tag{4.11}$$

The liberated sulfuric acid reacts with the scrap iron to form iron(II) sulfate, which is then also oxidized with air:

$$2\,\text{Fe} + 2\,\text{H}_2\text{SO}_4 \rightarrow 2\,\text{FeSO}_4 + 2\,\text{H}_2 \tag{4.12}$$

The reaction time for the Penniman process can vary from days to several weeks, depending on the reaction conditions and the desired pigment quality. At the end of the process, metal impurities and coarse particles are removed from the precipitate with sieves or hydrocyclones. In a downstream washing step, water soluble salts are removed. Drying is carried out with band or spray dryers. Disintegrators or jet mills are used for grinding. The main advantage of this process compared with the ordinary precipitation process is the small quantity of alkali and iron(II) sulfate required. Hydroxide in a relatively low quantity is used only to form the α-FeOOH nuclei. Iron(II) sulfate, on the other hand, is only required in small amounts for the seed formation and at the beginning of the main process step, where it is then continuously renewed by dissolving of iron by the reaction with the sulfuric acid liberated by hydrolysis (4.12). The process can, therefore, be considered as environmentally friendly.

Iron oxide red pigments can be produced by calcination of the iron oxide yellow from the Penniman process (4.4). Thermal treatment of α-Fe$_2$O$_3$ in reducing atmosphere, e.g., under hydrogen or hydrogen-containing gases, can be used for the production of black Fe$_3$O$_4$:

$$3\,\alpha\text{-Fe}_2\text{O}_3 + \text{H}_2 \rightarrow 2\,\text{Fe}_3\text{O}_4 + \text{H}_2\text{O} \tag{4.13}$$

The direct production of red α-Fe$_2$O$_3$ pigments using the Penniman process is likewise possible when suitable conditions are used. After filtration, washing, and drying, the pigments are ground using disintegrators or jet mills. The thus-obtained pigments have unsurpassed softness. Depending on the raw materials, they usually have a purer color than the harder red pigments produced by calcination [4].

4.2.4.3 The Laux process

The reduction of aromatic nitro compounds with iron under formation of iron oxide has been known for more than 150 years. It yields normally a black-gray iron oxide mud that cannot be transferred into an inorganic pigment [4]. Laux modified the process by adding iron(II) chloride or aluminum chloride solutions, sulfuric acid, and phosphoric acid. As a result, high-quality iron oxide pigments were achieved [28]. Various pigment types can be produced, dependent on the reaction conditions. The color range extends from yellow to red and black. Brown pigments are likewise possible by mixing of α-FeOOH, α-Fe$_2$O$_3$, and Fe$_3$O$_4$ in suitable ratios. The choice of additive is important for the color achieved. If the nitro compounds are reduced in the presence of aluminum chloride, high-quality yellow pigments are synthesized [29]:

$$2\,\text{Fe} + \text{C}_6\text{H}_5\text{NO}_2 + 2\,\text{H}_2\text{O} \rightarrow 2\,\alpha\text{-FeOOH} + \text{C}_6\text{H}_5\text{NH}_2 \tag{4.14}$$

The use of iron(II) chloride instead of aluminum chloride leads to black pigments with very high tinting strength [25]:

$$9\,\text{Fe} + 4\,\text{C}_6\text{H}_5\text{NO}_2 + 4\,\text{H}_2\text{O} \rightarrow 3\,\text{Fe}_3\text{O}_4 + 4\,\text{C}_6\text{H}_5\text{NH}_2 \tag{4.15}$$

Dark brown pigments with good tinting strength are accessible by the addition of phosphoric acid [30]. Calcination of thus-obtained black products results in light red to dark violet pigments. The Laux process as a whole is illustrated in Figure 4.3.

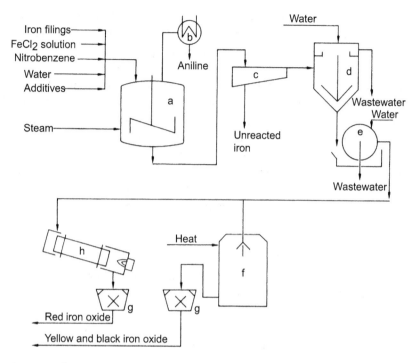

Fig. 4.3: Production of iron oxide pigments by the Laux process: (a) reactor; (b) condenser; (c) classifier; (d) thickener; (e) filter; (f) dryer; (g) mill; (h) rotary kiln [4].

The type and quality of the manufactured pigments is not only influenced by additives but also by the reaction rate. The rate depends especially on the following factors: composition and particle size of the iron used, addition rates of iron and nitrobenzene, and the pH value during the process. α-FeOOH and Fe_3O_4 are precipitated without the addition of hydroxides. The iron is oxidized by the nitrobenzene (or another suitable aromatic nitro compound) and iron ions are formed. Acid, which is generated in the vessel by hydrolysis at the same time, reacts with more metallic iron. The iron is dissolved, forming further iron ions for the advancement of the precipitation process accompanied by aniline formation, when nitrobenzene is used [4].

The iron used is ground material from iron casting or forging. It must be virtually free of oil and grease. The adjustment of the particle size takes place by grinding in edge runner mills and classification with vibratory sieves. Iron and the nitro compound are added gradually to a stirred reactor containing an aqueous solution of the other reactants (iron(II) chloride, aluminum chloride, sulfuric acid, and/or phosphoric acid). At temperatures of up to 100 °C, the nitro compound is reduced to form an amine (e.g., aniline from nitrobenzene), which is removed by steam distillation. Unreacted iron is also removed with the use of suitable classifiers. The iron oxide and the iron oxide hydroxide slurry is diluted with water and brought onto a filter. Soluble salts are removed by washing before the material is dried on suitable dryers to form yellow or black pigments. Calcination of the thus-obtained powders in rotary kilns in an oxidizing atmosphere at 700 to 800 °C allows the formation of red or brown pigments. Calcination in a nonoxidizing atmosphere at temperatures of 500 to 700 °C improves the tinting strength [31]. After thermal treatment, the pigments are ground in different types of mills to achieve the desired particle size.

The Laux process is a very attractive method for the manufacture of iron oxide pigments because of the simultaneously occurring production of aniline or other aromatic amines. It is notable that this process does not generate byproducts that harm the environment [4].

4.2.4.4 Other production processes
Iron oxide pigments of yellow, red, black, or brown color are industrially produced in large quantities using one of the processes described before. There are, however, some other chemical routes that are used on a small scale for the synthesis of iron oxide pigments for special applications:
1. Relatively pure iron(II) sulfate from the production of titanium dioxide by the sulfate process can be used for the oxidation with air in water and neutralization with sodium carbonate. Heavy metal impurities in the sulfate are isolated by a preprecipitation step. Pigments achieved using this route are powders with high analytical purity. They are suitable for the application in cosmetics as well as in the food and tobacco industries [6].
2. Chemically very pure iron oxide pigments are accessible by thermal decomposition of iron pentacarbonyl at 600 to 800 °C in an excess of oxygen. The size of the primary particles is in the nanometer range. The powders obtained using this process belong to the group of transparent iron oxide pigments [4–6, 3].
3. Platelet-shaped iron oxide red pigments are formed by hydrothermal dehydration and crystallization of iron oxide yellow, which is formed previously from iron(II) sulfate, in alkaline suspension [4, 6, 33, 34]. There is a certain use of the thus-synthesized α-Fe_2O_3 platelets as effect pigments in paints and coatings.

4.2.5 Pigment properties and uses

Figures 4.4, 4.5, and 4.6 show photographs of representative pigment samples of iron oxide yellow, iron oxide red, and iron oxide black. Depending on the morphological properties and exact composition, the powders may vary in terms of yellow, red, or black tint.

Scanning electron micrographs of an iron oxide yellow, an iron oxide red, and an iron oxide black pigment are shown in Figures 4.7, 4.8, and 4.9. The yellow pigment exhibits needle-shaped particles characteristic for many α-FeOOH powders. The red pigment is characterized by α-Fe_2O_3 particles with a tendencially spherical shape. The

Fig. 4.4: Photograph of an iron oxide yellow pigment.

Fig. 4.5: Photograph of an iron oxide red pigment.

Fig. 4.6: Photograph of an iron oxide black.

Fig. 4.7: Scanning electron micrograph of an iron oxide yellow pigment with needle-shaped particles.

Fig. 4.8: Scanning electron micrograph of an iron oxide red pigment with particles tending to a spherical shape.

Fig. 4.9: Scanning electron micrograph of an iron oxide black pigment with predominantly irregularly shaped particles.

Fe_3O_4 particles of the black pigment are rather irregularly shaped with a slight tendency to spheres and needles.

Iron oxide red and black pigments produced by the described processes exhibit Fe_2O_3 contents of 92 to 96 wt%. Very pure powders with Fe_2O_3 contents of 99.5 to 99.8 wt% are produced for special applications, e.g., for the manufacture of ferrites. Yellow and orange iron oxide pigments have a Fe_2O_3 content of 85 to 87 wt% corresponding to α-FeOOH contents of 96 to 97 wt%. Minor variations of 1 to 2% in the iron oxide content have nearly no influence on the technical quality of the pigments. On the other hand, the amount and nature of the water soluble salts as well as the particle size (effect on hue, tinge, and tinting strength) of the ground powders determine the pigment quality to a decisive extent. Thus, particle diameters of about 0.1 μm for red iron oxides lead to a yellow tinge and of about 1.0 μm to a violet tinge of the pigments [4].

The optical properties of the preferably needle-shaped iron oxide yellow pigments also depend on the particle size and on the length to width ratio of the particles. Typical lengths are in the range of 0.3 to 0.8 μm, diameters of 0.05 to 0.2 μm, and length:diameter ratios of 1.5 to 1.6. Some of the processes for α-FeOOH outlined above can be proceeded in such a way that spherical pigment particles are available. Black iron oxide pigments have particle diameters in the range of 0.1 to 0.6 μm [4].

The thermal stability of iron oxide pigments depends on the composition. α-Fe$_2$O$_3$ is stable up to temperatures of about 1200 °C in air, keeping its red color. Black Fe$_3$O$_4$ is transferred into brown γ-Fe$_2$O$_3$ at about 180 °C in the presence of oxygen. An increase of the temperature to 350 °C and higher leads to the formation of red α-Fe$_2$O$_3$. Yellow α-FeOOH is decomposed above 180 °C under water release to red α-Fe$_2$O$_3$. The thermal stability of iron oxide yellow can be increased to nearly 260 °C by incorporation of aluminum in the pigment. The thermal behavior of brown iron oxide pigments produced by mixing of yellow, red, and/or black oxides possesses different thermal behavior, depending on the composition of the mixtures.

Synthetic iron oxides are characterized by relatively good tinting strength and, depending on their particle size, excellent hiding power. They are lightfast, weather-resistant, and alkali-resistant. These properties are responsible for their versatility in different applications. Further important properties playing a role for their optical and application technical behavior are the refractive index (yellow 2.37, red 2.87, black 2.42), the specific surface area (yellow 10 to 18, red 3 to 15, black 3 to 15 m^2/g), and the density (yellow 4.0 to 4.2, red 5.0 to 5.2, black 4.7 to 4.9 g/cm^3).

Figure 4.10 shows the reflectance spectra of iron oxide yellow and iron oxide red pigments with their characteristic progression. For the α-Fe$_2$O$_3$ pigment, the green and the blue spectral range of the visible light are absorbed selectively. The nonabsorbed, i.e., the reflected part of the visible light (600 to 720 nm), generates the red color impression. In the case of the α-FeOOH pigment, the light reflection in the range of 520 to 720 nm is responsible for the yellow color. The color origin of the iron oxides can be plausibly explained based on the crystal lattice structure [35].

Iron oxide pigments have been used as colorants in many applications for a long time. More than 50% of natural and synthetic iron oxide pigments are produced for coloring of building materials [3]. They are used in concrete roof tiles, paving bricks, fibrous cement, bitumen, mortar, and rendering. Only small amounts of pigments that do not affect the setting time, compression strength, or tensile strength of the building materials are necessary in most application cases. Synthetic pigments show advantages compared with natural pigments due to their better tinting strength and purer hue.

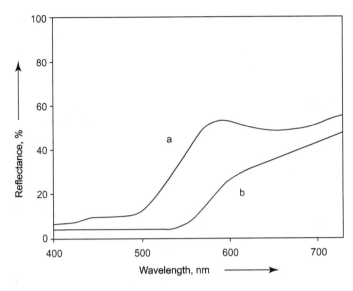

Fig. 4.10: Reflectance spectra of (a) iron oxide yellow and (b) iron oxide red.

Paints and coatings are other large application fields for iron oxide pigments. Approximately 30% of synthetically produced pigments are used here, especially in base coats, topcoats, dispersion paints, and wood protection systems [3]. In order to achieve a suitable hiding power in the paint and coating systems, iron oxide pigments with particle sizes of 0.1 to 1.0 μm are commonly used. The relationship of particle size and properties of the iron oxide red pigments is summarized in Table 4.5. With the exception of the transparent, extremely small-sized powders, e.g., iron oxide pigments are easy to incorporate in paint and coatings systems and exhibit good dispersion behavior.

Tab. 4.5: Relationship of particle size and selected properties of α-Fe$_2$O$_3$ pigments.

Particle size (μm)	0.001–0.01	0.1–1.0	10.0–100.0
Type of pigment	Transparent Iron oxide reds	Hiding Iron oxide reds	Micaceous Iron oxide
Color/appearance	Yellowish-red	Yellow-red to violet	Metallic brown to black
Hiding power	Low	High	Low
Specific surface area (m^2/g)	90–100	3–15	1–3
Dispersibility	Difficult	Easy	Very easy

Iron oxide red pigments are also widely used in plastics. The only small extent of migration and bleeding in polymer systems is of great advantage. The heat stability of red pigment types allows their application in plasticized polyvinyl chloride (plasticized PVC), stabilized rigid PVC, and other plastics. The relatively low thermal stability of iron oxide yellow pigments limits its use in thermoplastics for the coloration of PVC and LDPE. The heat resistance of the yellow pigment types can be increased up to 260 °C by coating the pigment particles with stable inorganic layers. Such an approach allows the use of these pigments also in polyolefins and polystyrene. Iron oxide black pigments can also be applied in plastics because of the reductive situation in most polymer melts, which enables their use at processing temperatures up to 280 °C. Pure grades of iron oxide pigments are permitted for the coloring of plastics used in food and commodity applications.

The pigmentation of rubber, especially of natural rubber, is usually done with iron oxides containing very low levels of copper and manganese in order to avoid deterioration effects. Both metals have a negative effect also in acrylonitrile butadiene styrene and polyolefins, where they should be kept at a very low level. Synthetic rubber is otherwise less sensitive. The high thermal resistance of iron oxide red pigments allows them to be used also in high-temperature applications. The pigments are consequently applied for coloring of ceramic bulks as well as for enamels and glazing.

In cosmetics and personal care products, synthetic iron oxide pigments are used in the formulation of a wide variety of product types including makeup and skin care preparations. The pigments are also allowed for use in pharmaceutical products [36].

Further applications of iron oxide pigments are found in the paper industry, amongst others for the fabrication of decor paper for furniture lamination. Application areas of some special iron oxides are magnetic coatings (audiotapes, videotapes, magnetic stripes), toner for photocopiers, friction pads (brake lining, clutch lining), and polishing agents for metal and glass surfaces.

4.2.6 Toxicology and occupational health

Iron oxide pigments are nontoxic unless they are contaminated with heavy metal compounds. This may be the case for some natural iron oxide types, which also contain crystalline silicon dioxide. The amount of crystalline silica in synthetic iron oxide pigments lies below the detection limit for X-ray diffraction analysis. In the cosmetics and pharmaceutical industries, only synthetic iron oxides of very high purity are allowed. The levels of heavy metals are defined by the national regularity authorities, and the small amounts that may eventually be in cosmetic or personal care products do not pose a risk to human health. Germany's workers' health authority for the chemical industry (BG Chemie) has recommended that all iron oxide pigments be classified as inert fine dusts with a MAK value (maximum allowable concentration, in Germany MAK) of 6 mg/m^3 air [37].

Iron oxides are approved for the use in feed and food [38]. Iron oxide yellow, iron oxide red and iron oxide black can thus also be used to color feedstuffs. The recommended concentrations are between 500 and 1.200 mg/kg. Iron oxides are excreted by the animals examined essentially unchanged in the feces. Iron absorption from the water insoluble iron oxides is low. It is unlikely that the use of iron oxides in animal feed would result in direct consumer exposure to the iron content of edible tissues and products from animals treated with iron oxides. Consequently, supplementation of feed with iron oxides for food-producing animals would not pose a risk to consumers [38].

Studies with Fe_3O_4 nanoparticles on rats showed that there were no pathological changes in the internal organs, except for a very weak pulmonary fibrosis that developed at the end of the first month in the treated animals. No mutagenic effect caused by the iron oxide samples was detected in the experiments. The results represent a useful extension of the knowledge on the safety of magnetite nanoparticles in view of their potential medical applications, e.g., in hyperthermia and magnetic resonance imaging [39].

4.3 Chromium oxide pigments

4.3.1 Fundamentals and properties

Chromium oxide pigments are representatives of the inorganic green and blue-green pigments [40–42]. They are also called chromium oxide green pigments and consist of chromium(III) oxide and chromium(III) oxide hydroxide. Both compounds crystallize in well-known structures:

- Cr_2O_3: corundum structure, pigment chromium oxide green C.I. Pigment Green 17, pigment particles with preferably spherical shape, olive green tint, color change with increasing particle size from brighter yellow-green to darker blue-green.
- CrOOH: 3 modifications with diaspore, indium oxide hydroxide, and delafossite structure, pigment chromium oxide hydrate green C.I. Pigment 18, attractive blue-green color.

There are no natural sources for chromium(III) oxide as opposed to corundum and natural iron oxides. It is also not possible to obtain chromium oxide hydroxide from natural deposits. Cr_2O_3 exhibits a high thermal resistance and also high hardness. Most of the synthesis routes lead to pigments with a Cr_2O_3 content of 99.0 to 99.5%. Other components with a very small content are silicon dioxide, aluminum oxide, and iron oxide. CrOOH is thermally not resistant. Heating leads to loss of crystal water with the subsequent deterioration of the application technical properties. This is the reason why CrOOH pigments have almost completely lost their technical importance.

Chromium(III) oxide is not only used for pigment purposes. Producers of Cr_2O_3 pigments usually also offer technical grades for applications other than coloration. These include the aluminothermic production of chromium metal by reaction of aluminum powder and Cr_2O_3, the production of thermally and chemically resistant bricks and lining materials, and the use of chromium(III) oxide as a grinding and polishing agent. Because of the hardness of chromium(III) oxide of ca. 9 on the Mohs scale, the abrasive properties of pigments with this composition must be considered in their applications [43].

4.3.2 Production of chromium oxide pigments

Chromium oxide pigments are produced starting from chromium(VI) compounds, mostly from alkali dichromates. These are available as industrial products in the required purity. High impurity contents are unfavorable for the hue of the final pigments [40].

4.3.2.1 Reduction of alkali dichromates

Alkali dichromates react with suitable reducing agents in solid mixtures or in aqueous medium to form chromium(III) oxide powders. Sulfur, charcoal, carbon compounds, or thiosulfates can act as reducing agents.

The reaction of solid alkali dichromate with sulfur or carbon compounds is strongly exothermic and proceeds with sodium dichromate and sulfur as follows:

$$Na_2Cr_2O_7 + S \rightarrow Cr_2O_3 + Na_2SO_4 \tag{4.16}$$

The water soluble sodium sulfate can be separated by washing the resulting product mixture after the reaction. If charcoal is used as reducing agent, Na_2CO_3 is formed as a byproduct [44].

The Cr_2O_3 synthesis with the use of sulfur was described for the first time in 1820 [45]. The use of $K_2Cr_2O_7$ instead of $Na_2Cr_2O_7$ goes back to 1927 [46].

In order to obtain Cr_2O_3 pigments with the desired properties, finely divided sodium dichromate dihydrate is mixed homogeneously with sulfur at the beginning of the process. The reaction of this mixture takes place in a furnace lined with refractory bricks at 750 to 900 °C. Sulfur is added in excess to ensure the completion of the reaction. After cooling down, the reaction mass formed is leached with water to remove all water soluble components such as sodium sulfate and the remaining sodium dichromate. The solid residue is dried, and ground. The use of $K_2Cr_2O_7$ instead of $Na_2Cr_2O_7$ as starting material leads to green pigments with a more bluish hue. The resulting chromium(III) oxide powders can be subjected to jet milling to achieve the required pigment properties necessary for the use in paints, lacquers, and other application systems.

The reduction of alkali dichromates can also be carried out in solution. Sulfur or thiosulfate can be used as reducing agents. If sulfur is applied, the reaction takes place in sodium hydroxide solution at boiling temperature:

$$4\,Na_2Cr_2O_7 + 12\,S + 4\,NaOH + 10\,H_2O \rightarrow 8\,Cr(OH)_3 + 6\,Na_2S_2O_3 \tag{4.17}$$

Chromium(III) hydroxide is formed in the alkaline medium besides sodium thiosulfate. After neutralization and addition of further sodium dichromate, even more $Cr(OH)_3$ is formed. The thiosulfate is transformed to sulfite during this reaction step:

$$Na_2Cr_2O_7 + 3\,Na_2S_2O_3 + 3\,H_2O \rightarrow 2Cr(OH)_3 + 6\,Na_2SO_3 \tag{4.18}$$

The separated reaction products are calcined and then washed with water to remove soluble components. Cr_2O_3 is formed during the calcination at 900 to 1100 °C [47, 48]:

$$2Cr(OH)_3 \rightarrow Cr_2O_3 + 3H_2O \tag{4.19}$$

The final particle size for the thus-obtained Cr_2O_3 pigments is adjusted by grinding and sieving steps.

The properties of chromium oxide pigments can be modified by special treatments, e.g., by deposition of titanium or aluminum-containing precipitates at the surface of the particles. Such treatments can change the color to yellow-green. They have also an influence on the surface properties with an impact on the application behavior, e.g., on the flocculation tendency. Organic compounds are also used for the aftertreatment of chromium oxide pigments [40].

4.3.2.2 Reduction of ammonium dichromate

Thermal decomposition of ammonium dichromate is an alternative route for the synthesis of chromium(III) oxide. Heating of the dichromate to temperatures above 200 °C initializes an exothermic reaction, leading to the formation of highly voluminous Cr_2O_3 under liberation of nitrogen:

$$(NH_4)_2Cr_2O_7 \rightarrow Cr_2O_3 + N_2 + 4H_2O \tag{4.20}$$

The thus-achieved chromium(III) oxide does not yet have pigment quality. Pigments are typically not obtained before addition of alkali salts, such as sodium sulfate, and subsequent calcination [49].

The suitable industrial process starts from a mixture of ammonium sulfate or chloride and sodium dichromate. When ammonium sulfate is used as the reducing agent, sodium sulfate is formed as a soluble byproduct [50]:

$$Na_2Cr_2O_7 + (NH_4)_2SO_4 \rightarrow Cr_2O_3 + N_2 + Na_2SO_4 + 4H_2O \tag{4.21}$$

Pigmentary chromium(III) oxide is isolated after washing the reaction products with water, drying, and if necessary grinding. Chromium(III) oxide pigments obtained by this process typically contain 99.0 to 99.5 wt% Cr_2O_3. Small amounts of SiO_2, Al_2O_3, Fe_2O_3, sulfur, and water are typically minor components found in the pigments.

4.3.2.3 Other production processes

Another route for the synthesis of Cr_2O_3 powders is the reaction of sodium dichromate with heating oil at 300 °C. Sodium carbonate is formed as a byproduct, which has to be removed by a washing step before calcination at 800 °C [51].

Shock heating of sodium dichromate in a flame at 900 to 1600 °C in the presence of hydrogen excess and chlorine can also be applied for the manufacture of chromium(III) oxide. Chlorine is necessary to bind the alkali as sodium chloride [52]. This method allows the preparation of high-purity chromium oxide pigments containing only very small sulfur amounts.

The synthesis of CrOOH pigments is possible by reaction of alkali dichromate with boric acid at 500 °C. $Cr_2(B_4O_7)_3$ is formed under liberation of oxygen. The subsequent hydrolysis in water leads to the formation of CrOOH and boric acid. An alternative route to form chromium(III) oxide hydrate is the reduction of alkali dichromates with formiate in the aqueous phase. The reaction takes place under pressure, and carbonate is formed as a byproduct [42].

4.3.3 Pigment properties and uses

Particle sizes of Cr_2O_3 pigments are in the range of 0.1 to 3 µm with mean values of 0.3 to 0.6 µm. Most of the particles are spherical. Coarser chromium oxides powders are manufactured for special applications (refractory industry, grinding, and polishing). A scanning electron micrograph of a Cr_2O_3 pigment is shown in Figure 4.11. The particles are characterized by a broad size distribution range from 100 to 500 nm.

00302863 ——— 300 nm 50.00 K X

Fig. 4.11: Scanning electron micrograph of a chromium oxide green pigment.

Chromium oxide has a refractive index of 2.5. Cr_2O_3 pigments possess an olive green tint. Figure 4.12 shows a representative photograph of a chromium(III) oxide pigment obtained by the reduction of an alkali dichromate. The photographed sample is characterized by the typical tint of chromium(III) pigments. Small particle sizes lead to lighter green with yellowish hues, larger ones to darker green with bluish tints. The darker pigments are the weaker colorants. Because of the strong light scattering property and the high absorption, chromium(III) oxide pigments show an excellent hiding power. They are very good UV absorbers and in organic binders contribute significantly to an enhancement of the weatherability for the overall system. There is a maximum in the reflectance curve in the green region of the spectrum at ca. 535 nm (Figure 4.13). A second maximum, which is weaker and lies in the violet region (ca. 410 nm) is caused by Cr–Cr interactions in the crystal lattice [40]. Chromium oxide green pigments also have a relatively high reflectance in the near infrared region (NIR). Especially products with larger particle size and more intense reflectance are, therefore, used in IR-reflecting camouflage coatings.

Fig. 4.12: Photograph of a chromium oxide green pigment.

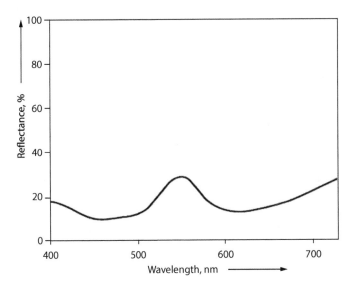

Fig. 4.13: Reflectance spectrum of chromium oxide green.

Cr_2O_3 is chemically inert. Consequently, chromium oxide green pigments are remarkably stable. They are insoluble in water, acids, and bases. In addition, they are extremely stable against sulfur dioxide and in concrete. Coating materials and plastics pigmented with chromium oxide green are lightfast and weather fast. The pigments are also temperature-resistant. A change of the color shade is observed only above 1000 °C due to particle growth.

Chromium oxide hydrate green shows a similar fastness to chromium oxide green. In contrast to Cr_2O_3, it is not stable against strong acids. The main reason for the low technical importance of CrOOH pigments is, however, the weak temperature stability.

Chromium(III) oxide pigments have a typical Cr_2O_3 content of 99.0 to 99.5 wt%. The minimum Cr_2O_3 content is determined with 96 wt% [40].

Chromium oxide green is a pigment, whose pigment particles exhibit high hardness. This means that pigmentary Cr_2O_3 has the highest abrasion wear of all inorganic pigments. It was found that the abrasion factor measured in printing inks is three to four times higher than for iron oxide pigments and 17 times higher than for anatase pigments [41]. This fact has to be accommodated for with the use of chromium(III) oxide pigments in several applications. It is necessary, therefore, to take care when grinding formulations containing chromium oxide green with steel balls. The abrasion of iron can lead to a significant shift of color shade. Attention is likewise necessary when Cr_2O_3-containing paints are sprayed because of a stronger wear of the spray nozzles.

A considerable part of the chromium oxide green pigments is used in the coatings industry (an estimated 45%) [41]. Excellent fastness properties and strong hiding power are the main criteria for the usage in this application field. Due to their very good light and weather stability, Cr_2O_3 pigments are often applied in dispersion and silicate paints in all grades of brightening. Examples for the use of the pigments are steel constructions (coil coating), facade coatings (emulsion paints), and automotive coatings. Chromium(III) oxide is the only green pigment that meets the high color stability requirements for building materials based on lime and cement besides the expensive cobalt green [53].

Chromium oxide green pigments can be used for the coloration of nearly all types of plastics. The main advantages for this application are the good temperature and migration stability and the excellent light and weather fastness of the pigments. The use of the pure pigments is only of minor importance because of their dull tint. More attractive colors are generated when Cr_2O_3 pigments are blended with other suitable colorants. Attractive and brilliant green colors can be achieved, for example, when the olive green chromium(III) oxide pigments are blended with stable yellow pigments. Examples of the use of Cr_2O_3 pigments in plastics are bottle cases and other large sized injection molded polyethylene articles of high density and stability.

4.3.4 Toxicology and occupational health

The use of dichromates as starting materials for the production of chromium(III) oxide or chromium(III) oxide hydrate pigments requires the compliance to occupational health regulations for the handling of hexavalent chromium compounds. Sulfur dioxide formed during the reduction of dichromates with an excess of sulfur must be removed from the flue gases, e.g., by oxidation to sulfur trioxide and conversion to H_2SO_4. Small amounts of unreacted dichromates or chromates in the wastewater are reduced with SO_2 or $NaHSO_3$ to low valent chromium and precipitated as chromium(III) hydroxide [54].

Chromium(III) oxide pigments do not exhibit acute toxicity (LD_{50} value rat oral: >5000 mg/kg). They are not irritating to the skin or mucous membranes. They are, therefore, not classified as hazardous materials and not subject to international transport regulations [42]. Toxicological or carcinogenic effects could not be detected in tests with rats fed up to 5% chromium(III) oxide in their feed [55].

Cr_2O_3 is not included in the MAK list (Germany), the TLV list (USA), or in the list of hazardous occupational materials of the European Community [56]. This means, in practice, that chromium(III) oxide can be considered as an inert fine dust with an MAK value of 3 mg/m^3 [40].

The use of chromium oxide pigments in toys, cosmetics, plastics, and paints that come in contact with food is permitted in correspondence with national and international regulations [40, 42]. The fulfillment of maximum limits for soluble heavy metals is a necessary prerequisite.

4.4 Mixed metal oxide pigments

4.4.1 Fundamentals and properties

An important family of inorganic pigments is termed mixed metal oxide pigments (MMO) or complex inorganic color pigments (CICP). These pigments are synthetic crystalline metal oxides with structures identical to those of naturally occurring minerals. They are called mixed or complex because they contain two or more different metals in the composition. The possibility of combining two or more metals in suitable structures gives a wide range of colors for this class of pigments for practical applications.

MMO pigments are colorants with outstanding fastness properties and are used in paints, plastics, building materials, glass coatings, and ceramics [57–60]. For ceramics, numerous other complex inorganic colored mixed oxides ("stains") exist.

A multitude of metals are typically found in MMO pigments as ions in the crystal lattices besides oxidic oxygen. Only a few of them are responsible for producing color: V, Cr, Mn, Fe, Co, Ni, and Cu. Other metal ions in the pigments do not really contribute to the color, but they are necessary for the crystal structure and are added as modifiers to balance the charge of the crystal lattice, or to slightly modify the shade produced by the coloring ions. Al, Si, Ti, Zn, Nb, Mo, Sb, W, and others belong to these ions.

MMO pigments are solid solutions of metal oxides, which are homogenously distributed in the crystal lattices and form new chemical compounds. These compounds have different types of crystal structures including rutile, spinel, inverse spinel, hematite, as well as the somewhat less common priderite and pseudobrookite (Table 4.6). MMOs have their own chemical character, which is different from that of a physical blend of their metal oxide components. The most decisive factor for the structure is the metal/oxygen (M/O) ratio. For rutile MO_2 it is 0.50, for hematite or corundum M_2O_3 it is 0.66, and for spinel M_3O_4 it is 0.75. The M/O ratio is especially important for the structure when the metal ions have radii of comparable size.

Rutile is the densest phase of titanium dioxide. Compositions with a M/O ratio of 0.50 containing metal ions of a radius similar to Ti(IV) often crystallize in the rutile structure. MMO phases with this structure contain a main fraction of TiO_2 as the base oxide, which

Tab. 4.6: Mixed metal oxide pigments with their chemical constituents, structures, and colors [57].

Color Index	Chemical constituents	Structure	Colors
Pigment Yellow 53	Ni(II), Sb(V), Ti(IV)	Rutile	Green shade yellow
Pigment Brown 24	Cr(III), Sb(V), Ti(IV)	Rutile	Ocher shade
Pigment Yellow 162	Cr(III), Nb(V), Ti(IV)	Rutile	Ocher shade
Pigment Yellow 164	Mn(II), Sb(V), Ti(IV)	Rutile	Brown
Pigment Yellow 119	Zn(II), Fe(II,III)	Spinel	Ocher-brown
Pigment Blue 28	Co(II), Al(III)	Spinel	Red shade blue
Pigment Blue 36	Co(II), Cr(III), Al(III)	Spinel	Green shade blue
Pigment Green 26	Co(II), Cr(III)	Spinel	Dark green
Pigment Green 50	Co(II), Ti(IV)	Inverse spinel	Green
Pigment Brown 29	Fe(II), Cr(III)	Hematite	Brown
Pigment Brown 35	Fe(II,III), Cr(III)	Spinel	Dark brown
Pigment Black 30	Ni(II), Fe(II,III), Cr(III)	Spinel	Black
Pigment Black 26	Mn(II), Fe(II,III)	Spinel	Black
Pigment Black 22	Cu(II), Cr(III)	Spinel	Black
Pigment Black 28	Mn(II), Cu(II), Cr(III)	Spinel	Black
Pigment Black 27	Co(II), Cr(III), Fe(II)	Spinel	Bluish black

acts as the host lattice. Consistently, rutile MMOs have several physical properties analogous to titanium dioxide in the rutile modification. Metal ions in the rutile structure are coordinated octahedrally by six oxygen ions. In MMO pigments, where titanium is partly substituted by other metal ions, these ions are equally distributed on Ti positions.

Typical examples of rutile type MMO pigments are nickel rutile yellow (nickel titanium yellow) and chromium rutile yellow (chromium titanium yellow). Nickel ions and chromium ions are the coloring components of these pigments located at titanium positions. With the valence states Ni(II) and Cr(III), both metals differ from Ti(IV). The necessary charge equalization is achieved by simultaneous placement of Sb(V) in the compositions. The general formulas of the pigments are, therefore, $(Ti,Ni,Sb)O_2$ for nickel rutile yellow and $(Ti,Cr,Sb)O_2$ for chromium rutile yellow. Another example is manganese rutile brown $(Ti,Mn,Sb)O_2$, where titanium is consequently substituted by manganese and antimony. Substitutions of antimony by tungsten or niobium, on the other hand, are also common. Relevant compositions with such substitutions are the yellows $(Ti,Ni,Nb)O_2$, $(Ti,Ni,W)O_2$, $(Ti,Cr,Nb)O_2$, and $(Ti,Cr,W)O_2$, as well as the browns $(Ti,Mn,Nb)O_2$ and $(Ti,Mn, W)O_2$ [57]. The amount of titanium dioxide as the base oxide is for all MMO pigments of the rutile type typically in the range of 70 to 90 wt% TiO_2. A broader overview on colored rutile MMO pigments and in addition on priderite $(BaNiTi_7O_{16})$ and pseudobrookite (Fe_2TiO_5) MMO pigments is given in [61].

The spinel structure AB_2O_4 (M_3O_4) is very common for many compositions containing mostly metal ions of a main group element (A) together with ions of a transition metal element from the row Ti to Zn (B) and oxide ions. Numerous transition metal oxides adopt this structure with the M/O ratio of 0.75 when the radii fit the spinel lattice. The structure is so preferred and stable that also stoichiometries different from M_3O_4 can crystallize in the spinel configuration. In such cases, the crystals typically contain metal ion vacancies.

Metal ions in the spinel structure are generally divalent (+2) and trivalent (+3), although differently charged ions can be accommodated. The structural coordination of the metal ions in the spinel lattice occurs in two different ways. The B ions in the AB_2O_4 structure are octahedrally coordinated by the oxide ions, while the A ions are tetrahedrally coordinated. Normal spinels have divalent metal ions in tetrahedral sites and trivalent metal ions in the octahedrally surrounded positions. Inverse spinels, on the other hand, also contain divalent metal ions in sixfold coordination and some trivalent metal ions in positions with the coordination number four. Table 4.6 contains only those spinel and inverse spinel compositions that are relevant mostly for industrial pigments.

Cobalt blue with the formula $CoAl_2O_4$ is one of the most important inorganic blue pigments. This red shaded blue pigment crystallizes in the normal spinel structure. The partial substitution of cobalt by chromium leads to green shaded blue pigments with spinel structure as well. Cobalt blue pigments contain up to 25 wt% Co and up to 27 wt% Cr [58].

The only industrially important representative with inverse spinel structure amongst the MMO pigments is cobalt titanate Co_2TiO_4. Due to its green color, it is colloquially called cobalt green. Pigments of this type contain up to 15 wt% Co. Cobalt can partially be substituted by zinc and nickel.

The ocher-brown zinc ferrite $ZnFe_2O_4$ is another pigment with spinel structure. Variation in the Zn: Fe ratio opens the spectrum of the shade from brighter to medium browns. The zinc content is typically at values of about 40 wt%.

Important spinel black pigments are copper chromites with the formula $CuCr_2O_4$. Chromium can be replaced partially by manganese and/or iron. Other black MMO pigments are iron chromium spinels, where cobalt or copper are introduced in the structure.

The replacement of iron(III) in the iron oxide spinel Fe_3O_4 by chromium(III) is a possible way to obtain dark brown pigments. The thus-achieved chromium iron brown contains up to 17 wt% Cr.

There are other products belonging to the complex inorganic color pigments besides the rutile and spinel type MMOs. These are pigments based on zirconium silicate (zircon, $ZrSiO_4$). The products are counted among the class of ceramic pigments (color stains for glazes) and are described in detail in Chapter 6.

Zirconium silicate stains are certainly used only in enamels and ceramics (high-temperature pigments). They combine sufficient color properties with extreme thermal and chemical stability and cover the color areas blue, yellow, and red. In addition, they allow the generation of all intermediate color tones [62–64]. The basis of all these pigments is the ZrO_2 lattice, in which transition metal ions are incorporated. The most important representatives of the zircon pigments are zircon vanadium blue and zircon praseodymium yellow. The proportion of vanadium or praseodymium in the pigments amounts to only a few percent.

The group of zirconium silicate color stains also covers the so-called inclusion pigments [65]. However, there is no incorporation of coloring ions in the zircon crystal lattice in this case. The principle is the envelopment of colorants by the thermally and chemically stable ZrO_2. The so included color-providing components are therewith protected and can be exposed to the extreme conditions of the ceramic glaze firing. Representatives of the zirconium silicate inclusion pigments are zircon iron pink (with up to 15 wt% Fe_2O_3) and products containing cadmium sulfoselenide included in $ZrSiO_4$. Tin dioxide and silicon dioxide can also act as enveloping substances besides zirconium silicate.

4.4.2 Production of MMO pigments

The synthesis of rutile and spinel MMO pigments is carried out using solid-state reactions of metal oxides, hydroxides, or carbonates at temperatures in the range of 800 to 1300 °C. The starting mixtures correspond to the desired compositions of the final pigments. The starting materials are deployed as fine grained powders. They are blended intimately before the high-temperature reaction starts. The solid-state reaction as the main step is a calcination process, where the mixture of the metal compounds is converted into a new chemical compound that already possesses preliminary color characteristics. Final color and other physical properties are optimized for the end use by finishing procedures. These involve micronizing, washing, and drying steps. Figure 4.14 schematically shows the manufacture of MMO pigments.

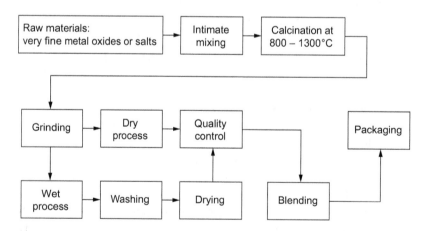

Fig. 4.14: Manufacturing scheme for mixed metal oxides pigments.

At relatively low temperatures, the raw ingredients decompose and the corresponding metal oxides are formed. The oxide mix becomes reactive at higher temperatures. Metal and oxide ions become more and more mobile and interdiffusion starts to form a new homogeneous compound. The ions in the so formed solid rearrange to a stable crystalline network, e. g., to a rutile or spinel lattice. The new structure is determined by the metal ions involved, the M/O ratio, and the calcination temperature.

4.4.2.1 Rutile MMO pigments

The formation of MMO pigments crystallizing in the rutile lattice occurs by calcination of homogeneous mixtures of titanium dioxide as the later host component and the color giving guest components at temperatures above 1000 °C in air. The guest components can be introduced in the form of their oxides but also as nonheat-resistant, decomposable compounds of the necessary coloring metal (hydroxides, oxide hydrates, carbonates, and acetates, nitrates). TiO_2 is mainly applied in the less dense and more reactive anatase modification. Dried titanium dioxide hydrate can be used for pigment synthesis instead of titanium dioxide. After the calcination step, the pigment brick is ground wet or dry to obtain the desired grain size for the rutile MMO pigment powder. It is possible at this point in time to equip the surface of the pigment particles with a suitable aftertreatment to enhance weather resistance and the behavior in the relevant binder systems.

The synthesis of nickel rutile yellow and chromium rutile yellow can be exemplary and summarily described as follows:

$$0.85\,TiO_2 + 0.05\,NiO + 0.05\,Sb_2O_5 \rightarrow (Ti_{0.85}Ni_{0.05}Sb_{0.10})O_2 \qquad (4.22)$$

$$0.90\,TiO_2 + 0.025\,Cr_2O_3 + 0.025\,Sb_2O_5 \rightarrow (Ti_{0.90}Cr_{0.05}Sb_{0.05})O_2 \qquad (4.23)$$

The Sb source for these syntheses is typically Sb_2O_3. Sb_2O_5, which is necessary for the pigment formation to achieve electroneutrality in the final composition, is formed during the reaction by oxidation of Sb_2O_3 with oxygen. Sb_2O_4 is formed intermediately, reacting further with anatase to $(Ti,Sb)O_2$, which already crystallizes in the rutile lattice. The reaction of $(Ti,Sb)O_2$ with the nickel or chromium oxide again proceeds with oxygen. Sb_2O_5, $NiSb_2O_6$, and $NiTiO_3$ can be detected as intermediates of these syntheses [61].

4.4.2.2 Spinel MMO pigments

MMO pigments with spinel structure are manufactured by intense mechanical blending of the fine particular powdered starting compounds followed by calcination at 800 to 1300 °C. Oxides or suitable decomposable metal compounds are used for the manufacture of spinel pigments. A very homogeneous starting mixture can be provided by simultaneous precipitation of the hydroxides or carbonates of the metals involved. The reaction temperature depends on the quality of the starting mixture, the reactivity of the raw materials, and the desired particle size of the final pigment. The color depth increases with the particle size. The hiding power decreases at the same time. The addition of mineralizers allows the lowering of the reaction temperature. Industrial production takes place in mineral crucibles in directly heated bogie hearth furnaces or tunnel kilns. The manufacture is likewise possible in indirectly heated metallic rotary kilns.

The so-obtained raw products are ground as suspension in ball mills, washed, filtrated, dried, and dry ground. Nearly dust-free and free flowing granulates are obtained by spray granulation of the washed and subsequently concentrated suspension.

The following two equations demonstrate the synthesis of cobalt blue and cobalt green:

$$CoO + Al_2O_3 \rightarrow CoAl_2O_4 \tag{4.24}$$

$$2\,CoO + TiO_2 \rightarrow Co_2TiO_4 \tag{4.25}$$

The production of MMO pigments with hematite, priderite, and pseudobrookite structures proceeds similarly to the described manufacture of spinel-type pigments.

4.4.2.3 Zirconium silicate pigments (ceramic colors)

The synthesis of zirconium silicate pigments proceeds in two steps. In the first step there is the formation of the zircon host lattice and in the second step the inclusion of the color providing component in the zircon structure formed. Generally, similar process steps are used as described for the synthesis of rutile or spinel MMO pigments. The control of solid-state reactions is essential to produce high quality pigments.

The presence of alkali metal halides, mainly of sodium fluoride, in the reaction mixture promotes the color formation. The color depends on a number of physical as well as chemical factors such as the proportions of vanadium pentoxide and sodium fluoride, the purity of the raw materials used, the particle sizes of the components, the distribution of the reactants and their initial degree of compaction, and the atmosphere used for the high-temperature reaction.

The following listing contains zirconium silicate color stains for selected pigments (percent values correspond to wt%):

- Zircon vanadium blue $(Zr,V)O_2$ [62]:
  Starting mixture: 62% ZrO_2, 30% SiO_2 (host lattice former), 5% NH_4VO_3, 3% NaF (mineralizer)
  Calcination conditions: 850 to 1000 °C, 2 to 3 h
- Zircon praseodymium yellow $(Zr,Pr)O_2$ [66]:
  Starting mixture: 62% ZrO_2, 30% SiO_2, 5% Pr_6O_{11}, 3% Na(F,Cl)
  Calcination conditions: 1250 °C, 2 to 3 h
- Zircon iron pink $(Fe_2O_3/ZrSiO_4)$ [62]:
  Starting mixture: 45% ZrO_2, 22% SiO_2 (envelop former), 21% $FeSO_4 \cdot 7\,H_2O$, 12% Na (F,Cl)
  Calcination conditions: 880 to 1100 °C, 1 h

It is necessary for the production of zircon-based colorants to work with very pure chemicals. Commonly, an expensive grade of ZrO_2 is required for the production of ceramic colors with the highest chroma. The high price of zirconia goes back to the processing of this material, where the separation from mineral zircon sand is necessary to obtain a suitable starting quality. Therefore, a new process has been developed to manufacture zircon-based colorants directly from zircon sand, which is comparably cheap and available. Equimolar proportions of ZrO_2 and SiO_2 are used to form $ZrSiO_4$, leading to intense colorants. In a special step, zircon is calcined with an alkali compound to form an alkaline silicozirconate. Acid treatment of this compound leads to decomposition and formation of a zirconia-silica mixture. After calcination, a zircon-based ceramic colorant is formed. This process can basically be used for the production of all important zircon colorants.

4.4.3 Pigment properties and uses

MMO pigments cover large areas of the color space. Blue, green, yellow, brown, and black pigments with partly intensive color strength are available. Only suitable reds are difficult to find amongst the MMO pigments.

High opacity, heat stability, special infrared properties, lightfastness, and weather fastness as well as chemical resistance are the main reasons for the use of these pigments in different application systems. Most MMO pigments possess an excellent heat stability due to the high temperature used during the manufacturing process.

Particle size and shape are important quality factors. MMO pigments have mean particle sizes ranging from 0.5 to 2.5 μm. Narrow particle size distributions are normally required to obtain beneficial color properties. The high-temperature calcination process promotes the formation of larger agglomerates of the MMO particles. These may exist as oversized particles in the calcined powders. Such large particles are the reason for longer grinding times to obtain pigments with a suitable particle size distribution that are easy to disperse. This situation does not only have an influence on the production costs but also on the color reproducibility. MMO, particularly nickel rutile yellow and chromium rutile yellow, exhibit a strong color dependency on the grinding time and intensity. Longer times and higher intensities can lead to an undesired stronger abrasion behavior of the pigment particles. More intense grinding steps combined with extensive sieving or sifting, on the other hand, also have a great impact on the particle shape, which is relatively irregular after the calcination step. A result of the grinding procedure for MMOs is that more spherical particles are generated. A spherical particle shape is favorable, particularly for the rutile yellows, and, therefore, desired. The advancement in the field has led to darker, redder, and more saturated chromium rutile pigments with a favorable spherical particle shape together with lower abrasion [57].

MMO pigments typically have low surface porosities. Specific surface areas measured using the BET method are in the range of 3 to 6 m^2/g for rutile-type MMOs and 3 to 10 m^2/g for spinel-type MMOs. Such low porosities and surface areas result in low oil absorption numbers. Consequently, high pigment loadings in color concentrates, such as pigment pastes, preparations, and master batches, are possible. MMO pigments are relatively easy to disperse in their application media. Shear stability problems, on the other hand, may create difficulties with some particular resins and applications [57].

Some MMO pigments, especially those with spinel and hematite structure find interest also because of their infrared reflectance, which is interesting for camouflage applications as well as for applications with a defined minimum total solar reflectance [67, 68]. Special pigments designed for infrared reflectance are the brownish to black pigments Pigment Brown 29, Pigment Brown 35, and Pigment Black 30 (Table 4.6).

Rutile MMO pigments exhibit purer color shades compared with iron oxide pigments. Therefore, they have broader coloristic application possibilities. The reflectance spectra of nickel rutile yellow and chromium rutile yellow are shown in Figure 4.15. Photographs of representative pigment samples of nickel rutile yellow and chromium rutile yellow are shown in Figures 4.16 and 4.17. Nickel rutile yellow is characterized by a warm yellow tone with a slightly green tint. Chromium rutile yellow, on the other hand, shows a darker yellow tone with a distinct reddish tint.

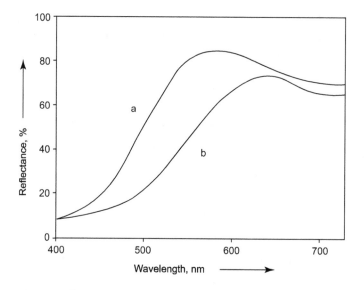

Fig. 4.15: Reflectance spectra of (a) the nickel rutile yellow pigment $(Ti_{0.85}Ni_{0.05}Sb_{0.10})O_2$ and (b) the chromium rutile yellow pigment $(Ti_{0.90}Cr_{0.05}Sb_{0.05})O_2$.

Scanning electron micrographs of a nickel rutile yellow pigment (Figure 4.18) and of a chromium rutile yellow pigment (Figure 4.19) show the large particle sizes known for MMO pigments. A considerable proportion of particles, especially in the case of the chromium-containing pigment, has sizes of more than 1 μm, which is attributable to the manufacturing conditions.

Insolubility, highest chemical and thermal stability as well as excellent lightfastness and weather fastness are fundamental advantages of the work with rutile MMOs. The photocatalytic activity known from titanium dioxide is significantly reduced because of the incorporation of foreign ions in the crystal lattice.

Fig. 4.16: Photograph of the nickel rutile yellow pigment $(Ti_{0.85}Ni_{0.05}Sb_{0.10})O_2$.

Fig. 4.18: Scanning electron micrograph of the nickel rutile yellow pigment $(Ti_{0.85}Ni_{0.05}Sb_{0.10})O_2$.

Fig. 4.17: Photograph of the chromium rutile yellow pigment $(Ti_{0.90}Cr_{0.05}Sb_{0.05})O_2$.

00302764 ——— 500 nm 20.00 K X

Fig. 4.19: Scanning electron micrograph of the chromium rutile yellow pigment $(Ti_{0.90}Cr_{0.05}Sb_{0.05})O_2$.

Rutile MMO pigments are often used in combination with high-performance organic pigments to formulate brilliant colors. Such combinations have the advantage of nontoxicity compared with lead chromate, lead molybdate, or cadmium-containing pigments. Nickel rutile yellow and chromium rutile yellow are used in formulations of many standard industry colors, e.g., in RAL tone colors, which are typically less clean than pure organic pigments. The use of rutile yellows in such formulations leads to an improvement of the weather stability due to the UV absorbing properties of the inorganic MMO pigments. Rutile yellows have also found an application as toners in pastel tone colors.

Spinel MMO pigments possess outstanding fastness properties. Their chemical stability and weather fastness are excellent, also in brightening blends with titanium dioxide. They outmatch organic green and blue pigments in bright pastel shades.

The reflectance spectra of cobalt blue and cobalt green can be seen in Figure 4.20. The broad bands in the spectra correspond to the colors of the two pigments. Photographs of representative pigment samples of cobalt blue, cobalt green, and spinel black are shown in Figures 4.21, 4.22, and 4.23.

Scanning electron micrographs of a cobalt blue pigment (Figure 4.24), cobalt green pigment (Figure 4.25), and spinel black pigment (Figure 4.26) once again show the large particle sizes known for MMO pigments. Many particles have sizes of more than 1 μm, especially in the cases of cobalt blue and spinel black, or consist of large units formed by sintering of smaller particles (cobalt green) at high temperatures.

Cobalt blue pigments cover the color area from red shaded to greenish turquoise-shaded blue. The spectra of these pigments exhibit a bathochromic shift in the maximum absorption when aluminum is substituted by chromium. There is an increase of the opacity at the same time. In a favorable way, the weather resistance of pigmented coatings also tends to be better from Pigment Blue 28 to Pigment Blue 36. Pigment Blue 28 is, nevertheless, applied in many systems because of its color and appearance (higher gloss). Efforts are being made to provide high strength cobalt blues, which enable improvement of UV opacity and weathering properties. The application of cobalt blue and green pigments in plastics is driven by their fastness properties. Pigment Green 26 has found special interest in camouflage applications [57].

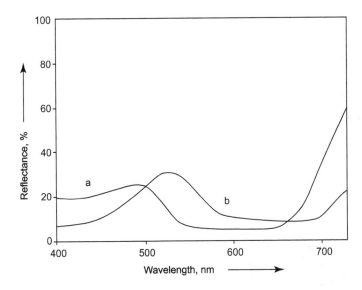

Fig. 4.20: Reflectance spectra of (a) the cobalt blue pigment $CoAl_2O_4$ and (b) the cobalt green pigment Co_2TiO_4.

Fig. 4.21: Photograph of the cobalt blue pigment $CoAl_2O_4$.

Fig. 4.22: Photograph of the cobalt green pigment Co_2TiO_4.

Fig. 4.23: Photograph of the spinel black pigment $CuCr_2O_4$.

Fig. 4.24: Scanning electron micrograph of the cobalt blue pigment $CoAl_2O_4$.

Fig. 4.25: Scanning electron micrograph of the cobalt green pigment Co_2TiO_4.

Zinc ferrite pigments are excellent representatives for the yellowish brown color area. In comparison to iron oxide yellow pigments, zinc ferrites offer higher heat stability and can be used in plastics and coatings at processing temperatures of above 120 °C. In terms of color, zinc ferrite pigments are not as clean as rutile yellows. The fact that zinc ferrites are slightly magnetic has to be taken into account.

00302864 —— 500 nm 20.00 K X

Fig. 4.26: Scanning electron micrograph of the spinel black pigment $CuCr_2O_4$.

Manganese titanate brown pigments are mainly used because of their excellent weathering properties. These iron-free browns have found application in plastics, particularly in rigid PVC (vinyl siding). The use of iron-containing pigments, particularly if highly soluble iron is involved, can degrade polyvinyl chloride catalytically under the influence of UV radiation [57].

Some of the spinel-type brown and black pigments, e.g., Pigment Brown 35 and Pigment Black 30, but also Pigment Brown 29 with its hematite structure, are predestinated for use as infrared reflecting materials. They are able to reflect in the NIR in contrast to other inorganic brown pigments like iron oxides or other black pigments like copper chromite or carbon black [69]. The nickel, iron, and chromium-containing Pigment Brown 30 reflects less of the infrared radiation compared with Pigment Brown 29 or other only iron and chromium-containing compositions, but offers a more bluish tint. The brown-colored iron chromium oxides are very suitable for use in PVC pipe and window profiles. They are deployed as a nontoxic alternative to color formulations containing lead chromates and lead molybdates.

Copper chromite black pigments like Pigment Black 22 are applied widely because of their outstanding durability and temperature resistance. Contrary to other spinel blacks, they do not exhibit infrared reflectance. Their main application is in coatings for black and gray colors. Cobalt chromites like Pigment Black 27 are preferably used in applications requiring higher heat stability. Manganese ferrites like Pigment Black 26 are less weather-resistant and acid stable than copper chromite blacks, but can be seen as a cost-effective black alternative if these properties are not required [57].

The requirements for high-temperature applications are fulfilled in a very effective manner by the zircon-based ceramic colorants vanadium zircon blue, praseodymium zircon yellow, and iron zircon coral. These colorants are stable up to temperatures of 1350 °C in all types of glazes. Cross mixing allows a wide range of additional high-temperature stable colors.

Vanadium zircon blue is most probably the best known ceramic colorant. Its color strength depends on the amount of vanadium in the zircon lattice and the incorporation of the vanadium on the mineralizer component used in the reaction mixture.

Praseodymium zircon yellow fulfills the demand for a clean, bright yellow for high-temperature applications. The addition of cerium oxide leads to orange shades. More intense colors are achieved by the use of lead compounds as mineralizers instead of alkali metal halides. The addition of ammonium nitrate to the alkali metal halide also leads to stronger colors. Praseodymium zircon yellow shows brighter and cleaner color properties than other high-temperature stable yellow colorants. The temperature stability surpasses that of cadmium sulfide and sulfoselenide pigments.

4.4.4 Toxicology and occupational health

MMO pigments with rutile and spinel structure do not exhibit acute toxicity (LD_{50} value rat oral: >5000 mg/kg). They are not irritating to the skin or mucous membranes. In the majority of available pigment chemistries, MMOs are highly inert chemical compounds, containing no relevant bioavailable or degradable substances. Therefore, most pigments, referring to manufacturers' statements on compliance, purity, and safe handling, are considered nontoxic and comply with food contact as well as toy safety regulations.

For the evaluation of chronic toxicology, it has to be considered that the incorporated metal oxides no longer exist as chemical individuals. In MMO pigments, they are fixed components of the rutile or spinel lattice. Several investigations in different countries with the focus on antimony, chromium, cobalt, copper, iron, manganese, and zinc have shown no indication of a relevant bioavailability of these heavy metals [59].

In the EU, there is a regulation that antimony compounds and preparations containing antimony above 0.25% (calculated as Sb metal) have to be marked as harmful to health. Due to the incorporation in the rutile lattice, the antimony loses its original chemical, physical, and physiological properties in the rutile-type MMO pigments. Antimony is not bioavailable in these pigments and, therefore, does not exhibit any toxic effects. Rutile MMOs as well as spinel MMOs are, therefore, not subjects to labeling as hazardous materials [59].

Zirconium silicate inclusion pigments containing cadmium sulfoselenide have to be considered separately and discussed with respect to toxicology and occupational health together with the pure cadmium-containing pigments.

4.5 Chromate and molybdate pigments

4.5.1 Fundamentals and properties

Chromate and molybdate pigments are representatives of the inorganic yellow, orange, and red pigments. Lead chromate (chrome yellow) and lead molybdate (molybdate orange and molybdate red), whose colors range from light lemon yellow to reds with a blue hue, are of the highest importance amongst the chromate and molybdate pigments [70–72]. The pigments crystallize in well-known structures. Chrome orange as a further pigment belonging to this group has lost its importance for coloring and is no longer produced. Mixtures of chromate pigments with iron blue or phthalocyanine pigments enlarge the potential up to green colors.

$PbCrO_4$, $Pb(Cr,S)O_4$: Chrome yellow pigments (C.I. Pigment Yellow 34) consist of pure lead chromate and lead sulfochromate (mixed phase pigment) (refractive index 2.30 to 2.65, density ca. 6 g/cm^3) [73]. Both pigments can crystallize in an orthorhombic or a monoclinic structure. The monoclinic structure is the more stable one [74]. The greenish yellow orthorhombic modification of lead chromate is metastable at room temperature and is readily transformed to the monoclinic modification under certain conditions (e.g., concentration, pH value, and temperature). The monoclinic modification exists naturally as crocoite. Partial replacement of chromate by sulfate in the mixed phase crystals causes a gradual reduction of the tinting strength and hiding power, but allows production of important chrome yellows with a greenish yellow hue.

$Pb(Cr,S,Mo)O_4$: Molybdate red and molybdate orange (C.I. Pigment Red 104), are mixed phase pigments, in which chromium is partially substituted by sulfur and molybdenum [73]. Most commercial products show an MoO_3 content of 4 to 6%, a refractive index of 2.3 to 2.65, and densities of about 5.4 to 6.3 g/cm^3. Their hue depends on the proportion of molybdate, crystal form, and particle size. Pure tetragonal lead molybdate, which is colorless, forms orange to red tetragonal mixed phase pigments with lead sulfochromate. The composition of molybdate red and molybdate orange pigments can be varied to give the required coloristic properties. Commercial products usually contain ca. 10% lead molybdate. Lead molybdate pigments have a thermodynamically unstable tetragonal crystal modification, which can be transformed into the undesirable stable yellow modification merely by dispersing [75]. This is especially true for the bluish varieties of molybdate red, which have larger particles whose color can be changed to yellow by shear forces. The tetragonal modification of lead molybdate pigments must, therefore, be stabilized after precipitation [76, 77]. Molybdate red and molybdate orange pigments are often combined with red organic pigments, providing a considerable extension and fine-tuning of the color range.

$PbCrO_4 \cdot PbO$: Chrome orange (C.I. Pigment Orange 21) is a basic lead chromate. It is, therefore, also described as $PbCrO_4 \cdot Pb(OH)_2$.

$Pb(S,Cr)O_4 + Fe_4^{III}[Fe^{II}(CN)_6]_3 \cdot x\,H_2O$: Chrome greens (C.I. Pigment Green 15) are pigment mixtures consisting of chrome yellow and iron blue pigments.

$Pb(S,Cr)O_4$ + phthalocyanine: Fast chrome greens (C.I. Pigment Green 48) are combinations of chrome yellow and phthalocyanine blue or phthalocyanine green. For high grade fast chrome greens, stabilized and highly stabilized chrome yellows are used in practice. The density and refractive index of chrome greens and fast chrome greens depend on the ratio of the components in the mixture. Their hues vary from light green to dark blue-green, again depending on the ratio of their components [70].

4.5.2 Production of chromate and molybdate pigments

Chrome yellow pigments are produced starting mostly from metallic lead, which reacts with nitric acid to give lead nitrate solution. The use of lead oxide instead of lead is possible, but not as common. Sodium dichromate solution is then added to the lead nitrate solution formed to obtain a yellow lead chromate precipitate.

The pH value is adjusted beforehand in order to obtain the chromate dichromate equilibrium on the side of the chromate:

$$Cr_2O_7^{2-} + H_2O \rightleftharpoons 2\,CrO_4^{2-} + 2H^+ \tag{4.26}$$

$$Pb(NO_3)_2 + K_2CrO_4 \rightarrow PbCrO_4 + 2\,KNO_3 \tag{4.27}$$

If the dichromate solution contains sulfate, lead sulfochromate is formed as a mixed phase precipitate. An example of this frequently practised case is demonstrated in the following equation:

$$Pb(NO_3)_2 + 0.5\,K_2CrO_4 + 0.5\,K_2SO_4 \rightarrow Pb(Cr_{0.5}S_{0.5})O_4 + 2\,KNO_3 \tag{4.28}$$

The thus-obtained precipitates are filtered off, washed, dried, and ground. The color of the final pigments depends on the ratio of the starting components and on the precipitation conditions (concentration, temperature, pH value, time). Orthorhombic crystals are generated, which are readily transformed into the more stable monoclinic crystals. Higher temperatures accelerate this conversion [78, 79]. Almost isometric particles are obtained by appropriate control of the process conditions [70].

When unstabilized, chrome yellow pigments exhibit poor lightfastness and darken due to possible redox reactions. An improvement in the fastness properties can be achieved by coating the pigment particles with a surface treatment containing mostly oxides of titanium, cerium, aluminum, antimony, and silicon [80–88].

Molybdate red and molybdate orange pigments are produced using the Sherwin–Williams process or the Bayer process. In the Sherwin–Williams process, lead nitrate solution reacts with a solution of sodium dichromate, ammonium molybdate, and sulfuric acid [89]. Instead of ammonium molybdate, the corresponding tungsten salt can be used, leading to a pigment based on lead tungstate. The pigment is stabilized by adding sodium silicate and aluminum sulfate to the suspension, which is then neutralized with sodium hydroxide or sodium carbonate to form the hydrated oxides of silicon and aluminum. The precipitate is filtered off, washed, dried, and ground. The stabilization improves the lightfastness and working properties of the final pigments and, in addition, leads to an increase of oil absorption.

In the Bayer process, molybdate red is formed starting from lead nitrate, potassium chromate, sodium sulfate, and ammonium molybdate [83]. The precipitate formed is stabilized by adding sodium silicate to the stirred suspension followed by solid antimony trifluoride and further addition of sodium silicate. The pH value is adjusted to 7 with diluted sulfuric acid and the precipitate is filtered off, washed free of electrolytes, dried, and ground to obtain the desired pigment properties.

To equip lead molybdate pigments with the necessary stability against light, weathering, chemical attack, and temperature, the same methods are used as those described for the stabilization of chrome yellow pigments [80–88].

Chrome orange is obtained by precipitating lead salts with alkali chromates in the alkaline pH range. The size of the pigment particles, and thus the hue, can be varied between orange and red by controlling the pH value and temperature. Chrome green and fast chrome green mixed pigments are produced by mixing chrome yellow with iron blue or phthalocyanine blue. Dry or wet mixing is used for the production process. The dry mixing process is executed in edge runner mills, high performance mixers, or mills allowing intimate contact between the particles of the inserted chrome yellow and the iron blue and/or the phthalocyanine pigments. Differences in the density and particle size of the components may result in segregation and floating of the pigment components in the coating systems. Therefore, wetting agents are added to avoid these effects [90].

Chrome green and fast chrome green pigments with brilliant colors, high color stability, very good hiding power, and good resistance to floating and flocculation are obtained by using the wet mixing process. In this case, one component is precipitated onto the other. Solutions of sodium silicate and aluminum sulfate or magnesium sulfate are finally added for further stabilization [89]. Alternatively, the pigment components are wet milled or mixed in suspension and then filtered. The pigment slurry is dried, and the pigment is ground to achieve the desired properties [70].

4.5.3 Pigment properties and uses

Chrome yellow, molybdate orange, molybdate red, and chrome green pigments are used in paints, coatings, and plastics. They are characterized by brilliant hues, good tinting strength, and good hiding power. Figure 4.27 shows typical reflectance spectra of chrome yellow and molybdate red pigments. Photographs of representative pigment samples of chrome yellow and molybdate red are shown in Figures 4.28 and 4.29. Chrome yellow is characterized by a brilliant yellow tone, and molybdate red shows an intensive red color.

Scanning electron micrographs of chromate and molybdate pigments are shown in Figure 4.30 (chrome yellow) and Figure 4.31 (molybdate red). The chrome yellow pigment is characterized by needle-shaped particles with lengths in the range from 0.3 to 1.2 µm. The molybdate red pigment consists of more spherical particles with a smaller percentage of needles.

Special surface treatment of the pigments permits an improvement of their resistance to light, weathering, chemicals, and temperature. All chromate and molybdate pigments are supplied as pigment powders, but also as low dust and dust-free preparations or as pastes. Chrome yellow pigments are insoluble in water. Solubility in

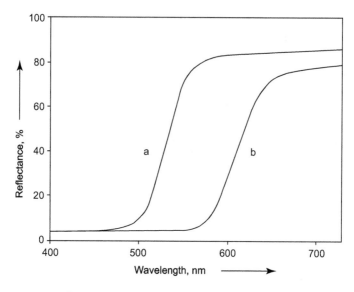

Fig. 4.27: Reflectance spectra of (a) chrome yellow and (b) molybdate red.

Fig. 4.28: Photograph of a chrome yellow pigment.

Fig. 4.29: Photograph of a molybdate red pigment.

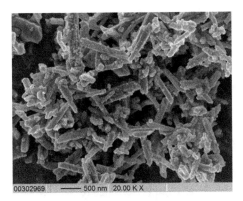

Fig. 4.30: Scanning electron micrograph of a chrome yellow pigment.

Fig. 4.31: Scanning electron micrograph of a molybdate red pigment.

acids and alkalis and discoloration by hydrogen sulfide and sulfur dioxide can be reduced to a minimum by precipitating inert metal oxides on the pigment particles.

Carefully controlled precipitation and stabilization provide chrome yellow pigments with exceptional fastness to light and weathering, and a very high resistance to chemical attack and temperature. The following qualities are commercially available [70]:

1. Unstabilized chrome yellows (limited importance)
2. Stabilized chrome yellows with higher color brilliance, stable to light, and weathering
3. Highly stabilized chrome yellow pigments:
   - very stable to light and weathering
   - very stable to light and weathering, and resistant to sulfur dioxide
   - very stable to high temperature, light, and weathering
   - very stable to high temperature, sulfur dioxide, light, and weathering
4. Low dust products (pastes or powders)

Chrome yellow pigments are mainly applied in paints, coil coatings, and plastics. They exhibit a low binder demand and favorable dispersibility as well as good hiding power, tinting strength, gloss, and gloss stability. Chrome yellows are not only used because of their beneficial pigment properties but also for economic reasons. They are important pigments for yellow colors in the production of automotive and industrial paints. Chrome yellow pigments stabilized with silicate at the surface are of major importance in the production of colored plastics (polyvinyl chloride, polyethylene, or polyesters) with high-temperature resistance. Incorporation into plastics also improves the chemical resistance of the polymer to alkali, acid, sulfur dioxide, and hydrogen sulfide [70].

Molybdate red and molybdate orange pigments possess fastness properties that are comparable with those of the chrome yellows. Similar to the lead chromates and sulfochromates, the pigment particles can be coated with metal oxides, metal phosphates, or silicates to obtain stabilized pigments with high color brilliance and good fastness properties. Highly stabilized grades with very good resistance to light, weathering, sulfur dioxide, and temperature, and with a very low content of acid soluble lead are also accessible.

The colors of lead molybdate pigments vary from red with a yellow hue to red with a blue hue. Molybdate orange has gained much more importance since the production of chrome orange was stopped. Molybdate orange and molybdate red are mainly used in paints, coil coatings, and for coloring plastics (polyethylene, polyesters, polystyrene). The temperature stable grades are the most suitable for coil coatings and plastics. Molybdate orange and molybdate red possess a low binder demand, good dispersibility, hiding power, and tinting strength, combined with very high lightfastness and weather resistance. Stabilization yields high grade types with good resistance to sulfur dioxide and high temperature as described for lead chromate and sulfochromate pigments. Like chrome yellows, molybdate reds are used to produce mixed pigments. Combinations with organic red pigments provide a considerably extended color range. Such combinations have very good stability properties because the lightfastness and weather resistance of many organic red pigments are positively affected by molybdate pigments.

Chrome greens provide excellent dispersibility, resistance to flocculation, bleeding, and floating. In addition, they have very good fastness properties. This applies especially to the fast chrome greens that are based on high grade phthalocyanine and highly stabilized chrome yellows. They are, therefore, used in the same applications as chrome yellow and molybdate red pigments, i.e., for the pigmentation of coating media and plastics.

4.5.4 Toxicology and occupational health

Chromate and molybdate pigments have to be handled carefully due to their components. Provisions have to be made and workplace concentration limits have to be kept when handling these pigments. Chromate and molybdate pigments do not exhibit acute toxicity (LD_{50} value rat oral: >5000 mg/kg).

They are not irritating to the skin or mucous membranes [72]. General regulations exist for all lead-containing materials. One of the concentration limits is the MAK value, which is <0.1 mg/m^3.

Lead chromate and molybdate pigments are hardly soluble compounds. It is, nevertheless, possible that at hydrochloric acid concentrations as available in gastric acid a partial liberation of lead can occur. The so-dissolved lead can be accumulated in the organism. This was confirmed by animal feeding studies with rats and dogs. The consequences include enzyme inhibitions and disturbances of the hemoglobin synthesis [72]. Lead chromate and molybdate pigments are, therefore, classified as chronically toxic. They are also qualified as toxic for reproduction and carcinogenic [91, 92].

Extensive epidemiological investigations have given no indication that the practically insoluble lead chromate pigments have any carcinogenic properties [93–95]. Such properties have been reported for the more soluble zinc chromate and strontium chromate pigments [70]. Chromate and molybdate pigments can be safely handled if the various rules and regulations regarding concentration limits, safe working practices, hygiene, and industrial medicine are complied with.

Environmental aspects involve particularly dust and water management. Dust emissions from approved manufacturing plants must not exceed a total mass flow of 5 g/h or a mass concentration of 1 mg/m^3 for the sum of lead and chromium [96]. Wastewater from inorganic pigment manufacturing processes in Germany discharging directly into public streams of water are limited for lead to 0.04 kg/t_{prod} and for chromium to 0.03 kg/t_{prod} [97]. Local or regional authorities might set lower limits, even for "nondirect" discharge into municipal sewer systems. In Germany, lead chromate and lead chromate pigments are assigned to Water Hazard Class (WGK) 2 (hazardous to water) [98], resulting in handling and storage restrictions and requirements on leak tightness. The European Community classifies lead chromate and lead chromate pigments as "very toxic to aquatic organisms" [99]. Waste containing lead chromate pigments that cannot be recycled must be transported to special waste disposal sites under appropriate control.

4.6 Bismuth vanadate pigments

4.6.1 Fundamentals and properties

Greenish yellow pigments based on bismuth orthovanadate, BiVO$_4$, represent a class of pigments with interesting coloristic properties [100–102]. They extend the range of yellow inorganic pigments, such as iron oxide yellow, chrome yellow, cadmium yellow, nickel rutile yellow, and chromium rutile yellow. In particular, they can substitute the greenish yellow lead chromate and cadmium sulfide pigments. In the Color Index, they are registered as C.I. Pigment Yellow 184.

Bismuth vanadate occurs naturally as the brown mineral pucherite (orthorhombic), as clinobisvanite (monoclinic), and as dreyerite (tetragonal). All these deposits are of no importance for the pigment industry. The development and market introduction of pigments based on BiVO$_4$ occurred later than for most of the other inorganic colored pigments [103, 104]. A driving force for this in the 1970s was the search for suitable nontoxic alternatives to existing yellow pigments such as lead chromate or cadmium sulfide. Pigments containing other phases besides $BiVO_4$, e.g., Bi_2XO_6 (X = Mo or W), have also been investigated and developed [105].

All commercial bismuth vanadate pigments are based on bismuth vanadate with monoclinic or tetragonal structure. The pigments are particularly brilliant in the greenish yellow range of the color space. The composition ranges from pure $BiVO_4$ to the mixed pigment $4BiVO_4 \cdot 3Bi_2MoO_6$, where molybdenum is incorporated in the structure. The bismuth vanadate component crystallizes here in the tetragonal scheelite structure, whereas bismuth molybdate occurs in the orthorhombic perovskite crystal lattice [102].

4.6.2 Production of bismuth vanadate pigments

Bismuth vanadate pigments can be synthesized either by solid-state processing using appropriate starting materials, e.g., Bi_2O_3 and V_2O_5, or by coprecipitation from aqueous solutions [103–109].

The solid-state reaction is performed at temperatures of about 500 °C:

$$Bi_2O_3 + V_2O_5 \rightarrow 2\,BiVO_4 \tag{4.29}$$

A much more suitable route for the synthesis of bismuth vanadate pigments is the precipitation process starting from solutions of bismuth nitrate and sodium vanadate or ammonium vanadate. The $Bi(NO_3)_2$ solution is obtained by reaction of bismuth metal with nitric acid. An amorphous precipitate is formed when sodium hydroxide is added to the strongly acidic nitrate and vanadate-containing solution. It consists of oxides and hydroxides and is washed salt free after the end of the reaction. The pigment formation can be summarized as follows:

$$Bi(NO_3)_3 + NaVO_3 + 2\,NaOH \rightarrow BiVO_4 + 3\,NaNO_3 + H_2O \tag{4.30}$$

By heating the suspension to reflux and controlling the pH-value, a transformation occurs, producing a fine crystalline product. Selective formation of particular $BiVO_4$ modifications is possible by controlling the coprecipitation conditions. Only two of the four bismuth vanadate polymorphs are of interest for pigments. These are the monoclinic and the tetragonal modifications, which exhibit brilliant yellow colors. It is, therefore, one of the targets to steer the process in such a way that these $BiVO_4$ crystal structures are formed. The color shade and brilliance strongly depend on the precipitation conditions, e.g., concentration, temperature, and pH-value.

The reaction for the formation of the pigment composition $4BiVO_4 \cdot 3Bi_2MoO_6$ requires ammonium molybdate in addition to the starting components described for pure $BiVO_4$:

$$10\,Bi(NO_3)_3 + 4\,NaVO_3 + 3\,(NH_4)_2MoO_4 + 20\,NaOH$$

$$\rightarrow 4BiVO_4 \cdot 3Bi_2MoO_6 + 24\,NaNO_3 + 6\,NH_4NO_3 + 10\,H_2O \qquad (4.31)$$

Improved pigment properties for all types of bismuth vanadate pigments can be achieved by an additional annealing step for the washed and dried precipitate at about 500 °C.

Bismuth vanadate pigments are often stabilized with additional layers to improve their properties, e.g., weathering and acid resistance. Surface treatments used for this purpose contain calcium, aluminum, or zinc phosphate, but also oxides like aluminum oxide. Specifically for the use in plastics, the pigments are coated with dense layers of silica or other silicon-containing components to increase the stability in polymers, such as polyamide, up to 300 °C [108]. The pigments are also offered in the form of fine granulates to avoid dust during handling. Such free flowing granulates are produced in spray tower dryers combined with an automatic packaging unit. In the final phase, the manufacture of pigment powders takes place by conventional drying using a belt dryer or other drying equipment followed by a grinding procedure.

4.6.3 Pigment properties and uses

Bismuth vanadate pigments are characterized by excellent brightness of shade, very good hiding power, high tinting strength, very good weather fastness, and high chemical resistance. They are easy to disperse and environmentally friendly. Compared with other inorganic yellow colorants, bismuth vanadate pigments are close to cadmium sulfide and lead chromate with respect to color properties. Their shade can be described as greenish yellow. Figure 4.32 shows a representative photograph of a bismuth vanadate pigment with the composition $BiVO_4$. Figure 4.33 demonstrates the coloristic situation for different yellow pigments by means of the remission curves of bismuth vanadate, cadmium sulfide, nickel rutile yellow, lead sulfochromate, and iron oxide yellow. The color saturation of bismuth vanadate pigments is significantly higher than for iron oxide yellow and nickel rutile yellow.

Unstabilized bismuth vanadate pigments show reversible color changes when irradiated intensely with light (photochromism). This effect can be reduced or nearly prevented by stabilization of the pigments using suitable surface treatment. The pigments have a density of 5.6 g/cm^3 and a refractive index of 2.45. The specific surface area (BET) reaches values of about 10 m^2/g. $BiVO_4$ pigments have a very good weather resistance in full shade and in combination with titanium dioxide.

Figure 4.34 shows a scanning electron micrograph of a pure bismuth vanadate pigment. The image shows a bimodal particle size distribution for the pigment. Particles with a size of about 300 nm are dominant. There are, however, particles with a size of less than 100 nm.

Fig. 4.32: Photograph of the bismuth vanadate pigment $BiVO_4$.

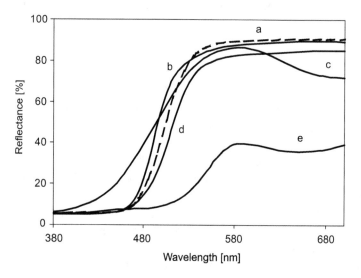

Fig. 4.33: Reflectance spectra of inorganic yellow pigments. (a) $BiVO_4$, (b) CdS, (c) $(Ti,Ni,Sb)O_2$, (d) $Pb(Cr,S)O_4$, and (e) α-FeOOH [100].

Bismuth vanadate pigments are used in the manufacture of lead-free, weather-resistant, brilliant yellow colors for automotive OEM and refinishes, industrial and decorative paints, powder coatings, and coil coatings. In combination with additional colorants, the pigments can be used as basis for a broad series of yellow, orange, red, and green color tones.

$BiVO_4$ pigments are available with a heat stability up to 300 °C. They exhibit very good lightfastness and weather resistance in plastics for outdoor use. The pigments have outstanding fastness to migration in plastics and are easily dispersible. With

00301688 ———— 300 nm 50.00 K X

Fig. 4.34: Scanning electron micrograph of the bismuth vanadate pigment $BiVO_4$.

their excellent heat stability, the thermostable types can be readily incorporated into polyolefins and ABS at 260 to 280 °C and even in polyamide at 280 to 320 °C [100].

4.6.4 Toxicology and occupational health

Bismuth vanadate pigments do not exhibit acute toxicity (LD_{50} value rat oral: >5000 mg/kg). They are not irritating to the skin or mucous membranes. Animal studies gave some indication of inhalation toxicity, which may have been due to the vanadium content in the pigments [100, 102]. Toxic effects are observable for rats only when the concentration in the lungs reaches levels that do not occur under the usual conditions of industrial hygiene. For risk reduction, some producers supply bismuth vanadate pigments in a free flowing, low dust form. Such preparations make the pigments inaccessible to the lungs. Therefore, the dust-free pigments can be handled under usual hygienic working conditions.

4.7 Ultramarine pigments

4.7.1 Fundamentals and properties

Ultramarine blue is the synthetic equivalent to the naturally occurring mineral lapis lazuli, the quality of which is found in Afghanistan and Chile. These semiprecious stones have been used since prehistoric times for decorative purposes. Lapis lazuli was imported already in the thirteenth century from the Far East to Europe. A blue pigment was obtained out of the mineral in a circumstantial way by grinding and separation of the lighter rock using a kneading process. The so-obtained blue pigment powder was named "ultramarine", which means "across the sea".

The deep blue color was used by many artists and can be seen already in paintings of Giotto and Jan van Eyck. The pigment was very soon an indispensable colorant for artists' colors. Excellent lightfastness and color purity of the pigment powder were decisive for the use in art. Increasing transportation costs, depletion of mines, and subsequent processing led to the situation that the pigment was more expensive than gold at the end of the eighteenth century. As a result, scientists in Europe started activities in the 1820s to find a synthetic and economic process for the manufacture of ultramarine blue. The chemical composition was determined already in 1806 by French chemists, who detected the presence of sodium, silica, alumina, and sulfur in ultramarine [110–112].

Today it is assumed that Guimet in France and Gmelin in Germany independently developed similar processes for the synthetic preparation of ultramarine in 1828. Guimet, who won an award for this development, is, however, frequently referred to as the first person to produce ultramarine blue on a commercial scale [113].

Synthetic ultramarines as inorganic pigments are today commercially available in three colors:
- reddish blue, C.I. Pigment Blue 29
- violet, C.I. Pigment Violet 15
- pink, C.I. Pigment Red 259

It has been known since the determination of the chemical composition that ultramarine is a sulfur-containing sodium aluminosilicate. However, it was only much later that it was possible by means of X-ray diffraction to elucidate the basic structure. Accordingly, it is a sodalite crystal lattice corresponding to the formula $Na_6Al_6Si_6O_{24}$ (NaS_n) with n values from 2 to 4. In the meantime, the use of modern spectroscopic methods allowed a profound insight into the nature of the color giving components [114–117]. Electron paramagnetic resonance spectroscopy was able to detect the attendance of free polysulfide radicals in the crystal structure. These radicals are normally very unstable, but incorporated as negatively charged polysulfide ions, e.g., as S_3^-, in the aluminosilicate lattice together with sodium counterions, they are stabilized. The crystal structure of ultramarine pigments is shown in Figure 4.35.

In the simplest case of the three-dimensional basic structure, equal numbers of silicon and aluminum ions are present, and the basic lattice unit is $Na_6Al_6Si_6O_{24}$ or $(Na^+)_6$ $(Al^{3+})_6(Si^{4+})_6(O^{2-})_{24}$. The nature of the incorporated polysulfide groups is responsible for the color of the pigments derived from this structure [118–122]. The basic lattice $(Na^+)_6$ $(Al^{3+})_6(Si^{4+})_6(O^{2-})_{24}$ is derived from $Si_{12}O_{24}$ by substituting six of the silicon ions by aluminum ions. Every Al^{3+} ion must be incorporated together with a Na^+ ion so that the overall ionic charge for the structure is zero. Hence, six of the eight available sodium sites in the structure are always occupied by sodium ions required for lattice stability, and the remaining two sites are filled with sodium ions associated with polysulfide groups. This means that only one S_3^{2-} polysulfide ion can be inserted into the lattice (as Na_2S_3), even though subsequent oxidation to S_3^- leads to loss of one of the accompanying sodium ions. This gives a basic ultramarine lattice formula of $Na_7Al_6Si_6O_{24}S_3$ [110].

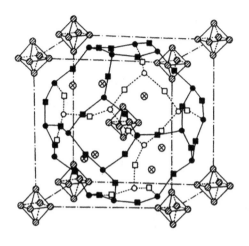

□ O
O Si oder Al
⊗ Na
⊘ S

Fig. 4.35: Crystal structure of ultramarine pigments.

There are two types of sulfur groups in ultramarine blue: S_3^- and S_2^-. Both species are free radicals stabilized by entrapment in the crystal lattice. The spacing between the three sulfur atoms in the predominant S_3^- radicals is 0.2 nm, and the angle between them is 103°. S_3^- absorbs a broad energy band in the visible green-yellow-orange region centered at 600 nm, whilst S_2^- absorbs in the violet-ultraviolet region at 380 nm [110].

An increase of the sulfur content improves the color quality. To achieve this, the lattice aluminum content can be decreased by using high silicon feldspar in the manufacturing process. This approach also reduces the number of sodium ions needed for lattice stabilization and leaves more for polysulfide group equivalence. A typical product resulting from this procedure is $Na_{6.9}Al_{5.6}Si_{6.4}O_{24}S_{4.2}$, which has a stronger, redder shade of blue than the simpler composition.

The reflectance spectra of ultramarine blue and ultramarine violet are shown in Figure 4.36. In the case of ultramarine violet, and also of ultramarine pink, the lattice structure is slightly modified, but the polysulfide color centers are further oxidized, possibly to S_3Cl^-, S_4, or S_4^- [110].

Synthetic and natural ultramarines with their sodalite-related crystal structure belong to the clathrates. Although they have a system of very small cavities (cages), their lattice paths are limited to 0.4 nm diameter channels. The sodium ions can be substituted by other metal ions, e.g., silver, potassium, lithium, or copper. The exchange with potassium ions leads to slightly redder shades of ultramarine blue [123]. Under specific reaction conditions, an ultramarine green pigment can also be prepared

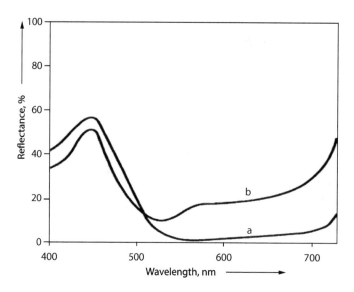

Fig. 4.36: Reflectance spectra of (a) ultramarine blue and (b) ultramarine violet.

[124, 125]. However, this product has not found any technical use due to insufficient color and application properties.

4.7.2 Production of ultramarine pigments

The production of ultramarine pigments is carried out in the four steps of clay activation, blending and heating of the raw materials, oxidation, and purification and refinement. The raw materials used are china clay (kaolinitic clay), feldspar, anhydrous sodium carbonate, sulfur, and a reducing agent (oil, pitch, or coal). In order to improve and simplify the complex and time-consuming technical synthesis of ultramarine pigments, a wide range of investigations have been carried out, but in the end consequence these have not led to any fundamental change in the manufacturing process [126–131].

The clay activation takes place continuously in a rotary or a tunnel kiln or in a batch process in crucibles in a tiled stove. Kaolinite $(Al_4(OH)_8Si_4O_{10})$ is transformed to metakaolinite $(Al_4(OH)_{8-2x}O_xSi_4O_{10})$ by heating the clay to about 700 °C. The kaolinite structure is destabilized by water removal and metakaolinite is formed:

$$Al_4(OH)_8Si_4O_{10} \rightarrow Al_4(OH)_{8-2x}O_xSi_4O_{10} + x\,H_2O \tag{4.32}$$

The thus-activated clay is blended with the other raw materials and dry ground. Batch or continuous ball mills are typically used to obtain mean particle sizes of about 15 μm. Table 4.7 contains typical recipes for ultramarine blue pigments.

Tab. 4.7: Typical recipes for ultramarine blue pigments (in wt%) [110].

Component	Green shade type	Red shade type
Calcined clay	32.0	30.0
Feldspar		7.0
Sodium carbonate	29.0	27.0
Sulfur	34.5	33.0
Reducing agent	4.5	3.0

The ground powder mixture is heated to temperatures of 750 to 800 °C under reducing conditions, normally in a batch process. Directly fired kilns with the blend in lidded crucibles of controlled porosity or muffle kilns are traditionally used for this reaction step. The ground powder mixture is in practice often densified to form bricks, with the aim to improve the throughput. The bricks are stacked in a defined pattern and fired in chambers, which are indirectly heated using gas-fired burners [110]. Sodium carbonate reacts in a very complicated firing process under reducing conditions (sulfur dioxide atmosphere) with the sulfur and the reducing agent at 300 °C to form sodium polysulfide [132]:

$$Na_2CO_3 + xS + SO_2 \rightarrow Na_2S_x + CO_2 + SO_3 \tag{4.33}$$

At higher temperatures of 700 to 800 °C and over a period of several days, the meta-kaolinite reacts with sodium carbonate and is transformed into a framework with a three-dimensional sodalite structure:

$$3\,Al_4(OH)_{8-2x}O_xSi_4O_{10} + 6\,Na_2CO_3 \rightarrow 2\,Na_6Al_6Si_6O_{24} + 6\,CO_2 + (12-3x)\,H_2O \tag{4.34}$$

Na_2S_x formed according to (4.33) is also involved in this chemical conversion. It reacts with the $Na_6Al_6Si_6O_{24}$ to pre-ultramarine, which is still colorless:

$$Na_6Al_6Si_6O_{24} + Na_2S_x \rightarrow Na_6Al_6Si_6O_{24} \cdot Na_2S_x \tag{4.35}$$

The thus-formed pre-ultramarine contains entrapped sodium and polysulfide ions in the structure.

In the subsequent oxidation step, the furnace is cooled to 500 °C and air is admitted in controlled amounts. The oxygen reacts slowly in up to 20 days with excess sulfur to form sulfur dioxide, which exothermically oxidizes the diatomic and triatomic polysulfide ions to S_2^- and S_3^- free radicals, leaving sodium-sulfur compounds and sulfur as by-products:

$$2\,Na_6Al_6Si_6O_{24} \cdot Na_2S_x + O_2 + SO_2 \rightarrow 2\,Na_7Al_6Si_6O_{24} \cdot S_x + Na_2SO_4 \tag{4.36}$$

The furnace is cooled down and unloaded when oxidation is finished. A full kiln cycle takes 3 to 4 weeks. The thus-produced "raw" ultramarine product typically contains 75 wt% blue ultramarine, 23 wt% sodium sulfoxides, and 2 wt% sulfur [110].

Purification and refinement procedures can take place batchwise or continuously. The raw ultramarine blue is crushed and ground, slurried in warm water, then filtered, and washed to remove the sulfur compounds. Repeated elutriating and wet grinding release the sulfurous impurities and reduce the ultramarine particle size to a range of 0.1 to 10.0 μm. The impurities are floated off by boiling or cold froth flotation.

Discrete particle size fractions are achieved by gravity or centrifugal separation processes. Flocculation and filtration are used to remove residual fine particles. The separated fractions are dried and disintegrated to give pigment grades of different particle size distribution. Blending to sales grade standard, adjusting hue, brightness, and color strength to achieve specified color tolerance are the final operations before the pigments can be sold [110].

Violet ultramarine pigments can be produced by heating a mid-range blue grade with ammonium chloride at about 240 °C in the presence of air. Treating the violet pigment with hydrogen chloride gas at 140 °C leads to the pink derivative.

4.7.3 Pigment properties and uses

Ultramarine pigments exhibit intensive shades of blue and relatively weak violet and pink color tones. The basic blue ultramarine color is a rich, bright reddish blue. The blue color tone of the final pigments is adjusted by varying the chemical composition and the particle size distribution. The violet and pink representatives have significantly weaker and less saturated colors.

The final color quality of commercial pigments is developed by grinding to reduce the particle size to desired distribution and thus enhance tinting strength. Mean particle sizes for most of the pigments are in the range from 0.7 to 5.0 μm. Fine powders are lighter in shade and rather greener than coarser grades. When used together with white pigments, their color is brighter and more saturated. Figures 4.37 and 4.38 show photographs of representative pigment samples of ultramarine blue and ultramarine violet.

Fig. 4.37: Photograph of an ultramarine blue pigment.

Fig. 4.38: Photograph of an ultramarine violet pigment.

Figure 4.39 shows a scanning electron micrograph of an ultramarine blue pigment. As a result of the manufacturing conditions, a larger proportion of the particles stick together. The particles are characterized by an irregular shape and by sizes in the range from 50 to more than 500 nm.

Fig. 4.39: Scanning electron micrograph of an ultramarine blue pigment.

Ultramarine pigments give a transparent blue in gloss paints and clear plastics. The reason for this is the refractive index, which is close to 1.5 and, therefore, similar to that of paints and plastics. Opacity can be achieved by the addition of small quantities of a white pigment. Increasing quantities of white create paler shades, and a trace of ultramarine added to a white enhances the whiteness and acceptability [110].

Ultramarine blue is thermally stable in different applications to around 400 °C, violet to 280 °C, and pink to 220 °C. All three pigments have an excellent lightfastness. Ultramarines are not stable against acids. The pigments can completely be decomposed by the action of acids. The color is lost in this case and silica, sulfur, sodium, and aluminum salts, as well as hydrogen sulfide, are formed. Evolution of hydrogen sulfide under the influence of acids is a useful test for ultramarine.

A short-term stability against acids can be achieved by protecting the pigment particles with a coating of impervious silica. Blue and violet pigment grades are stable in diluted alkaline solutions, whereas pink changes the color to a violet shade.

Ultramarine pigments are insoluble in water and organic solvents. Therefore, the color does not bleed or migrate from paints or polymers. As a consequence, ultramarines are approved for a wide range of food contact applications.

The fine particles of ultramarine pigments have a high surface energy and are cohesive. Finer grades with their larger surface area are, therefore, less easy to disperse than the coarser types. There are also pigments with specifically treated surfaces available, where the energy is reduced and the dispersibility is improved.

Ultramarine pigments absorb moisture on the external particle surfaces and at the internal surfaces of the sodalite structure. External surface moisture of up to 2% according to particle size is desorbed at about 100 °C, but the additional 1% of internal moisture needs 235 °C for complete removal [110]. Particles of ultramarine pigments are relatively hard. They are known to cause abrasion in equipment handling either as dry or slurried pigment.

The specific surface area of ultramarine powders varies with the particle size distribution and has values in the range of 1 to 3 m²/g. The stability and safety of ultramarine pigments are good prerequisites for their use in many applications. The pigments are applied among others in plastics, paints and powder coatings, printing inks, paper and paper coatings, rubber and thermoplastic elastomers, latex products, detergents, cosmetics and soaps, artists' colors, toys and educational equipment, and leather finishes.

Blue ultramarine pigments can be used in any polymer. Violet ultramarine grades have maximum processing temperatures of 280 °C. Pink ultramarines can be used in polymers up to maximum processing temperature of 220 °C. Acid-resistant grades are commonly used in PVC, where color can fade during processing. Surface-treated grades are applied, when enhanced dispersibility is necessary.

Ultramarine pigments are used in many paint systems. Decorative paints, stoving finishes, transparent lacquers, industrial paints, and powder coatings are typical applications. The transparent nature of the pigments makes them attractive for some impressive coatings in combination with mica-based special effect pigments [110]. Ultramarine pigments can be used in inks for most of the printing processes. Letterpress as well as flexographic and gravure printing need high strength grades. Lithography, on the other hand, needs water-repellent types. Screen printing, fabric printing, and hot foil stamping can work with any grade of ultramarine pigment. Improved strength grades were developed for high solid aqueous dispersions, which are finding increasing use in flexographic printing applications [110]. Ultramarine blue pigments are used to enhance the hue of white paper and to correct the yellow tinge. They are added directly to the paper pulp. Alternatively, they are used in applied coatings. Blue ultramarines are also widely used to enhance the effects of optical brightening agents in improving whiteness of laundered fabrics. In addition, ultramarine pigments have found a wide use in cosmetics.

They are regarded as completely safe, nonstaining, and conformal to all major regulations in this application field.

Artists' colors belong to the traditional uses for ultramarine pigments. They are still an important application for all ultramarine colors. Unique color properties, stability, and safety are important for the use of the pigments in this area of application. Plastics and surface coatings for toys, children's paints and finger paints, modeling compositions, colored paper, and crayons are additional applications for ultramarines. They are compliant with major regulations and standards for these uses. Ultramarines are categorized as laundry grades, which are low strength and sometimes low-purity materials, and as industrial/technical grades, which are high strength, high-purity pigments [110].

4.7.4 Toxicology and occupational health

Ultramarine pigments do not exhibit acute toxicity (LD_{50} value rat oral: >5000 mg/kg). They are not irritating to the skin or mucous membranes. The pigments have had a long term and widespread human and environmental exposure without any reported instances of ill effects. The worldwide application of ultramarine as an additive for sugar, to produce a whitening effect, and as a whitening agent for clothes for use in the household washing have not had any adverse health effects. The use of ultramarine for whitening of clothes is still widespread.

The only known hazard of ultramarines is the evolution of hydrogen sulfide on contact with acid. This is the reason why ultramarine pigments must not be mixed with acids. It must be ensured that no contamination with acids is possible during storage and disposal. Ultramarine pigments are not combustible, but in the case of a surrounding fire, sulfur dioxide can be formed. When handling the pigments, excessive dust formation should be avoided. There is no hazard designation or transport classification for ultramarine pigments.

The manufacturing process releases larger amounts of sulfur dioxide and of water soluble sodium sulfur compounds. These must be disposed of in accordance with environmental regulations. Reduction of the sulfur dioxide emissions is achieved by conversion of the SO_2 to SO_3 using a vanadium pentoxide catalyst and the subsequent condensation of the sulfur trioxide to produce very concentrated and pure sulfuric acid. The soluble sulfur salts can be fully oxidized to sulfates, which can be disposed using a suitable sewage system [110].

4.8 Iron blue pigments

4.8.1 Fundamentals and properties

The term "iron blue pigment" has replaced a number of older names (e.g., Paris blue, Prussian blue, Berlin blue, Milori blue, and Turnbull's blue) to a large extent. These names were used for insoluble pigments based on microcrystalline Fe(II)Fe(III) cyano complexes. Many of these were associated with specific hues. A standardization of the naming system has led to a reduction in the number of varieties [133–136].

Iron blue with the nomenclature C.I. Pigment Blue 27:77,510 (so-called "insoluble iron blue") was prepared for the first time in 1706 by the color maker Diesbach and the alchemist Johann Conrad Dippel in Berlin, who used a precipitation reaction. The resulting product can be regarded as the oldest synthetic coordination compound. Milori was the first to produce it as a pigment on an industrial scale in the early nineteenth century [133]. There is also a pigment called "soluble iron blue" having the nomenclature C.I. Pigment Blue 27:77520. In technical applications, the pigment is today usually referred to as iron blue.

X-ray and infrared spectroscopic measurements show that "soluble" iron blue pigments can be described by the formula $M^I[Fe^{III}Fe^{II}(CN)_6] \cdot x\ H_2O$ [137]. M^I represents sodium, potassium, or ammonium ions. Potassium and ammonium ions are preferred in industrial manufacturing processes because they produce excellent hues. The ammonium compound is the most common one [133].

$M^I[Fe^{III}Fe^{II}(CN)_6] \cdot x\ H_2O$ is called "soluble iron blue" because it consists of colloidal particles, which are colloidally solved after the synthesis. The addition of excessive Fe^{2+} and Fe^{3+} ions leads to the "insoluble iron blue". The composition of this precipitate can be described with the formula $Fe^{III}_4[Fe^{II}(CN)_6]_3 \cdot x\ H_2O$ or $Fe^{III}[Fe^{III}Fe^{II}(CN)_6]_3 \cdot x\ H_2O$ ($x = 14$–16).

Figure 4.40 shows the crystal structure of $M^I[Fe^{III}Fe^{II}(CN)_6] \cdot x\ H_2O$, particular the $Fe^{III}Fe^{II}(CN)_6$ lattice [138, 139]. Fe^{2+} ions (Fe^{II}, ferrous ions) form a face-centered cubic lattice, which is interlocked with another face-centered cubic lattice of Fe^{3+} ions (Fe^{III}, ferric ions), to give a three-dimensional cubic lattice with the corners and face centers occupied by ferrous ions. The CN groups are located at the edges of the cubes, each between a ferrous ion and the neighboring ferric ion. The carbon atom of the CN group is bonded to the Fe^{II} ion, whereas the nitrogen atom is linked to the Fe^{III} ion. The M^+ ions (M^I) and the water molecules are arranged inside the cubes formed by the iron ions. The intensive color of iron blue can be explained by a charge transfer between Fe^{2+} and Fe^{3+} ions.

The crystal structure of $Fe^{III}_4[Fe^{II}(CN)_6]_3 \cdot x\ H_2O$ also consists of a three-dimensional cubic Fe^{III}–NC–Fe^{II} framework, but in this case, one-fourth of the Fe^{II} sites (and cyanide positions as well) are vacant. Water molecules at empty nitrogen positions complete the coordination sphere of the Fe^{3+} ions [140]. Coordinative water as expressed in the formula $M^I[Fe^{III}Fe^{II}(CN)_6] \cdot x\ H_2O$ is absolutely necessary for the stabilization of the crystal structure. Iron blue loses its pigment character when this water is removed. Additional studies on the detailed structure of iron blue pigments used methods like neutron diffraction and X-ray photoelectron spectroscopy [141–144].

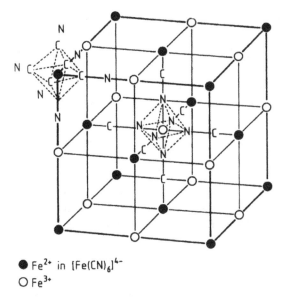

● Fe²⁺ in [Fe(CN)₆]⁴⁻
○ Fe³⁺

Fig. 4.40: Section of the crystal structure of iron blue [138, 139].

4.8.2 Production of iron blue pigments

Iron blue pigments ("insoluble iron blue") are manufactured by a precipitation process in aqueous solution starting from potassium or sodium hexacyanoferrate(II) and iron(II) sulfate or iron(II) chloride. A white precipitate consisting of iron(II) hexacyanoferrate(II) ($M^I_2 Fe^{II}[Fe^{II}(CN)_6]$, Berlin white) is formed:

$$K_4\left[Fe^{II}(CN)_6\right] + FeSO_4 \rightarrow K_2 Fe^{II}\left[Fe^{II}(CN)_6\right] + K_2SO_4 \qquad (4.37)$$

Soluble reaction products are separated by decantation. In a next step, the precipitate is aged by heating the aqueous suspension and then oxidized with alkali chlorate, hydrogen peroxide, or alkali dichromates in the presence of hydrochloric acid to give a blue pigment with the composition $Fe^{III}_4[Fe^{II}(CN)_6]_3$. When pure sodium hexacyanoferrate(II) solution is used at the beginning of the process, the pigment properties are adjusted by adding a potassium or ammonium salt during the precipitation of the white paste product or prior to the oxidation stage.

The industrial precipitation is executed batchwise in stirred tanks by simultaneous or sequential addition of aqueous solutions of alkali hexacyanoferrate(II) and the iron(II) compound to a diluted acid. The reaction conditions, in particular, temperature and concentration of the starting solutions, have a significant influence on size and shape of the precipitated particles. The aging period for the white precipitate, which follows after decantation, varies in duration and temperature, depending on the properties required of the finished pigment. The oxidation step to form the blue pigment by adding, e.g., the oxidizing agent together with hydrochloric acid can be done in the same reactor. Finally, the suspension of the blue pigment is pumped into filter presses. After filtering, the resulting cake is washed until there are no acids and salts present. The washed filter cake is dried using tunnel or belt driers as well as spray or spin flash driers. The thus-obtained solid, which is ground in dependence on the desired product quality, is packed into bags, or stored in silos. There is also the possibility of forming rods or pellets by extruding the washed filter cake with a granulator. In this case, a dust-free iron blue pigment is obtained after drying [133].

The dispersibility of iron blue pigments can be improved by adding organic compounds to the pigment suspension before filtering to prevent the particles from agglomeration during drying [133]. "Soluble iron blue" pigments are accessible by the reaction of potassium hexacyanoferrate(II) (yellow prussiate of potash) with an iron(III) salt or by the reaction of potassium hexacyanoferrate(III) (red prussiate of potash) with an iron(II) salt in aqueous solution:

$$K_4\left[Fe^{II}(CN)_6\right] + Fe^{3+} \rightarrow K\left[Fe^{III}Fe^{II}(CN)_6\right] + 3\,K^+ \tag{4.38}$$

$$K_3\left[Fe^{III}(CN)_6\right] + Fe^{2+} \rightarrow K\left[Fe^{III}Fe^{II}(CN)_6\right] + 2\,K^+ \tag{4.39}$$

In both cases, colloidal iron blue is formed in when the mole ratio 1:1 is used. Manufacture of the insoluble $Fe^{III}{}_4[Fe^{II}(CN)_6]_3$ out of $K[Fe^{III}Fe^{II}(CN)_6]$ by the addition of excessive Fe^{2+} and Fe^{3+} ions is basically possible, but not favorable for an industrial process [145].

4.8.3 Pigment properties and uses

Iron blue pigments have several properties of special practical relevance, such as hue, relative tinting strength, dispersibility, and rheological behavior. Pure pigments cause a black color impression. Their chroma is recognizable to the human eye only by addition of light scattering colorants. The reflectance spectrum of iron blue is shown in Figure 4.41 together with the curve of copper phthalocyanine blue. Figure 4.42 shows a photograph of a representative pigment sample of iron blue.

Iron blue pigments consist of extremely fine particles (primary particles from 0.01 to 0.1 μm). Due to their small sizes, the pigments are difficult to disperse. Figure 4.43 shows a scanning electron micrograph of an iron blue pigment. It gives a clear picture of the small particle sizes and of the tendency to agglomeration respectively aggregation.

The lightfastness and weather resistance of the pigments are excellent, especially when used in their pure form. When mixed with white pigments, these properties can be affected negatively [146]. A suitable topcoat, as commonly applied in automobile manufacture, can overcome this problem [133]. Iron blue pigments are stable against diluted mineral acids and oxidizing agents. Concentrated acids and hydroxides decompose the pigments. They do not bleed in common application systems.

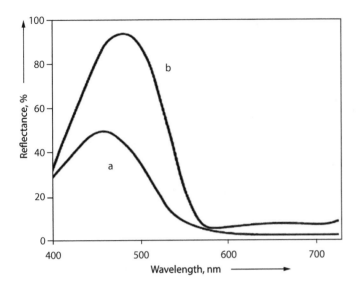

Fig. 4.41: Reflectance spectrum of iron blue (a) in comparison with the spectrum of copper phthalocyanine blue (b).

Fig. 4.42: Photograph of an iron blue pigment.

Iron blue pigments are mostly used in its pure form, e.g., in printing inks, and do not need any additives to improve them. Well-dispersed iron blue pigments impart printing inks a pure black tone. Iron blue pigments are thermally stable for short periods up to temperatures of 180 °C. They can, therefore, be used in stoving finishes. The pigment powders have an explosion hazard with an ignition temperature in the range of 600 to 625 °C. They are combustible with an ignition temperature in air above 140 °C [138]. The main applications for iron blue pigments are the printing ink

00301806 ———— 300 nm 50.00 K X

Fig. 4.43: Scanning electron micrograph of an iron blue pigment.

industry and agriculture. Micronized iron blue pigments are used for coloring of fungicides.

Iron blue pigments are very interesting for printing applications, especially for rotogravure printing. Their deep hue, good hiding power, and economic cost/performance basis are decisive for their use in this printing method. For multicolor printing, iron blue is often mixed with phthalocyanine pigments. Controlling the shade of black printing inks is another relevant application. Typical amounts used in practice are 5 to 8% for full shade rotogravure inks and 2 to 8% for toning black gravure and offset inks [133]. Iron blue pigments have even been used in the printing design of blue postage stamps [147].

The pigments are also applied in the production of single and multiple use carbon papers and blue copying papers. In both cases, the pigments are used for toning carbon black [138]. Toning of black gravure inks and black offset printing inks is described in detail in [133].

Blue inorganic fungicides based on copper compounds are mostly used for treating vines, olives, or citrus fruits and have been utilized already for a long time in Mediterranean countries, but they have now been replaced by colorless organic compounds. Micronized iron blue pigments are used to color these fungicides. Even small amounts become visible due to the high color intensity of the pigment and, therefore, precise control of the fungicide application is possible. The fungicide is usually mixed with a micronized iron blue pigment [148].

Iron blue pigments are used only to a limited extent in the paint industry, e.g., for full, dark blue colors for automotive finishes. Typical pigment concentrations to obtain a full shade with good hiding power are 4 to 8 wt% [133]. "Water soluble" iron blue pigments are used directly in the aqueous phase for manufacturing of blue paper. An alternative approach is the use of a suitable iron blue pigment, which is ground together with a water soluble binder, applied to the paper, then dried and glazed. The pigment concentration used here is 8 wt% [133].

Iron blue has also become important as an agent for medical applications. It is used for decontamination of persons or animals having ingested radioactive material. The isotope ^{137}Cs, which would otherwise be freely absorbed via the human or animal digestive tract, is exchanged with the iron(II) of the iron blue and is then excreted through the intestines [133, 149–151]. $Fe^{III}_4[Fe^{II}(CN)_6]_3$ is also used as an effective antidote for thallium intoxications. It interferes with the entero systemic circulation of thallium ions and enhances their fecal excretion [152].

4.8.4 Toxicology and occupational health

Iron blue pigments do not exhibit acute toxicity (LD_{50} value rat oral: >5000 mg/kg). They are not irritating to the skin or mucous membranes. Additional long-term tests have shown that there are no harmful effects of iron blue pigments on humans and animals. The adsorption of iron blue pigments in living beings is very low. It has been shown that the $Fe^{II}(CN)_6$ units of an intravenously injected ^{59}Fe radio labeled iron blue pigment are rapidly and virtually completely excreted with the urine. After oral administration of $Fe^{III}_4[Fe^{II}(CN)_6]_3$, only 2% of the labeled hexacyanoferrate ions (^{59}Fe) were adsorbed by the gastrointestinal tract [153]. Most of the substance is excreted through the intestines. There was no evidence for the decomposition of the hexacyanoferrate unit [133].

The decomposition of iron blue to toxic cyanide in aqueous media is found to be very low. The hydrogen cyanide release from $K[Fe^{III}Fe^{II}(CN)_6]$ in artificial gastric and intestinal juice was 141 and 26 µg/g after 5 h. The measured value for water after the same time was 37 µg/g. The investigation of $Fe^{III}_4[Fe^{II}(CN)_6]_3$ under comparable conditions showed values of 64, 15, and 22 µg/g [133]. Primary irritation tests did not detect any or only slight effects on the skin or in the eyes of treated rabbits [133]. The medical application of $Fe^{III}_4[Fe^{II}(CN)_6]_3$ in doses of up to 20 g/d for the decontamination of persons exposed to radiocesium was not accompanied with any toxicological effects [154]. The effect of iron blue pigments on fishes was found to be harmless. There are, however, slight toxic effects on bacteria when iron blue is in the water [133].

4.9 Cadmium sulfide/selenide pigments

4.9.1 Fundamentals and properties

Cadmium sulfide/selenide pigments, also referred to as cadmium pigments, consist of cadmium sulfides and sulfoselenides as well as zinc-containing sulfides of cadmium. Cadmium pigments have the broadest range amongst the inorganic pigments with respect to brilliant yellow, orange, and red colors. The yellows vary from pale primrose to deep golden yellow, the reds from light orange via deep orange, light red, crimson to maroon [155–158].

Cadmium sulfide occurs in nature as cadmium blende or greenockite. It crystallizes in the hexagonal wurtzite structure. Natural cadmium sulfide sources, however, are not of importance for the pigment producing industry. The color of cadmium pigments can be adjusted by the composition and the size of the primary particles.

All cadmium pigments are based on cadmium sulfide and crystallize in a wurtzite lattice. Cadmium ions fill half of the tetrahedral vacancies in a hexagonal close-packed arrangement of sulfide ions. In this lattice, the cations as well as the anions can be replaced within certain limits by chemically related elements. Only zinc and mercury as cations and selenide as anion have gained importance as exchangeable ions. It was also tried to substitute selenide by telluride, but the so-obtained pigments have inferior coloristic properties and were, therefore, not introduced onto the market [156]. Incorporation of zinc leads to greenish yellow pigments, mercury, and selenide change the color shade to orange and red.

The following four cadmium pigments have achieved technical importance:

- (Cd,Zn)S (cadmium zinc sulfide): pigment cadmium yellow C.I. Pigment Yellow 35, greenish yellow
- CdS (cadmium sulfide): pigment cadmium yellow C.I. Pigment Yellow 37, reddish yellow
- Cd(S,Se) (cadmium sulfoselenide): pigment cadmium orange C.I. Pigment Orange 20, orange
- Cd(S,Se) (cadmium sulfoselenide): pigment cadmium red C.I. Pigment Red 108, red

Cadmium zinc sulfide pigments contain 59 to 77 wt% Cd, 0.2 to 13 wt% Zn and in addition 1 to 2 wt% Al_2O_3 for lattice stabilization. Cadmium orange and cadmium red have 66 to 76 wt% Cd and 1 to 24 wt% Se in the composition [158].

Cadmium sulfides and selenides are semiconductors. Their color can be explained using the band model by the distance between the valence and conduction bands in the crystal lattice. Light with wavelengths in the visible range of the electromagnetic spectrum is sufficient to lift electrons from the valence to the conduction band. The pigments exhibit an extremely high color purity.

Cadmium pigments were first produced on industrial scale around 1900. They were used for paints, artists' colors, and ceramics. The relatively high costs for raw materials and production caused high pigment prices and were limiting factors for a broad application. The upcoming demand for colored plastics led to an increased use of cadmium pigments in this application. The pigments exhibited excellent properties in coloring of polymers, offering a broad range of bright, intermixable, dispersible, and lightfast shades. They withstand the rigorous processing temperatures necessary for engineering polymers much better than organic pigments.

The beginning cadmium discussion and the development of substitute products, especially of high value organic, temperature stable yellow and red pigments, have led to a significant reduction of the cadmium pigment volumes.

4.9.2 Production of cadmium sulfide/selenide pigments

Cadmium pigments must be produced from pure grades of chemicals because the final colorants must be free of transition metal compounds, which form deeply colored sulfides (e.g., copper, iron, nickel, cobalt, and lead) [158, 159]. The manufacturing process starts with the dissolution of high-purity cadmium metal (99.99%) in sulfuric, hydrochloric, or nitric acid. Mixtures of mineral acids are likewise possible. A mixture of sulfuric and nitric acid is preferred in order to achieve controllable reaction rates and to avoid producing noxious fumes. In an alternative process, the cadmium metal is first oxidized to cadmium oxide, which is then diluted in mineral acids. The oxide can be produced by melting the metal, vaporizing it at about 800 °C, and oxidizing the vapor with air [159]. Zinc salt is added in dependence on the desired color tone of the final pigments. After dissolution of the cadmium metal, cadmium sulfide is precipitated by addition of sodium sulfide solution. A very fine crystalline cadmium or cadmium zinc sulfide is precipitated, which has a cubic crystal form and does not yet show pigment properties. Parameters such as concentration, temperature, pH value, and mixing conditions have an effect on the particle size and the shape of the precipitate. The following equation demonstrates the CdS precipitation using a $CdSO_4$ solution as cadmium provider:

$$CdSO_4 + Na_2S \rightarrow CdS + Na_2SO_4 \qquad (4.40)$$

The precipitated particles are separated from the suspension by a filter press or a belt filter. Sodium sulfate and unreacted cadmium salts have to be washed out carefully. Remaining salts would have a detrimental effect on the pigment properties. Moreover, any soluble cadmium must be removed in order to fulfill the regulatory requirements. The removal is typically achieved by acidification with diluted mineral acids followed by decantation and washing or by filtration and washing.

The filter cake is dried in an oven and then crushed into small lumps prior to calcination. Rotary, tunnel, and static kilns are used for the calcination step at about 600 °C. The cubic crystal lattice of the precipitated and dried material is transformed into the hexagonal structure during calcination, and the particles grow to a size that makes the products suitable for pigments. Sulfur dioxide and in the case of the sulfoselenide pigments also selenium and selenium dioxide are emitted. Careful fume removal through a gas scrubber or similar equipment is, therefore, required. The calcination temperature and residence time as well as atmosphere affect color and texture of the resulting pigments.

In a second process, the so-called "powder process" finely divided cadmium carbonate or cadmium oxide is subjected to intensive mechanical mixing with sulfur and mineralizers and then calcined in the absence of oxygen. The addition of zinc produces the same color effects as in the case of the precipitation process. The so-obtained products are washed, dried, and calcined at about 600 °C.

The manufacture of cadmium red and cadmium orange pigments is carried out with sodium sulfide solutions containing the amount of selenium required for the final pigment. Selenium is added as powder to the sodium sulfide solution and is dissolved in it. The sulfide ions react with the cadmium ions first until the sodium sulfide has been used up [159]. Cadmium sulfoselenide is then formed during the calcination step. An example of the formation of a sulfoselenide pigment is shown in the following summarized equation:

$$CdSO_4 + 0.9\,Na_2S + 0.1\,Na_2Se \rightarrow Cd(S_{0.9}Se_{0.1}) + Na_2SO_4 \tag{4.41}$$

The subsequent processing is similar to that described for pure cadmium sulfide pigments. A manufacturing scheme for cadmium sulfide and sulfoselenide pigments is shown in Figure 4.44.

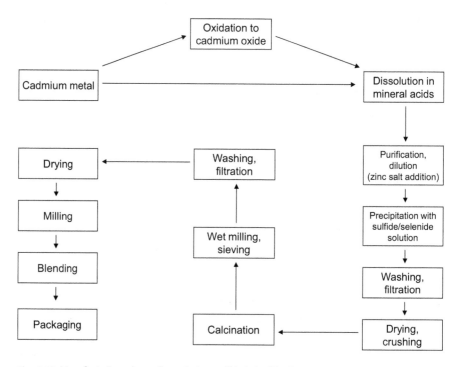

Fig. 4.44: Manufacturing scheme for cadmium sulfide/selenide pigments.

4.9.3 Pigment properties and uses

Cadmium sulfide and sulfoselenide pigments are intensely colored, lightfast, stable at high temperatures, and migration-resistant. They are practically insoluble in water, alkaline solutions, and organic solvents, but are attacked and decomposed by acids. The weather fastness of cadmium pigments is limited. Cadmium sulfide is oxidized slowly by atmospheric oxygen and under the influence of light, especially sunlight, to cadmium sulfate.

Fig. 4.45: Scanning electron micrograph of a cadmium yellow pigment.

Fig. 4.46: Scanning electron micrograph of a cadmium orange pigment.

Fig. 4.47: Scanning electron micrograph of a cadmium red pigment.

The density of cadmium pigments is between 4.2 and 5.6 g/cm^3. The average particle size of the powders lies in the range of 0.2 to 0.5 µm. Scanning electron micrographs of cadmium pigments are shown in Figure 4.45 (cadmium yellow), Figure 4.46 (cadmium orange), and Figure 4.47 (cadmium red). The primary particles of all three pigments are strongly sintered due to the manufacturing process and form larger units with sizes of up to 0.5 µm.

Cadmium pigments are stable up to temperatures of 600 °C. Their hiding power in most application systems is very good. The high color purity of the pigments is due to the steep reflectance spectra, which is typical for a semiconductor related pigment (Figure 4.48). Figures 4.49, 4.50, and 4.51 show photographs of representative pigment samples of cadmium yellow, cadmium orange, and cadmium red.

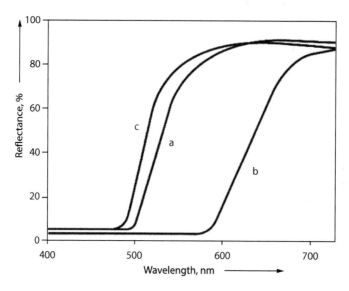

Fig. 4.48: Reflectance spectra of the pigments CdS (a), $Cd(S_{0.53}Se_{0.47})$ (b), and $(Cd_{0.82}Zn_{0.18})S$ (c).

Fig. 4.49: Photograph of a cadmium yellow pigment.

Cadmium pigments are offered as pure powders but also as pigment preparations, where the color strength is adjusted by addition of barium sulfate. They are also sold in a dust-free and dispersed form as highly concentrated plastic granulates (masterbatch

Fig. 4.50: Photograph of a cadmium orange pigment.

Fig. 4.51: Photograph of a cadmium red pigment.

pellets), as pasty concentrates, and as liquid colorants. All these dosage forms can be added to polymers in different stages of their processing chain.

The largest application fields of cadmium pigments are plastics with 90% and ceramics with 5%. Coating materials are a further application area with minor importance. The use of the pigments in polymers is almost universal. Polyethylene, polystyrene, polypropylene, and ABS are the most important polymers for the application of cadmium sulfides and sulfoselenides with the aim of coloration.

Glazes and enamels are typical applications for the pigments in ceramics. High temperatures are used in both cases for the burning process. For stability reasons, organic pigments and some of the inorganic pigments cannot be used here. There are some inorganic alternatives for the yellow color range. In the red area, however, only cadmium sulfoselenides are suitable for the generation of brilliant colors. Care must be taken during application of the pigments concerning the sensitivity to friction. Especially red cadmium sulfoselenides have the problem that excessive shear forces can cause a color change of the pigments.

4.9.4 Toxicology and occupational health

Cadmium sulfide and sulfoselenide pigments do not exhibit acute toxicity (LD_{50} value rat oral: >5000 mg/kg). They are not irritating to the skin or mucous membranes. The pigments are compounds with low solubility. Small amounts of cadmium are, however, solved in diluted hydrochloric acid comparable with gastric acid of similar concentration. Animal feeding studies over three months with dogs and rats show a cadmium intake and accumulation in the internal organs, especially in the kidneys [160]. There was no indication of carcinogenic effects after a long-term animal feeding study [156]. In general, cadmium ions are immunologically active. They are to some extent transferred to the fetus. Synergistic effects with other toxins have been found [161].

Seafoods contain relatively high amounts of cadmium. The concentrations of cadmium and other heavy metals in sewage sludge and feces are substantial. Cadmium is also present in some dental materials with substantial exposure to appliances such as silver solder alloys. However, acute cadmium poisoning is found to be rare [161].

The use of cadmium pigments is controversial and no longer considered optimal due to the principal possibility of uncontrolled solution of at least small amounts of the pigments and the potential danger that cadmium compounds can unintendedly get into the cycle of nature. Recycling of cadmium pigment-containing materials demands special care and advanced processing technologies. Cadmium is encountered across a wide range of contemporary polymer products, mainly because of the unregulated recycling of electronic waste and polyvinyl chloride. However, concentrations are generally low, conforming with current limits and posing minimal risk to consumers. Of greater concern is the relatively high concentration of cadmium pigments (up to 2 wt%) in old products, and in particular children's toys that remain in circulation [162].

A relatively great consumer risk seems to be the use of cadmium pigments in the enamels of decorated drinking glasses. Any pigmented enamel within the lip area is subject to ready attack from acidic beverages because the pigments are neither encapsulated nor overglazed. Glass bottles decorated with cadmium-containing enamel do not appear to represent a direct health hazard but have the propensity to contaminate recycled glass products. It is recommended, therefore, that decorated glassware is better regulated and that old, brightly colored toys are treated cautiously [162].

The legal requirements for the use of cadmium sulfide and sulfoselenide pigments in the relevant scopes of application must be adhered to. Waste materials coated with cadmium pigments should never be incinerated in fire places, ovens, or on open fire. They have to be treated as special refuse and disposed by a certified waste management company. Modern incineration plants for domestic waste are equipped with effective filter systems that can withhold contained cadmium particles completely.

Cadmium present in the environment is primarily due to contaminated sites from former times. Human beings and animals accumulate cadmium in their kidneys from where it is released again only very slowly. A verifiable cadmium exposure for human beings takes place via the lungs by cigarette smoke and car exhaust gases and, to a lesser degree, via food. The insoluble cadmium pigments cannot be absorbed through the skin. They are less toxic than soluble cadmium compounds. To avoid chronic effects, all safety measures for work with cadmium sulfide and sulfoselenide pigments have to be adhered to. Inhalation of pigment dust and skin contact must be avoided. Mercury-containing cadmium colors from older production may no longer be used.

4.10 Cerium sulfide pigments

4.10.1 Fundamentals and properties

Cerium sulfide pigments are orange and red inorganic pigments consisting of more than 90% of Ce_2S_3. They were introduced onto the market as supposed candidates for inorganic red and orange pigments. Due to high raw material costs, the prices of cerium sulfide pigments are in the upper range of inorganic colorants. The aim of pigment developments on the basis of rare earth metal sulfides was the replacement of other red and orange-colored pigments that are in public discussion because of their toxicity and environmental problems (cadmium sulfide, sulfoselenide, and selenide, lead molybdate). Cerium sulfide pigments have been application wise primarily developed for the coloration of plastics.

Due to the good thermal stability, lightfastness, dispersibility, and the strong hiding power, cerium sulfide pigments would basically be promising candidates for the substitution of the cadmium and lead-containing pigments [163–166]. The following pigments belong to the family of cerium sulfide pigments [165]:

- Ce_2S_3/La_2S_3 (cerium lanthanum sulfide): pigment cerium sulfide light orange C.I. Pigment Orange 78, light orange.
- Ce_2S_3 (cerium sulfide): pigment cerium sulfide orange C.I. Pigment Orange 75, orange.
- Ce_2S_3 (cerium sulfide): pigment cerium sulfide red C.I. Pigment Red 265, red.
- Ce_2S_3 (cerium sulfide): pigment cerium sulfide burgundy C.I. Pigment Red 275, burgundy red.

Cerium(III) sulfide exists in three allotropic modifications: α, β, and γ; all forms are different with reference to their thermal stability and their color. The only phase suitable for the use as a pigment is γ-Ce_2S_3. This compound is colored dark red. γ-Ce_2S_3 is formed only at temperatures of about 1000 °C. Therefore, its industrial production is very complicated. γ-Ce_2S_3 is isomorphic with Ce_3S_4 and has the ability to accommodate other cations at cerium vacancies in the lattice, e.g., alkaline earth metal ions or noncerium lanthanides [167–169]. Its formula can be expressed as $Ce_{3-x}S_4$, where x stands for cationic vacancies. The dark red color of some cerium sulfide pigments is due to electronic transitions from the CE 4f level into the Ce 5d conduction band corresponding to an energy gap of about 1.9 eV [165].

4.10.2 Production of cerium sulfide pigments

Cerium sulfide is not accessible by precipitation from cerium compounds with hydrogen sulfide because it is highly sensitive for hydrolysis. It is also difficult to stabilize cerium in the oxidation state +3. Therefore, only solid–gas reactions are suitable for the manufacture of cerium sulfide pigments [165]. Most synthesizing methods for lanthanide sulfides involve high-temperature gas–solid reactions between a sulfurizing agent (H_2S and/or CS_2) and lanthanide precursors, sometimes even under high pressure [170]. Various lanthanide compounds can be used as precursors, e.g., oxides, salts, alkoxides, oxalates, tartrates, or malonates [171, 172].

The solid-state reaction of a precursor with elemental sulfur is also possible. Lanthanide metals can act as precursors in this case. Such processes have several disadvantages and are difficult to perform because they require excessive temperatures and complicated pressure conditions. The typical production method for cerium sulfide pigments starts with the synthesis of a cerium containing precursor [165]. The precursor is synthesized from a solution of a cerium compound by precipitation with hydroxides. Alkali metal or alkaline earth metal salts can be added to affect the desired color shade of the final pigment. The colorless, mainly cerium hydroxide-containing precipitate is filtrated, dried, and then transformed at temperatures of about 1000 °C in a sulfurizing atmosphere into the colored γ-cerium(III) sulfide:

$$Ce(OH)_3 + 3\,H_2S \longrightarrow \gamma\text{-}Ce_2S_3 + 6\,H_2O \tag{4.42}$$

The cooled reaction product is ground in order to destroy agglomerates that may have been formed during the sulfurization process. Deagglomeration is decisive for the achievement of a high pigment quality. The γ-Ce_2S_3 powder is now surface treated using a wet process. This step increases the stability of the pigment and optimizes its compatibility with the application media. The surface treatment can be of inorganic and/or organic nature. Finally, the treated pigment is filtered, dried, and packaged.

4.10.3 Pigment properties and uses

Cerium sulfide pigments have a density of about 5 g/cm³ and a refractive index of 2.7. Their average particle diameter is in the magnitude of 1 µm. The color strength reaches 50 to 70% of a cadmium sulfoselenide pigment and is comparable with lead molybdate. The hiding power of cerium sulfide pigments is similar to that of the cadmium and lead-containing pigments. The reflectance spectra of a cerium sulfide red pigment in comparison with a cadmium red and an iron oxide red are shown in Figure 4.52. Cerium sulfide reflects less than cadmium red but significantly more strongly than iron oxide red. The course of the reflection curves demonstrates clearly that the cadmium pigment exhibits the purest red color (the steepest curve without any additional band) followed by the cerium sulfide pigment (a relatively steep curve bending earlier than the curve of the cadmium pigment). With the broad and humped reflection band, the iron oxide curve demonstrates why it is the less attractive one of the three red pigments with respect to color. A photograph of a representative sample of a cerium sulfide pigment is shown in Figure 4.53.

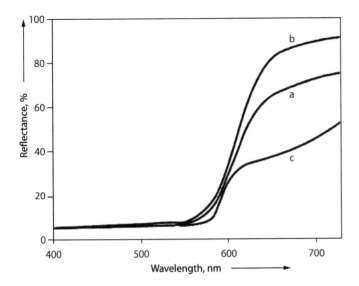

Fig. 4.52: Reflectance spectrum of cerium sulfide red (a) in comparison with the spectra of cadmium red (b) and iron oxide red (c).

Fig. 4.53: Photograph of a cerium sulfide pigment.

The heat stability of cerium sulfide pigments in plastics is very good. Their use is possible in high-density polyethylene (310 °C), in polypropylene, polyamide and polycarbonate (each at 320 °C). Light and weather fastness is also good. Cerium sulfide pigments are easily dispersible and migration fast. They show dimension stability also at large area spraying parts. The main disadvantage of γ-Ce_2S_3 pigments is their limited stability in water and water-containing systems and in humid atmosphere. The pigments decompose very slowly in cold water. They degrade in hot water and generally in humid atmosphere of more than 100 °C or in acids to form hydrogen sulfide. Coloration of plastics with cerium sulfide pigments can lead to odor problems due to reaction with residual water, especially in systems capable of swelling. This behavior is problematic and limits the broader application of the pigments in various media. The use of a suitable surface treatment for stabilization can improve the situation but is obviously not able to solve the problem fully. Cerium sulfide pigments can basically be used in several coatings formulations such as automotive refinish (acrylic-isocyanate binders), general industrial solvent-based systems (polyester-melamine binders), and oil-based architectural coatings.

4.10.4 Toxicology and occupational health

Cerium sulfide pigments do not exhibit acute toxicity (LD_{50} value rat oral: >5000 mg/kg). They are not irritating to the skin or mucous membranes [165, 173]. Genotoxic or carcinogenicity effects have not been found. The pigments are nearly insoluble and cerium ions, moreover, show effectively no toxicity (acute and chronic) [165]. So, there is practically no risk to humans or environment when cerium sulfide pigments are used in accordance with the standard safety measures.

There is, however, the limited stability against water and especially acids associated with the liberation of hydrogen sulfide gas. Producers must, therefore, ensure that packages for cerium sulfide pigments are clearly and legibly labeled with a warning of the formation of toxic gases in the case of contact with acids. The production process must be performed in such a way that the reaction of the pigment or preliminary stages thereof with moisture and acids under hydrogen sulfide formation is avoided. A properly working exhaust air system is necessary for the manufacture because of the use of hydrogen sulfide gas for the cerium sulfide formation.

4.11 Oxonitride pigments

4.11.1 Fundamentals and properties

Oxonitride pigments, also called oxynitride pigments, are basically interesting for the yellow, orange, and red color range. The most important and investigated compositions are $LaTaON_2$, $CaTaO_2N$, and $SrTaO_2N$. The development of these pigments started in the 1990s. The intention was the replacement of cadmium containing pigments by high chromatic yellow, orange, and red pigments. Especially $CaTaO_2N$ and $LaTaON_2$ are still seen as promising candidates in this sense [174, 175].

The color can be adjusted from yellow via orange to red in dependence on the composition. Oxonitride pigments are in chemical terms solid solutions and exhibit semiconducting properties. Decisive for the color is the band gap between valence and conductivity band. The color within the spectrum can be determined from yellow to red only by the variation of the amount and nature of the cations in the oxonitride. Successive substitution of a cation by another cation with a higher valency allows the increase of the ratio of nitrogen to oxygen in the composition. This ion exchange can be used to shift the color toward the region of longer wavelengths. On the contrary, the substitution of a cation by another cation with a one unit lower valency leads to a shortwave shift of the color. The compositions that are relevant for oxonitride pigments predominantly crystallize in the perovskite, chalcolamprite, and spinel structures [175–177].

Many binary nitrides are colored or black substances: Mg_3N_2 is yellow, Li_3N is red-brown, and Zn_3N_2 is black. Transition metal nitrides are generally black. Tantalum nitrides make an exception: Ta_3N_5 is red-orange and $TaON$ is yellow [178–180]. The color of these nitride-type compounds can be explained by a decrease in the band gap energy value, as confirmed by band structure calculations [181, 182]. It is possible to maintain the highest oxidation state of tantalum and other transition metal elements in ternary oxonitrides by taking advantage of the inductive effect of an appropriate counter-cation such as calcium, strontium, or lanthanum. Solid solutions of this oxonitride-type are basically excellent candidates for colored pigments. It could be shown that there is the possibility to adjust the color of oxonitride pigments in the perovskite-type system $Ca_{1-x}La_xTaO_{2-x}N_{1+x}$ in a systematic manner [174]. In dependence on the nitrogen content from $x = 0$ to 1, the hue varies from yellow ($CaTaO_2N$) to brown ($LaTaON_2$) through orange and red intermediate colors. Comparable colors could be achieved using the system $AZr_{1-x}Ta_xO_{3-x}N_x$ (A = Ca, Sr, Ba); from white $AZrO_3$ to yellow, red, or brown $ATaO_2N$ [183].

4.11.2 Production of oxonitride pigments

The synthesis of oxonitride pigments is carried out using a high-temperature solid-state–gas reaction [174, 184, 185]. In the case of $LaTaON_2$, an equimolar blend of Ta_2O_5 and La_2O_3 is ground with the threefold quantity of an NaCl/KCl mixture (mineralizer) and afterward thermally treated for several hours at 900 °C in an ammonia stream:

$$Ta_2O_5 + La_2O_3 + 4\,NH_3 \rightarrow 2\,LaTaON_2 + 6\,H_2O \qquad (4.43)$$

Alkaline earth metal-containing oxonitrides are produced with the addition of $CaCO_3$ or $SrCO_3$ to the starting mixture. In the case of $CaTaO_2N$, a mixture of Ta_2O_5 and $CaCO_3$ is used:

$$Ta_2O_5 + 2\,CaCO_3 + 2\,NH_3 \rightarrow 2\,CaTaO_2N + 3\,H_2O + 2\,CO_2 \qquad (4.44)$$

The resulting products are washed to remove soluble compounds, filtrated, and dried.

The entire production process for oxonitride pigments is difficult to conduct on a larger scale. High production costs must, therefore, be considered. Moreover, the costs for the pigments are comparatively high due to the lanthanum and tantalum content. These facts are the main reasons why oxonitride pigments have not been manufactured on production scale so far. They belong to the coloristically interesting but not commercialized inorganic pigments.

4.11.3 Pigment properties and uses

Oxonitride pigments are representatives of the yellow, orange, and red colorants. The important properties of these pigments are their good brilliance and color strength, strong hiding power, good dispersibility, lightfastness, and temperature stability [174]. A color impression of oxonitrides with the composition $Ca_{1-x}La_xTaO_{2-x}N_{1+x}$ is given in Figure 4.54. A broad color variety from yellow ($x = 0.05$) to red ($x = 0.90$) via orange tones is possible [174]. The average particle size of developing products is in the range of 1 µm. Oxonitride pigments are stable against acids and hydroxide solutions. The pigments are basically suitable for the manufacture of easy to burn glazes and vitrifiable colors. They can also be used for the coloration of plastics, coatings, and printing inks as well as for cosmetic formulations. The high thermal stability allows the mass pigmentation of plastics with following extrusion as well as the use in annealing lacquers [184, 185]. The firing range of the pigments for vitrifiable and enamel color is at 500 to 680 °C.

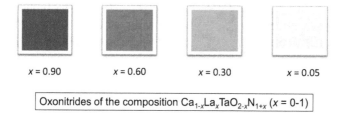

$x = 0.90$ $x = 0.60$ $x = 0.30$ $x = 0.05$

Oxonitrides of the composition $Ca_{1-x}La_xTaO_{2-x}N_{1+x}$ ($x = 0$-1)

Fig. 4.54: Color impression of oxonitrides of the composition $Ca_{1-x}La_xTaO_{2-x}N_{1+x}$ ($x = 0$–1) [174].

4.11.4 Toxicology and occupational health

Usable studies on toxicological properties of oxonitride pigments have not yet been documented. Because all constituent elements are harmless, the pigments basically seem to be promising replacements for lead and cadmium-containing compositions. The chemical and thermal stability of the oxonitride pigments makes it possible to expect them to be problem free in various application media. The production process must be performed in such a way that a release of gaseous ammonia during the solid-state–gas reaction is prevented. A properly working exhaust air system is absolutely necessary for the manufacture of the described oxonitride compositions.

4.12 Yttrium indium manganese oxide pigments

4.12.1 Fundamentals and properties

Yttrium indium manganese oxide with the composition $YIn_{1-x}Mn_xO_3$ ($x = 0$–1), also referred to as YInMn Blue and commercially known as Blue 10G513, represents a product series with the potential for colored inorganic pigments in the blue range of the color space. Compositions of this type have a unique crystal structure in which trivalent manganese ions (Mn^{3+}) in the trigonal bipyramidal coordination are responsible for the observed intense blue color [186, 187]. The development of YInMn Blue goes back to investigations on oxidic solid solutions between $YInO_3$, a ferroelectric material, and $YMnO_3$, an antiferromagnetic material, at temperatures above 1000 °C. It was found during these studies that certain compositions in the system $YIn_{1-x}Mn_xO_3$ have an intense blue color [186–188]. Corresponding to their chemical composition, yttrium indium manganese oxide pigments can also be classified as MMO pigments or complex inorganic color pigments (CICP).

The intense color of YInMn Blue can be explained by trivalent manganese that it is introduced at dilution in trigonal bipyramidal coordination. Density functional theory calculations indicate that the blue color results from an intense absorption in the red/green region. This absorption is due in turn to a symmetry allowed optical transition between the valence band maximum, composed of Mn $3d(x^2 - y^2, xy)$ states strongly hybridized with O $2p(xy)$ states, and the narrow Mn $3d(z^2)$-based conduction band minimum. Starting with the hexagonal $YMnO_3$-$YInO_3$ solid solution, it could be shown that the behavior is a general feature of diluted Mn^{3+} in this coordination environment [186]. Certain pigments with the composition $YIn_{1-x}Mn_xO_3$ are characterized not only by a vibrant blue color but also by unusually high NIR reflectance [186–188]. The color may be adjusted by the variation of the In/Mn ratio in the pigment composition. The pigment with the strongest blue hue was found to be $YIn_{0.8}Mn_{0.2}O_3$. It has a color comparable to that of cobalt blue pigments of the formula $CoAl_2O_4$ [187].

4.12.2 Production of yttrium indium manganese oxide pigments

Pigments with the composition $YIn_{1-x}Mn_xO_3$ can be produced by solid-state reaction of the oxides of yttrium, indium, and manganese in stoichiometric amounts at temperatures in the range of 1100 to 1200 °C. The synthesis with the composition $YIn_{0.8}Mn_{0.2}O_3$ can be exemplary and summarily described as follows:

$$Y_2O_3 + 0.8\,In_2O_3 + 0.2\,Mn_2O_3 \rightarrow 2\,YIn_{0.8}Mn_{0.2}O_3 \tag{4.45}$$

Before the start of the high-temperature reaction, the oxides have to be mixed intensely to obtain homogeneous products. After cooling of the reaction product, suitable steps of powder processing follow to achieve the necessary properties of the final pigment. It is basically also possible to execute the solid-state reaction with decomposable yttrium, indium, and manganese compounds such as hydroxides, carbonates, or nitrates. A wet chemical route, e.g., via the coprecipitation of the metal hydroxides followed by a high-temperature reaction, is likewise possible.

4.12.3 Pigment properties and uses

YInMn Blue pigments offer stable and long lasting blue color tones for many applications. They have high weatherability, heat and chemical resistance, are nonwarping, and easy to disperse. The pigments are based on a new structural chemistry that expands the range of colors that stay cooler when exposed to the sun, allowing building material manufacturers to meet regulatory requirements and building owners to potentially save energy. The high-temperature calcination production process makes YInMn Blue pigments highly inert. While they are highly infrared reflective, they are extremely opaque in the visible and UV regions of the solar spectrum. The inertness of the pigments means that they can be used in a wide range of coating and plastic applications. The special infrared reflecting properties allow a broad use in building materials. The pigments are also offered for the application in anticounterfeiting features, glass enamels, and artists' colors. In dependence on the composition, YInMn Blue pigments can show color tones from light to dark blue. An example of a strong blue pigment of the $YIn_{1-x}Mn_xO_3$ series is shown in Figure 4.55.

Fig. 4.55: Photograph of a compacted YInMn Blue pigment.

Because of the expensive raw materials required, e.g., for yttrium and indium compounds, relatively high production costs for YInMn Blue pigments must be assumed. The pigments belong, however, to the most promising candidates among the new developed colored inorganic pigments.

The special reflective properties in the infrared range probably extend the possibilities of their application considerably.

4.12.4 Toxicology and occupational health

Toxicological studies on YInMn Blue pigments found no toxic components in the investigated samples. There was no effect of the products themselves that would require acute or chronic hazard labeling to conform with ASTM D4236, the Labeling for Hazardous Art Materials Act, LHAMA regulations 16 CFR 1500.14(B)(8), or the Federal Hazardous Substances Act. The pigments are, therefore, classified as not being toxic [189]. Further investigations have shown that the pigments are not corrosive, skin and eye irritant, or combustible/flammable as defined in 16 CFR 1500.3(b)(5) and 1500.3(b)(7)–(10) of the US Federal Hazardous Substances Act regulations. They do not contain hazardous substances when evaluated under ASTM F963.8.2 of the Standard Consumer Safety Specification for Toy Safety [189].

4.13 New approaches for the development of inorganic colored pigments

4.13.1 Intention for new developments

The search for novel inorganic colored pigments, which was also the intention for the development of the cerium sulfide, oxonitride, and YInMn Blue pigments, has led to further new pigmentary products. Almost all of these have not left the laboratory stage. The new developments cover a white range of colors from blue via violet, yellow, and green to orange. An additional important aspect in the search for new pigment systems is the replacement of technical pigments containing toxic or cancerogenic components, e.g., chromate, cadmium, and lead-containing types. A key challenge for the acceptance in the pigment industry is also that new colorants are environmentally benign, earth abundant, and stable against weather influences, chemicals, and high temperatures. Finally, the new pigments must fulfill the criteria of compatibility with the components of the relevant application systems. A few selected of these new pigment developments are presented here.

4.13.2 Blue pigments

Blue pigments with the composition $Ca_6Ba(P_{1-x}Mn_x)_4O_{17}$ ($x = 0.01–0.13$) were synthesized in a single phase form by a solid-state reaction method [190]. The pigments exhibit strong optical absorption in the UV and the yellow-red spectral regions due to a charge transfer transition in MnO_4 tetrahedra and transitions of Mn^{5+}, which is the cause of the blue color. The optimization with the composition resulted in the attainment of the highest blue hue for the composition $Ca_6Ba(P_{0.99}Mn_{0.01})_4O_{17}$. Pigments of this type show a low chemical stability against acidic and alkaline solutions, but an excellent color durability against heat and light. Although they are attractive from the coloristic point of view, they will hardly find an application without an improvement of the chemical stability [190].

New blue oxide-based materials with the composition $LaGa_{1-x}Mn_xGe_2O_7$ ($x = 0.1-0.4$) were developed by a solid-state reaction process [191]. The substitution of Ga^{3+} by Mn^{3+} in $LaGaGe_2O_7$ results in pigmentary blue products. The color is based on Mn^{3+} in the trigonal bipyramidal coordination with longer apical bond lengths. The light absorption corresponds to the energy region of 1.7 to 2.5 eV [191].

Another structural type of compounds usable for blue pigments with tunable hue is based on solid solutions of the composition $CaAl_{12-2x}Co_xTi_xO_{19}$ ($x = 0.2-1.0$) [192]. Such solid solutions crystallize in the hexagonal structure of the mineral hibonite ($CaM_{12}O_{19}$) with five distinct crystallographic sites for M cations (M = Al^{3+}, Co^{2+}, and Ti^{4+}). The intense blue color is attributed to a synergistic effect of allowed d–d transitions involving the chromophore Co^{2+} in both tetrahedral and trigonal bipyramidal crystal fields. Compared with commercially available cobalt blue pigments, these hibonite-structured blue materials possess a more reddish hue. Pigments like $CaAl_{10.6}Co_{0.7}Ti_{0.7}O_{19}$ ($x = 0.7$) and $CaAl_{10}CoTiO_{19}$ ($x = 1.0$) can be synthesized using the solid-state reaction process. Suitable raw materials are $CaCO_3$, Al_2O_3, $CoCO_3$, and TiO_2 [192].

Blue pigments of the composition $Sr_{1-x}La_xCu_{1-y}Li_ySi_4O_{10}$ ($x = y = 0-0.5$) were prepared by the solid-state reaction of suitable compounds such as La_2O_3, $SrCO_3$, CuO, Li_2CO_3, and SiO_2 [193]. These raw materials were mixed intensely in the desired stoichiometric proportions with acetone as wetting medium and dried in an air oven. The process of mixing and drying was repeated thrice to obtain a homogeneous mixture for the high-temperature solid-state reaction. The thus-synthesized blue pigments show a high reflectance in the NIR. The investigation of this property on a roof of a building has shown that the pigments have a potential application as so-called "cool pigments" [193].

4.13.3 Violet pigments

Violet pigments with the composition $LaAl_{1-x}Mn_xGe_2O_7$ ($x = 0.01-0.20$) can be synthesized using a solid-state reaction process [194]. Al^{3+} is substituted by Mn^{3+} in $LaAlGe_2O_7$ to obtain colored pigments. These exhibit a strong optical absorption in the yellow-red light region due to the Mn^{3+} d–d transitions that explain the violet color. Among the investigated compositions, $LaAl_{0.90}Mn_{0.10}Ge_2O_7$ shows the most vivid violet color [194].

4.13.4 Green pigments

A green-colored nanopigment with the composition Y_2BaCuO_5 was synthesized by a nanoemulsion method [195]. The pigment exhibits an impressive NIR reflectance (61% at 1100 nm). Structural investigations of pigment samples reveal the orthorhombic crystal structure for Y_2BaCuO_5, where yttrium is coordinated by seven oxygen atoms with the local symmetry of a distorted trigonal prism, barium is coordinated by eleven oxygen atoms, and the coordination polyhedron of copper is a distorted square pyramid [CuO$_5$]. The spectra in the UV and in the visible range show an intense d–d transition associated with the CuO$_5$ chromophore between 2.1 and 2.5 eV in the visible domain. The potential utility of the pigment as a "cool pigment" could be demonstrated on a pigmented roof of a building [195].

4.13.5 Yellow pigments

(Bi$_2$O$_3$)$_{1-x}$ – (Nb$_2$O$_5$)$_x$ solid solutions derived from the binary system Bi$_2$O$_3$-Nb$_2$O$_5$ were synthesized by solid-state reaction between Bi$_2$O$_3$ and Nb$_2$O$_5$ at 800 °C (x = 0–0.20) [196]. Powder samples show colors from light yellow to brilliant reddish yellow depending on the composition. The most brilliant reddish yellow color was obtained for the composition (Bi$_2$O$_3$)$_{0.94}$(Nb$_2$O$_5$)$_{0.06}$ with x = 0.06. Other inorganic reddish yellow and orange pigments with Bi$_2$O$_3$ as the main component are lanthanide-doped Bi$_2$O$_3$ solid solutions such as CeO$_2$-ZrO$_2$-Bi$_2$O$_3$, Er$_2$O$_3$-Bi$_2$O$_3$, and Ho$_2$O$_3$-Bi$_2$O$_3$ [197–199].

4.13.6 Orange pigments

Compositions of the series Bi$_{1-x}$La$_x$FeWO$_6$ (x = 0–0.30), derived from BiFeWO$_6$ by substitution of bismuth by lanthanum in the crystal structure, were synthesized by a coprecipitation method as new inorganic orange pigments [200]. Strong optical absorption was observed below 550 nm in all investigated compositions. The most vibrant orange hue was obtained for Bi$_{0.90}$La$_{0.10}$FeWO$_6$ (x = 0.10). Although the yellowness value of this pigment is a little smaller than that of commercially available orange pigments, the redness value is significantly larger. Pigments like Bi$_{0.90}$La$_{0.10}$FeWO$_6$ are characterized by a high stability against chemicals [200].

Compositions of the series La$_3$LiMn$_{1-x}$Ti$_x$O$_7$ (x = 0–0.05), derived from La$_3$LiMnO$_7$ by incorporation of titanium at manganese positions in the crystal structure, were synthesized by solid-state reaction at 900 °C as new inorganic orange pigments [201]. Single-phase compositions were obtained for x ≤ 0.03. Strong optical absorption is observed at wavelengths between 400 and 550 nm. A shoulder absorption peak appears around 690 nm. Among the compositions synthesized, the most brilliant orange color was obtained for La$_3$LiMn$_{0.97}$Ti$_{0.03}$O$_7$ (x = 0.03). The redness and yellowness values of this orange pigment are higher than those of commercially available orange pigments [201].

New orange oxide-based materials with the composition Ca$_{14}$(Al$_{1-x}$Fe$_x$)$_{10}$Zn$_6$O$_{35}$ (x = 0–0.30) were developed using a solid-state reaction method [202]. The substitution of Al^{3+} by Fe^{3+} in Ca$_{14}$Al$_{10}$Zn$_6$O$_{35}$ results in pigmentary orange products. The iron-containing pigments strongly absorb visible light at wavelengths from 380 to 490 nm and around

600 nm due to d–d transitions of Fe^{3+}. By increasing the Fe^{3+} content, it is possible to modify the color of the pigments from yellow to orange [202].

$Bi_{2-x}Zr_{3x/4}O_3$ solid solutions derived from the binary system Bi_2O_3-ZrO_2 were synthesized by solid-state reaction at 800 °C ($x = 0.3$–05) [203]. Single-phased $Bi_{2-x}Zr_{3x/4}O_3$ solid solutions with tetragonal crystal structure are formed in the x-range of 0.3 to 0.5. A very brilliant orange powder is obtained for the composition with $x = 0.5$. The stability of the thus-prepared $Bi_{2-x}Zr_{3x/4}O_3$ compositions against acids is not sufficient. The treatment of the orange powders, e.g., with diluted nitric acid, leads to a significant discoloration [203]. Without a suitable stabilization, the powders cannot be used for pigment purposes.

Bismuth oxyhalide solid solutions of the composition $BiOBr_{1-x}I_x$ offer a broad color spectrum from bright reddish yellow to reddish orange by increasing the iodine amount [204]. Conversely, the physical mixture of the colorless bismuth oxybromide (BiOBr) with the brick red bismuth oxyiodide (BiOI) leads to duller brick red-colored powders with unchanged hue of the color compared to BiOI. The morphology of the pigments is characterized by platelets of 0.2 to 0.5 µm in diameter and a couple of nanometers in thickness. $BiOBr_{1-x}I_x$ pigments are obtained by a wet chemical precipitation process. The chemical stability and the outdoor durability of the pigments are relatively low. It is, however, possible to improve these properties by using doping agents or by stabilizing the pigments with layers of metal oxides. Another option is the improvement by the use of an encapsulation technology. The thus-obtained pigments are more heat and weather stable and show clean shades from the orange yellow to the reddish orange [205]. Applications for stabilized bismuth oxyhalide pigments are solventborne and waterborne coatings as well as powder coatings [204].

4.14 Conclusions

Inorganic colored pigments are manufactured today mainly by modern chemical processes on industrial scale production. They are among the most important representatives of colorants. Their advantages over organic colored pigments are their broad applicability, chemical and thermal stability, lightfastness as well as value in use. Organic pigments, however, exhibit partly higher color strength and purer colors. Their disadvantages are the significantly lower stability in various application systems and the high manufacturing costs.

Inorganic colored pigments are applied in all pigment relevant systems such as coatings, paints, plastics, artists' paints, cosmetics, printing inks, leather, construction materials, paper, glass, and ceramics. They are produced mostly by precipitation and by solid-state processes. Natural pigments, with the exception of the iron oxide types, are no longer relevant. Chemical purity, particle size distribution, and surface quality determine the coloristic and application technical properties of the individual pigments. Surface treatments are used in many cases to improve the quality of pigments with respect to stability and compatibility with the application system.

Major efforts have been made to develop new colored pigments and to improve existing ones. Particle management, sophisticated surface treatment, and provision of pigment preparations are suitable ways for the improvement of the pigments and the optimization of the dosage form. Further work is necessary to develop an appropriate replacement for chromate, cadmium, and lead-containing pigments, which are used more and more only rarely nowadays.

Study questions

4.1   What are colored pigments?
4.2   Which colors can be achieved by using of inorganic colored pigments?
4.3   Which iron oxides and iron oxide hydroxides are used for pigment purposes?
4.4   What are the processes used for the production of synthetic iron oxide pigments?
4.5   How are chromium oxide pigments produced?
4.6   What are main representatives of MMO pigments based on the rutile and the spinel structures?
4.7   How can the color of ultramarine pigments be explained?
4.8   What is the reason for the decreasing use of chromates and lead and cadmium compounds for pigments?

Bibliography

[1]   Endriss H. Aktuelle Anorganische Buntpigmente, Zorll U, Curt R ed. Vincentz Verlag, Hannover, 1997, 17.
[2]   Spille J, in Kittel – Lehrbuch der Lacke Und Beschichtungen, Vol. 5, Pigmente, Füllstoffe Und Farbmetrik, 2nd edn. Kittel H ed. S. Hirzel Verlag, Stuttgart, Leipzig, 2003, 20.
[3]   Pfaff G Estimation based on data from professional journals, reports, and conference information, 2022.
[4]   Buxbaum G, Wiese J. Industrial Inorganic Pigments, 3rd edn. Buxbaum G, Pfaff G ed. Wiley-VCH Verlag, Weinheim, 2005, 99.
[5]   Buxbaum G, Kresse P, in Kittel – Lehrbuch der Lacke Und Beschichtungen, Vol. 5, Pigmente, Füllstoffe Und Farbmetrik, 2nd edn. Spille J ed. S. Hirzel Verlag, Stuttgart, Leipzig, 2003, 71.
[6]   Endriss H. Aktuelle Anorganische Buntpigmente, Zorll U ed. Vincentz Verlag, Hannover, 1997, 45.
[7]   Schwertmann U, Cornell RM. Iron Oxides in the Laboratory, Preparation, and Characterization, VCH Verlagsgesellschaft, Weinheim, 1991.
[8]   Cornell RM, Schwertmann U. The Iron Oxides, 2nd edn. Wiley-VCH Verlag, Weinheim, 2003.
[9]   Jolly JLW, Collins CT. Iron Oxide Pigments (In Two Parts), Part 2: Natural Iron Oxide Pigments – Location, Production, and Geological Description, Dept. of the Interior, Bureau of Mines, Washington, 1980.
[10]   Ack E. Farben Ztg. 1922/23,28,493.
[11]   Bouchonnet MA. Bull. Soc. Chim. Fr. 1912,9,345.
[12]   Patent DE 38 20 499 (Bayer AG) 1988.
[13]   Carter EV, Laundon RD. J. Oil Colour Chem. Ass. 1990,1,7.
[14]   Patent DE 26 53 765 (Bayer AG) 1976.
[15]   Patent DE 11 91 063 (Bayer AG) 1963.
[16]   Patent DE 25 17 713 (BASF) 1975.
[17]   Patent US 6,689,206 (Bayer AG) 2004.
[18]   Patent US 2,388,659 (Interchem. Corp.) 1943.

[19]   Patent US 3,133,267 (C. K. Williams & Co.) 1934.

[20]   Patent US 2,631,085 (Reconstruction Finance Corp.) 1947.

[21]   Patent GB 668929 (Reymers Holms Gamla Ind.) 1950.

[22]   Patent DE 704 295 (O. Glemser) 1937.

[23]   Patent DE 22 12 435 (Pfizer Inc.) 1972.

[24]   Patent DD 26901 (M. Hollnagel, E. Kühn) 1960.

[25]   Patent US 2,620,261 (C. K. Williams & Co.) 1947.

[26]   Patent US 1,327,061 (West Coast Kalsomine Co.) 1917.

[27]   Patent US 1,368,748 (National Ferrite Co.) 1920.

[28]   Patent DE 463 773 (IG Farbenindustrie AG) 1925.

[29]   Patent DE 515 758 (IG Farbenindustrie AG) 1925.

[30]   Patent DE 551 255 (IG Farbenindustrie AG) 1930.

[31]   Patent EP 0 249 843 (Bayer AG) 1987.

[32]   Patent DE 466 463 (IG Farbenindustrie AG) 1926.

[33]   Patent EP 0 014 382 (Bayer AG) 1980.

[34]   Pfaff G, in Special Effect Pigments, 2nd edn. Pfaff G ed. Vincentz Network GmbH & Co., KG, Hannover, 2008, 38.

[35]   Köhler P, Schneider V, Kischkewitz J. Farbe + Lack. 2014,120,32.

[36]   Rowe RC. Pharm. Int. 1984,9,221.

[37]   Meyer AH. Römpp Lexikon Lebensmittelchemie, 2nd edn. Georg Thieme Verlag, Stuttgart, New York, 2006, 290.

[38]   Aquilina G, Azimonti G, Bampidis V, De Lourdes Bastos M, Bories G, Chesson A, Cocconcelli PS, Flachowsky G, Gropp J, Kolar B, Kouba M, Lopez Puente S, Marta L, Lopez-Alonso M, Mantovani A, Mayo B, Ramos F, Rychen G, Saarela M, Villa RE, Wallace RJ, Wester P. Safety and efficacy of iron oxide black, red and yellow for all animal species, European Food Safety Authority EFSA Journal 2016, ISSN. 1831–4732, John Wiley and Sons Ltd.

[39]   Szalay B, Tatrai E, Nyiro G, Vezer T, Dura GJ. Appl. Toxicol. 2012,32,446.

[40]   Rieck H, in Industrial Inorganic Pigments, 3rd edn. Buxbaum G, Pfaff G ed. Wiley-VCH Verlag, Weinheim, 2005, 111.

[41]   Räde D, in Kittel – Lehrbuch der Lacke Und Beschichtungen, Vol. 5, Pigmente, Füllstoffe Und Farbmetrik, 2nd edn. Spille J ed. S. Hirzel Verlag, Stuttgart, Leipzig, 2003, 89.

[42]   Endriss H. Aktuelle Anorganische Buntpigmente, Zorll U ed. Vincentz Verlag, Hannover, 1997, 77.

[43]   Keifer S, Wingen A. Farbe + Lack. 1973,79,866.

[44]   Patent US 4,127,643 (PPG Industries) 1977.

[45]   Lassaigne J-L. Ann. Chim. Phys., Ser. 2. 1820,14,299.

[46]   Patent US 1,728,510 (Roth) 1927.

[47]   Patent US 2,560,338 (C. K. Williams & Co) 1950.

[48]   Patent US 2,695,215 (C. K. Williams & Co) 1950.

[49]   Patent EP 0 068 787 (Pfizer Inc.) 1982.

[50]   Patent US 4,040,860 (Bayer AG) 1976.

[51]   Patent ES 438129 (Colores Hispania S. A.) 1975.

[52]   Patent US 3,723,611 (Bayer AG) 1971.

[53]   Püttbach E. Betonwerk + Fertigteiltechnik. 1987,2,124.

[54]   Patent DE 31 23 361 (Bayer AG) 1981.

[55]   Ivankovich S, Preussmann R Ed. Cosmet. Toxicol. 1975,13,347.

[56]   Regulation (EC) No 1272/2008 – Classification, labelling and packaging of substances and mixtures (CLP) of the European Parliament and of the Council. Latest update: 19/03/2021.

[57]   Brussaard H, in Industrial Inorganic Pigments, 3rd edn. Buxbaum G, Pfaff G ed. Wiley-VCH Verlag, Weinheim, 2005, 116.

[58] Etzrodt G, in Kittel – Lehrbuch der Lacke Und Beschichtungen, Vol. 5, Pigmente, Füllstoffe Und Farbmetrik, 2nd edn. Spille J ed. S. Hirzel Verlag, Stuttgart, Leipzig, 2003, 96.

[59] Endriss H. Aktuelle Anorganische Buntpigmente, Zorll U ed. Vincentz Verlag, Hannover, 1997, 94.

[60] White J. High Performance Pigments, 2nd edn. Faulkner EB, Schwartz RJ ed. Wiley-VCH Verlag, Weinheim, 2009, 41.

[61] Maloney J, in High Performance Pigments, 2nd edn. Faulkner EB, Schwartz R ed. Wiley-VCH Verlag, Weinheim, 2009, 53.

[62] Batchelor RW. TransBr. Ceram. Soc. 1974,73,297.

[63] Kleinschmidt P. Chem. Unserer Zeit. 1986,20,182.

[64] Pfaff G, in Winnacker–Küchler: Chemische Technik, Prozesse Und Produkte, Vol. 7, Industrieprodukte, 5th edn. Dittmeyer R, Keim W, Kreysa G, Oberholz A ed. Wiley-VCH Verlag, Weinheim, 2004, 334.

[65] Broll A, Beyer HH, Kleinschmidt P. Chem. Ztg. 1977,101,319.

[66] Stefanov S. Sprechsaal. 1968,101,404.

[67] White JP. Paint Coat. Ind. 2000,16,54.

[68] Bendiganavale AK, Malshe VC. Recent Pat. Chem. Eng. 2008,1,67.

[69] Burkhart G, Detrie T, Swiler D. Paint Coat. Ind. 2001,1,30.

[70] Brandt K, in Industrial Inorganic Pigments, 3rd edn. Buxbaum G, Pfaff G ed. Wiley-VCH Verlag, Weinheim, 2005, 128.

[71] Etzrodt G, in Kittel – Lehrbuch der Lacke Und Beschichtungen, Vol. 5, Pigmente, Füllstoffe Und Farbmetrik, 2nd edn. Spille J ed. S. Hirzel Verlag, Stuttgart, Leipzig, 2003, 118.

[72] Endriss H. Aktuelle Anorganische Buntpigmente, Zorll U ed. Vincentz Verlag, Hannover, 1997, 57.

[73] Hund F. Farbe + Lack. 1967,73,111.

[74] Algra GP, Erkens LJH, Kok DM. J. Oil Colour Chem. Assoc. 1988,71,71.

[75] Schäfer H. Farbe + Lack. 1971,77,1081.

[76] Patent DE 21 27 279 (Hoechst) 1971.

[77] Patent DE 20 62 775 (Hoechst) 1970.

[78] Wagner H. Z. Anorg. Allg. Chem. 1931,208,249.

[79] Wagner H. Farben-Ztg. 1933,38,932.

[80] Lesche H. Farbe + Lack. 1959,65,79.

[81] Patent US 2,808,339 (DuPont) 1957.

[82] Patent DE 18 07 891 (DuPont) 1969.

[83] Patent DE 19 52 538 (Bayer AG) 1969.

[84] Patent DE 20 49 519 (ICI) 1970.

[85] Patent DE 26 00 365 (Ten Horn Pigment) 1976.

[86] Patent DE 33 23 247 (BASF) 1983.

[87] Patent DE 38 06 214 (Heubach) 1988.

[88] Patent DE 39 06 670 (Heubach) 1989.

[89] Patent US 2,237,104 (Sherwin-Williams) 1938.

[90] Newkirk AE, Horning SC. Ind. Eng. Chem. Ind. Ed. 1941,33,1402.

[91] TRGS 900 Grenzwerte in der Luft am Arbeitsplatz (Luftgrenzwerte), October 2000 (BArbBl. 10/2000, p. 34), revised August 1, 2003 (BArbBl. 9/2003, p. 42 (48)), *MAK-Werte-Liste* (11/2003).

[92] TRGS 905 Verzeichnis krebserzeugender, erbgutverändernder oder fortpflanzungs-gefährdender Stoffe, März 2001 (BArbBl. 3/2001, p. 94 (97)), revised August 1, 2003 (BArbBl. 9/2003, p. 42 (48)).

[93] Davies JM. Lancet. 1978,2,18.

[94] Davies JM. Br. J. Ind. Med. 1984,41,158.

[95] Cooper WC. Dry Colour Manufacturers' Association, Arlington, 1983.

[96] 1. BImSchVwV: TA Luft – Technische Anleitung zur Reinhaltung der Luft, July 24, 2002 (GMBl. Nr. 25 – 29 vom 30.07.2002, p. 511).

[97]   § 7a Wasserhaushaltsgesetz (WHG) – Abwasserverordnung (AbwV); Anhang 37: Herstellung anorganischer Pigmente, October 15, 2002 (BGBl. I Nr. 74 vom 23.10.2002, p. 4047), revised December 16, 2002 (BGBl. I Nr. 85 vom 19.12.2002,p. 4550).

[98]   § 19g through § 19l WHG – Verwaltungsvorschrift wassergefährdende Stoffe (VwVwS),May 17, 1999 (BAnz. Nr. 98a vom 29.05.1999).

[99]   Council Directive 67/548/EEC, June 27, 1967 (ABl. EG vom 16.08.1967 Nr. L 196, p. 1) together with the 28th adaption Council Directive 2001/59/EC, August 6, 2001, (ABl. EG vom 21.08.2001 Nr. L 225, p. 1).

[100]   Seeger O, Wienand H, in Industrial Inorganic Pigments, 3rd edn. Buxbaum G, Pfaff G ed. Wiley-VCH Verlag, Weinheim, 2005, 123.

[101]   Etzrodt G, in Kittel – Lehrbuch der Lacke Und Beschichtungen, Vol. 5, Pigmente, Füllstoffe Und Farbmetrik, 2nd edn. Spille J ed. S. Hirzel Verlag, Stuttgart, Leipzig, 2003, 93.

[102]   Endriss H, in Aktuelle Anorganische Buntpigmente, Zorll U ed. Vincentz Verlag, Hannover, 1997, 139.

[103]   Patent US 4,026,722 (DuPont) 1976; Patent US 4,063,956 (DuPont) 1976; Patent US 4,115,141 (DuPont) 1977; Patent US 4,115,142(DuPont) 1977.

[104]   Patent US 4,251,283 (Montedison) 1978; Patent US 4,230,500 (Montedison) 1978; Patent US 4,272,296 (Montedison) 1979; Patent US 4,316,746 (Montedison) 1980.

[105]   Patent EP 0 074 049 (BASF) 1981; Patent EP 0 271 813 (BASF) 1986.

[106]   Patent DE 33 15 850 (Bayer AG) 1983; Patent DE 33 15 851 (Bayer AG) 1983; Patent EP 0 492 224 (Bayer AG) 1991; Patent EP 0 723 998 (Bayer AG) 1995.

[107]   Patent EP 0 239 526 (Ciba) 1986; Patent EP 0 304 399(Ciba) 1987; Patent EP 0 430 888, (Ciba) 1989; Patent US 5,123,965 (Ciba) 1989.

[108]   Patent EP 0 551 637 (BASF) 1992; Patent EP 0 640 566 (BASF) 1993.

[109]   Patent US 5,399,197 (Colour Research Company) 1990; Patent EP 0 650 509 (Colour Research Company) 1992.

[110]   Calvert D, in Industrial Inorganic Pigments, 3rd edn. Buxbaum G, Pfaff G ed. Wiley-VCH Verlag, Weinheim, 2005, 136.

[111]   Wehner M, in Kittel – Lehrbuch der Lacke Und Beschichtungen, Vol. 5, Pigmente, Füllstoffe Und Farbmetrik, 2nd edn. Spille J ed. S. Hirzel Verlag, Stuttgart, Leipzig, 2003, 125.

[112]   Endriss H, in Aktuelle Anorganische Buntpigmente, Zorll U ed. Vincentz Verlag, Hannover, 1997, 85.

[113]   Ball P. Bright Earth – The Invention of Color, Penguin, 2001.

[114]   Podschus E, Hofmann U, Leschewski K. Anorg. Allg. Chem. 1936,228,305.

[115]   Tarling SE, Barnes P, Mackay AL. J. Appl. Cryst. 1984,17,96.

[116]   Tarling SE, Barnes P, Klinowsky J. Acta Cryst. B. 1988,44,128.

[117]   Booth DG, Dann SE, Weller MT. Dyes Pigm. 2003,58,73.

[118]   Clark RJH, Dines TJ, Kurmoo M. Inorg. Chem. 1983,22,2766.

[119]   Clark RJH, Franks ML. Chem. Phys. Lett. 1975,34,69.

[120]   Clark RJH, Cobbold DG. Inorg. Chem. 1978,17,3169.

[121]   Gobeltz N, Demortier A, Lelieur JP, Duhayon C. J. Chem. Soc. Faraday Trans. 1998,94,2257.

[122]   Gobeltz-Hautecoeur N, Demortier A, Lede B, Lelieur JP, Duhayon C. Inorg. Chem. 2002,41,2848.

[123]   Kendrick E, Dann SE, Hellgardt K, Weller MT, The Effect of Different Precursors on the Synthesis of Ultramarine Blue using a Modified Test Furnace, International Zeolite Conference, 2004.

[124]   Seel F, Güttler H-J, Wieckowski AB, Wolf BZ. Naturforsch. B. 1979,34,1671.

[125]   Gobeltz N, Demortier A, Lelieur JP, Lorriaux A, Duhayon C. New J. Chem. 1996,20,19.

[126]   Kowalak S, Jankowska A. Catal. Today. 2004,90,167.

[127]   Arieli D, Vaughan DEW, Goldfarb D. J. Am. Chem. Soc. 2004,126,5776.

[128]   Rahman AU, Khan F, Riaz M, Latif A. Pak. J. Sci. Ind. Res. 2009,52,15.

[129]   Goncalves L, Moronta D, Ocanto F, Linares C. Rev. de la Fac. de Ing. 2010,25,25.

[130]   Kajinebaf VT, Rezaeian F, Rajabi M, Baghshahi S. Pigment Resin Technol. 2013,43,1.

[131]   Borhade A, Arun GD, Sanjay GW, Dipak RT. Res. J. Chem. Environ. 2013,17,14.

[132] Gobeltz N, Demortier A, Lelieur JP, Duhayon C. Inorg. Chem. 1998,37,136.

[133] Böhland T, in Industrial Inorganic Pigments, 3rd edn. Buxbaum G, Pfaff G ed. Wiley-VCH Verlag, Weinheim, 2005, 145.

[134] Winkeler H, in Kittel – Lehrbuch der Lacke Und Beschichtungen, Vol. 5, Pigmente, Füllstoffe Und Farbmetrik, 2nd edn. Spille J ed. S. Hirzel Verlag, Stuttgart, Leipzig, 2003, 122.

[135] Endriss H. Aktuelle Anorganische Buntpigmente, Zorll U ed. Vincentz Verlag, Hannover, 1997, 114.

[136] Ferch H, in Ullmann's Encyclopedia of Industrial Chemistry, vol. A 20, 5th edn. VCH-Verlag, Weinheim, 1992, 326.

[137] Dix MF, Rae AD. J. Oil Colour Chem. Assoc. 1978,61,69.

[138] Wertheim GK, Rosencwaig A. J. Chem. Phys. 1971,54,3235.

[139] Buser HJ, Ludi A, Petter W, Schwarzenbach D. J. Chem. Soc. Chem. Commun. 1972,23,1299.

[140] Buser HJ, Schwarzenbach D, Petter W, Ludi A. Inorg. Chem. 1977,16,2704.

[141] Ivanov VD. Ionics. 2020,26,531.

[142] Emrich RJ, Traynor L, Gambogi W, Buhks E. J. Vac. Sci. Techn. A. 1987,5,1307.

[143] Herren F, Fischer P, Ludi A, Haelg W. Inorg. Chem. 1980,19,956.

[144] Ludi A. Chem. Unserer Zeit. 1988,22,123.

[145] Pfaff G, in Encyclopedia of Color, Dyes, Pigments, vol. 2, Pfaff G ed. Walter de Gruyter GmbH, Berlin, Boston, 2022, 777.

[146] Müller-Fokken L. Farbe + Lack. 1978,84,489.

[147] Groß U. ChemTexts. 2018,4(18).https://doi.org/10.1007/s40828-018-0070

[148] Vossen-Blau zur Färbung von Fungiziden. Schriftenreihe Pigmente, Nr. 50, Degussa AG, Frankfurt/M, 1985.

[149] Nigrovic V. Int. J. Rad. Biol. 1963,7,307.

[150] Nigrovic V. Phys. Med. Biol. 1965,10,81.

[151] Tananayev IV. Zh. Neorg. Khim. 1956,1,66.

[152] Pai V. West Indian Med. J. 1987,36,256.

[153] Dvorak P. Z. Ges. Exp. Med. 1969,151,89.

[154] Verzijl JM, Joore HCA, van Dijk A, Wierckx FCJ, Savelkoul TFJ, Glerum JH. Clin. Toxicol. 1993,31,553.

[155] Etzrodt G, in Industrial Inorganic Pigments, 3rd edn. Buxbaum G, Pfaff G ed. Wiley-VCH Verlag, Weinheim, 2005, 121.

[156] Endriss H, in Aktuelle Anorganische Bunt-Pigmente, Zorll U ed. Vincentz Verlag, Hannover, 1997, 121.

[157] Etzrodt G, in Kittel – Lehrbuch der Lacke und Beschichtungen, Vol. 5, Pigmente, Füllstoffe Und Farbmetrik, 2nd edn. Spille J ed. S. Hirzel Verlag, Stuttgart, Leipzig, 2003, 116.

[158] Pfaff G, in Winnacker–Küchler: Chemische Technik, Prozesse und Produkte, Vol. 7, Industrieprodukte, 5th edn. Dittmeyer R, Keim W, Kreysa G, Oberholz A ed. Wiley-VCH Verlag, Weinheim, 2004, 336.

[159] Dunning P, in High Performance Pigments, 2nd edn. Faulkner EB, Schwartz RJ ed. Wiley-VCH Verlag, Weinheim, 2009, 13.

[160] Klimisch HJ. Toxic. 1993,84,103.

[161] Lilljequist R, Kauppi M. Heavy Met. Bull. 1997,2,12.

[162] Turner A. Sci. Total Environ. 2019,657,1409.

[163] Maestro P, Huguenin D. J. Alloys Comp. 1995,225,520.

[164] Pfaff G, in Winnacker–Küchler: Chemische Technik, Prozesse Und Produkte, Vol. 7, Industrieprodukte, 5th edn. Dittmeyer R, Keim W, Kreysa G, Oberholz A ed. Wiley-VCH Verlag, Weinheim, 2004, 354.

[165] Berte JN, in High Performance Pigments, 2nd edn. Faulkner EB, Schwartz RJ ed. Wiley-VCH Verlag, Weinheim, 2009, 27.

[166] Pfaff G, in Encyclopedia of Color, Dyes, Pigments, vol. 1, Pfaff G ed. Walter de Gruyter GmbH, Berlin, Boston, 2022, 103.

[167] Julien-Pouzol M, Guittard M. Ann. Chim. 1972,7,253.

[168] Mauricot R, Gressier P, Evain M, Brec R. J. Alloys Comp. 1995,223,130.

[169] Laronze H, Demourges A, Tressaud A, Lozano L, Grannec J, Guillen F, Macaudiere P, Maestro P. J. Alloys Comp. 1998,275,113.

[170] Cutler M, Leavy JF. Phys. Rev. 1966,133,1153.

[171] Henderson JR, Murato M, Loh M, Gruber JB. J. Chem. Phys. 1967,47,3347.

[172] Kumta PN, Risbud SH. J. Mater. Sci. 1994,29,1135.

[173] Endriss H. Aktuelle Anorganische Buntpigmente, Vincentz Verlag, Hannover, 1997, 150.

[174] Jansen M, Letschert HP. Nature. 2000,404,980.

[175] Tessier F, Maillard P, Cheviré F, Domen K, Kikkawa S. J. Ceram. Soc. Jap. 2009,117,1.

[176] Günther E, Hagenmayer R, Jansen M. Z. Anorg. Allg. Chem. 2000,626,1519.

[177] Pfaff G, in Winnacker–Küchler: Chemische Technik, Prozesse Und Produkte, Vol. 7, Industrieprodukte, 5th edn. Dittmeyer R, Keim W, Kreysa G, Oberholz A ed. Wiley-VCH Verlag, Weinheim, 2004, 357.

[178] Brese NE, O'Keeffe M, Rauch P, DiSalvo FJ. Acta Cryst. C. 1991,47,2291.

[179] Fang CM, Orhan E, De Wijs GA, Hintzen HT, De Groot RA, Marchand R, Saillard JY, de With G. J. Mater. Chem. 2001,11,1248.

[180] Orhan E, Tessier F, Marchand R. Solid State Sci. 2002,4,1071.

[181] Diot N Thesis 2222, Université de Rennes 1, France, 1999.

[182] Orhan E, Jobic S, Brec R, Marchand R, Saillard JY. J. Mater. Chem. 2002,12,2475.

[183] Grins J, Svensson G. Mater. Res. Bull. 1994,29,801.

[184] Patent EP 0 697 373 (dmc^2 Degussa Metals Catalysts Cerdec AG) 1995.

[185] Patent EP 0 627 382 (Cerdec AG) 1994.

[186] Smith AE, Mizoguchi H, Delaney K, Spaldin NA, Sleight AW, Subramanian MA. J. Am. Chem. Soc. 2009,191,17084.

[187] Mizoguchi H, Sleight AW, Subramanian MA. Inorg. Chem. 2011,50,10.

[188] Smith AE, Comstock MC, Subramanian MA. Dyes Pigm. 2016,133,214.

[189] The Shepherd Color Company. Website information, last access 2022-09-03.

[190] Kim SW, Sim GE, Ock JY, Son JH, Hasegawa T, Toda K, Bae DS. Dyes Pigm. 2017,139,344.

[191] Saraswathy D, Rao PP, Sameera S, Jamesa V, Raj AKV. RSC Adv. 2015,5,27278.

[192] Duell BA, Li J, Subramanian MA https://pubs.acs.org/doi/10.1021/acsomega.9b03255.

[193] Jose S, Reddy ML. Dyes Pigm. 2013,98,540.

[194] Kim SW, Saito Y, Hasegawa T, Toda K, Uematsu K, Sato M. Dyes Pigm. 2017,136,243.

[195] Jose S, Prakash A, Laha S, Natarajan S, Reddy ML. Dyes Pigm. 2014,107,118.

[196] Kusumoto K. J. Ceram. Soc. Jap. 2016,124,926.

[197] Masui T, Shiraishi A, Furukawa S, Nunotani N, Imanaka N. J. Jpn. Soc. Colour Mat. 2012,85,913.

[198] Sulcova P, Trojan M. J. Therm. Anal. Calorim. 2006,83,557.

[199] Sulcova P, Trojan M. J. Therm. Anal. Calorim. 2007,88,111.

[200] Oka R, Takemura A, Shobu Y, Minagawa K, Masui T. J. Ceram. Soc. Jap. 2022,130,39.

[201] Oka R, Koyama J, Morimoto T, Masui T. Molecules. 2021,26,6243.

[202] Oka R, Kosaya T, Masui T. Chem. Lett. 2018,47,1522.

[203] Kusumoto K. J. Ceram. Soc. Jap. 2017,125,396.

[204] Devreux V, Clabaux E. PCI Paint Coatings Industry 2021,10,7.

[205] Patent WO 002017102735 (CapellePigments NV) 2017.

5 Black pigments (carbon black)

5.1 Fundamentals and properties

The most important inorganic black pigments are carbon black, iron oxide black, and spinel blacks (Table 5.1). Carbon black pigments are of greatest importance by far among these three blacks.

The term "carbon black" stands for a group of well-defined industrially manufactured products. These are produced under exactly controlled conditions. The physical and chemical properties of each carbon black type are kept within narrow limits.

Tab. 5.1: Summary of inorganic black pigments.

Name/Color index	Chemical composition	Structure
Carbon Black	C	amorphous
Magnetite	Fe_3O_4	spinel
(C.I. Pigment Black 11)		
Pigment Black 30	Ni(II), Fe(II,III), Cr(III) oxide	spinel
Pigment Black 26	Mn(II), Fe(II,III) oxide	spinel
Pigment Black 22 / ⎱	Cu(II), Cr(III) oxide	spinel
Pigment Black 28 ⎰	Mn(II), Cu(II), Cr(III) oxide	spinel
Pigment Black 27	Co(II), Cr(III), Fe(II) oxide	spinel

Carbon black is a form of highly dispersed, virtually pure elemental carbon (diamond and graphite are other forms of nearly pure carbon) with extremely small, almost spherical colloidal particles that are produced by incomplete combustion or thermal decomposition of gaseous or liquid hydrocarbons [1–5]. Carbon blacks still contain considerable amounts of chemically bound hydrogen, oxygen, nitrogen, and sulfur, depending on the manufacturing conditions and the quality of the raw materials. Carbon black possesses outstanding pigment characteristics such as high light stability and full insolubility in different media. It is not only applied as a pigment, but also as active filler material in rubber. A major application area here is in car tires. The use of carbon black as filler is much more effective when compared to zinc oxide with respect to abrasion resistance. Today, more than 35 types of carbon black are applied as fillers in rubber and nearly 80 carbon black types as pigments in various special applications. More than 90% of the carbon black produced worldwide is used in the rubber industry [4].

Carbon black pigments consist of extremely small, mostly spherically shaped primary particles, which form aggregates with each other, which in turn consist of chains and clusters. The degree of aggregation is called the "structure" of carbon black. There is a high agglomeration tendency of the aggregates with each other. The primary particles have diameters ranging from 5 to 500 nm. The structure of these particles contains mainly amorphous but also microcrystalline subregions. Diffraction patterns obtained using high resolution transition electron microscopy show that the spherical primary particles consist of relatively disordered nuclei surrounded by concentrically deposited carbon layers [6]. It has been found that the degree of order increases from the center of the particles to their peripheral areas. This structural phenomenon is relevant for the comprehension of the chemical reactivity of carbon black. The mean diameter of the primary particles, particle size distribution, and degree of aggregation can be affected within relatively wide ranges by parameter variation during the production process. Figure 5.1 shows a scanning electron micrograph of a carbon black pigment produced according to the gas black process. Nearly all primary particles have a similar size of about 30 nm. The small size associated with a very high surface area leads to the formation of larger aggregates.

00301816 ——— 300 nm 50.00 K X

Fig. 5.1: Scanning electron micrograph of a carbon black pigment produced according to the gas black process.

Carbon black contains structural motifs that are very similar to those of graphite, which crystallizes in a layer structure of hexagonally bound carbon atoms. The tightly bound layers are connected to each other only by loose van der Waals forces. The carbon atoms within the layers of carbon black are arranged almost in the same way as in graphite. The carbon layers are oriented nearly parallel to each other [7, 8]. The relative position of these layers in carbon black is arbitrary however, unlike in the case of graphite. Thus, there is no order in the structural c direction for carbon black. Crystalline regions in carbon black are, therefore, only of small size, typically 1.5 to 2 nm in length and 1.2 to 1.5 nm in height, corresponding to four to five carbon layers [4]. The simplified structures of carbon black and graphite are shown in Figure 5.2. The fact that both carbon species – graphite and carbon black, are electrically conductive (unlike diamond) can easily be explained by their structural relationship. The surface chemistry of carbon black is closely related to that of graphitic and of turbostratic carbons through the predominant hexagonal carbon lattice from graphene fragments forming its basic structural units [9].

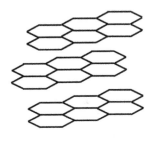

Fig. 5.2: Simplified layer structure of carbon black (left) and graphite (right).

The internal structure of carbon black particles considerably influences the optical behavior, apart from the shape and agglomeration state of the primary particles [10]. It could be shown that the UV $\pi - \pi^*$ absorption feature of carbon black varies in position between 196 and 265 nm depending on the state of bending of the graphene layers in the subunits of particles and/or the dimensions of the plane graphitic microcrystallites, and the incorporation of hydrogen. In the mid-infrared region of the electromagnetic spectrum, the absolute value of the absorption coefficient κ is dominated by continuous absorption due to free charge carriers, which are also influenced by the ratio of sp^2/sp^3 hybridized carbon in the primary particles. The appearance of prominent bands is related to the existence of functional groups, like $C-H_n$, $C = O$, and/or $C-O-C$ [10].

Carbon black is provided in the form of pellets or finely divided powders. There is clear correlation between the properties of the different carbon black types, e.g., color, particle size distribution, specific surface area, or electrical conductivity and their use in tires, rubber, plastics, printing inks, and coatings.

The worldwide production range for carbon black comprises six types of carbon black, which are produced by different manufacturing processes: furnace black, channel black, gas black, lamp black, thermal black, and acetylene black. Chemical and physical characteristics can differ considerably, depending on the production method.

5.2 Production of carbon black pigments

5.2.1 Raw materials

Table 5.2 contains a compilation of the most important processes for the manufacture of carbon black. The processes are basically classified in two groups, those with partial combustion of the raw materials (thermal oxidative decomposition) and those that are based on pure pyrolysis (thermal decomposition). The two processes differ significantly in the conduct of the thermal decomposition step. The thermal oxidative decomposition uses air for the combustion of part of the starting materials (the energy for the pyrolysis is provided at the same time). Pure thermal decomposition uses an external power supply to achieve the necessary high temperatures.

Tab. 5.2: Production methods for carbon black pigments.

Chemical process	Manufacturing methods	Main raw materials
Thermal oxidative decomposition	Furnace black process	Aromatic oils on coal tar basis or mineral oil, natural gas
	Gas black and channel black process	Coal tar distillates
	Lamp black process	Aromatic oils on coal tar basis or mineral oil
Thermal decomposition	Thermal black process	Natural gas or mineral oils
	Acetylene black process	Acetylene

The raw materials preferentially used for most carbon black production processes, especially the furnace black process, are heavy oils, which consist mainly of aromatic hydrocarbons. These carbon compounds deliver the highest carbon-to-hydrogen ratio, thus maximizing the available carbon for the carbon black synthesis. The use of aromatic hydrocarbons is, therefore, the most efficient way to achieve high yields of carbon black. Hydrocarbons with higher numbers of combined rings would be beneficial, but their use is limited because they are solid and difficult to decompose thermally. Preferred raw materials are, therefore, those organic compounds in which the majority of the carbon atoms are bound in three- or four-membered rings.

Other sources of raw materials for carbon black are oily distillates from coal tar and residual oils that are created by catalytic cracking of mineral oil fractions and olefin manufacture. Hydrocarbons produced by thermal cracking of naphtha or petrochemical oil can also be used as a source of feedstock. Chemical and physical properties of the raw materials are important for the process management and for the achievement of suitable final products: density, distillate residue, viscosity, carbon/hydrogen ratio, asphaltene content, specified impurities, etc. Feedstock specifications must take these properties into account.

Further quality requirements concern impurities from foreign matter. Alkaline metals can play a significant role in the production process because they have a direct influence on specific carbon black characteristics. The sulfur content of the used feedstock is important for production, since emission of sulfurous gases formed during the combustion process is restricted by law in many countries. Carbon black types with high sulfur contents are moreover prohibitive for various applications.

5.2.2 Furnace black process

The development of the furnace black process as one of the most important production routes for carbon black goes back to an increasing demand from different industries. Carbon black was especially used for tires and for black pigmented plastics, paints, and printing inks. As a result, a new and better manufacturing process, the furnace black process, was developed for rubber grades and also for new carbon black pigments for refined applications. This production process can be used for nearly all types of carbon black required today. It also meets the high demands concerning economy and ecology.

The furnace black process has been greatly refined and improved since its development in the 1920s. The process is carried out continuously in closed reactors. Most of the semireinforcing rubber blacks with specific surface areas of 20 to 60 m^2/g and the active reinforcing blacks with specific surface areas of 65 to 150 m^2/g are manufactured by this process. A large amount of pigment-grade carbon blacks with much greater specific surface areas and smaller particle sizes is also produced using the furnace black process [4]. The process allows the control of several product properties, such as specific surface area, particle size distribution, structure, and absorption. Application relevant properties of carbon black-containing rubber, such as abrasion resistance, tear strength, jetness, and tinting strength can also be varied in the furnace black process by adjusting the operating parameters. This flexibility is important to achieve the narrow specifications required by customers.

The central unit of a furnace black production plant is the special furnace in which the carbon black is formed. The starting components are injected mostly as an atomized spray into the high-temperature zone of the furnace. High temperatures are achieved by burning a fuel (natural gas or oil) with air. The fuel is burned in an excess of oxygen, which is not sufficient for the complete combustion of the feedstock. Most of the feedstock is pyrolyzed to form carbon black at temperatures of 1200 to 1900 °C. The reaction products are quenched with water and cooled further in heat exchangers. The resulting carbon black is collected from the tail gas by using suitable filter systems.

The feedstock for the process, preferably petrochemical or carbochemical heavy oils, which usually begin to crystallize near the ambient temperature, is stored in heated tanks equipped with circulation pumps to maintain a homogeneous mixture. Rotary pumps transport the oil to the reactor via heated pipes and a heat exchanger. There the oil is heated to 150 to 250 °C and obtains a viscosity appropriate for atomization. Various spraying devices are used to introduce the oil and the fuel into the reaction zone. Frequently used devices are axial oil injectors with a spraying nozzle at its tip (hollow cone spray shape) and one or two-component atomizing nozzles (with air and steam as the preferred atomizing agents) [4, 11].

Alkali metal compounds, preferably aqueous solutions of potassium hydroxide or potassium chloride, are often added to the oil in the oil injector to have an influence on the carbon black structure [12, 13]. The additives may also be sprayed separately into the combustion chamber. Other additives, e.g., alkaline earth metal compounds, which lead to an increase of the specific surface area of the formed carbon black powders, can be introduced in a similar manner.

Natural gas is the fuel of choice in most cases to obtain the necessary high pyrolysis temperatures, but other gases, e.g., coke oven gases or vaporized liquid gas, can also be used. Various oils including the feedstock are also used as fuel because of their favorable economic balance. Fast and complete combustion is achieved by the use of special burners adapted to the oil type (Figure 5.3) [4].

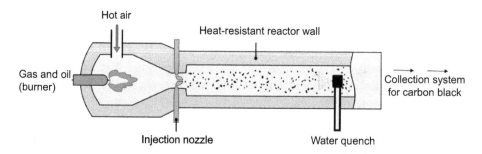

Fig. 5.3: Scheme of furnace black production process (source: MagentaGreen, Wikimedia Commons).

Rotating piston compressors or turbo blowers ensure that the required air is compressed before combustion. Heat exchangers are used to preheat the air using hot gases leaving the furnace. Energy is conserved in this way, and the carbon black yield is improved, at the same time. Temperatures of 500 to 700 °C are typical for the preheated air [4].

Modern combustion units in carbon black plants vary significantly in internal geometry, flow characteristics, and the way in which fuel and feedstock are introduced. On the other hand, they all have the same basic process steps, in particular the provision of hot combustion gases in a combustion chamber, the injection of the feedstock and its rapid mixing with the combustion gases, vaporizing of the oil, pyrolyzing it in the reaction zone, and rapid cooling of the reaction mixture in the quenching zone to temperatures of 500 to 800 °C.

The main pyrolysis reaction can be described in a simplified way with the following equation:

$$C_n H_m + energy \rightarrow n\,C + m/2\,H_2 \tag{5.1}$$

The morphology of carbon black primary particles formed during the pyrolysis reaction indicates that during formation of carbon black, the first nuclei of pyrolyzed hydrocarbons condense from the gas phase. Further carbon layers or their precursors are then adsorbed onto the surface of the growing particles. New layers are always orientated parallel to the existing surface. Aggregates are created by further carbon deposits on initially formed loose agglomerates.

The carbon layers undergo a conversion to a graphitic arrangement beginning at the particle surface at temperatures above 1200 °C. Graphite crystallites are formed at 3000 °C. At the same time the carbon black particles show a polyhedral shape [4].

The furnace black process is typically performed in horizontally arranged reactors. Such reactors can have a length of up to 18 m and an outer diameter of up to 2 m. In some cases, especially for the manufacture of certain semireinforcing blacks, vertical reactors are used [14].

The ratio of fuel, feedstock, and air has a strong influence on the properties of the resulting carbon black powders. Increasing amounts of excess air relative to the amount needed for the complete combustion of the fuel lead in most cases to smaller particle sizes of the formed carbon black. Rising temperatures together with higher oil combustion rates are achieved when larger amounts of air in the reaction zone are used. The nucleation velocity and the number of particles formed increase under such conditions, but the mass of each particle and the total yield decrease [4].

Further parameters having an influence on the carbon black quality are the way in which the oil is injected, atomized, and mixed with the combustion gases, the type and amount of additives, the preheating temperature of the air, and the quench position. Several reactions can take place at the surface of the freshly formed carbon black particles in the reactor with participation of the hot surrounding gases (e.g., Boudouard reaction, water gas reaction). As a consequence, the chemical nature of the carbon black surface is modified with increasing residence time. These reactions are stopped by quenching the reaction products to temperatures below 900 °C and a certain state of surface activity for the carbon black is fixed. Pelletizing and drying conditions are other possibilities to adjust the surface properties of carbon black powders [4].

When the mixture of carbon black and gas leaves the furnace reactor, it is cooled down in heat exchangers to temperatures of 250 to 350 °C and conducted into a collection system. This system consists typically of only one high performance bag filter with several chambers. These are periodically purged by counterflow of filtered gas or by pulsejets. The filter may contain hundreds of bags with a total filter area of several thousand square meters, depending on the capacity of the production unit [4]. The carbon black is now pneumatically transferred from the filter into a first storage tank. Small amounts of impurities of iron, rust, or coke particles are either removed by magnets and classifiers or milled to an appropriate consistency [4].

The bulk density of freshly synthesized carbon black is extremely low at 20 to 60 g/l. The way to facilitate handling and further processing is compaction of the powdered material. Weak densification of the powder can be done using a process by which the carbon black is conducted over porous, evacuated drums. Thus, treated carbon black retains its powdery state [15, 16]. This form of compacting is favorable for most of the carbon black types used in paints, inks, and plastics because a good dispersibility of the pigment in the application system must be maintained.

Carbon black obtained in this way contains ca. 50 wt% water. In the next step, the material is dried in dryer drums, indirectly heated by burning tail gas. The dried powder is transported via conveyor belts and elevators to storage tanks or packing stations. The bulk density of wet, pelletized carbon black reaches values of 250 to 500 g/l [4].

5.2.3 Channel black and gas black processes

The oldest method for producing small particle size carbon blacks on an industrial scale is the channel black process. It has been used in the USA since the late nineteenth century. Low profitability and environmental regulations led to the closure of the last production plant in the USA in 1976. Natural gas was used as the feedstock and the yield of carbon black was only 3% to 6% [4].

The gas black process, which is similar to the channel black process, was developed in the 1930s. This process uses coal tar oils instead of natural gas. Much higher yields and production rates are achievable using oil based raw materials. Degussa started the manufacture of carbon black with the gas black process on an industrial scale in Germany in 1935. Today carbon blacks of various qualities for a diversity of applications are produced using the gas black process [4].

As shown in Figure 5.4, the raw materials are partially vaporized in the reactor. The residual oil is continuously removed. Combustible carrier gases, such as hydrogen, coke oven gas, or methane, transport the oil vapor to the production vessel. Very small carbon black particles are formed when air is added to the oil–gas mixture. The

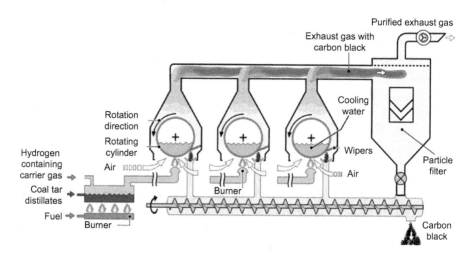

Fig. 5.4: Scheme of gas black production process (source: MagentaGreen, Wikimedia Commons).

gas black process is technically not as flexible as the furnace black process. However, it is possible to produce a variety of different gas black types by varying the relative amounts of carrier gas, oil, and air [4].

The manufacturing facility for the gas black process consists of a burner pipe approximately 5 m long, which carries 30 to 50 diffusion burners. The flames are orientated in such a way that they are in contact with a water-cooled drum, where about half of the carbon black formed during the process is deposited. The deposited carbon black is scraped off from the drum and transported by a screw to a pneumatic conveying system. At the bottom, the gas black equipment is surrounded by a steel box. The off-gas is sucked by fans at the top of the apparatus into filters, which collect the carbon black suspended in the gas [4].

The amount of air used in the process can be regulated by valves in the exhaust pipes. A combination of several gas black apparatuses can form one production unit. One oil vaporizer feeds the whole unit in this case. Typically, the yield of carbon black produced using the gas black process is 60%, while the production rate is 7 to 9 kg/h. The yield of special high-quality blacks is only 10% to 30% [4].

The use of air in the gas black process leads to the contact of oxygen with the formed carbon black particles at high temperatures. As a result, acidic oxides are formed on the surface of the particles. Gas blacks react, therefore, unlike furnace blacks, acidic when suspended in water [4].

5.2.4 Lamp black process

The oldest commercially used production process for carbon black is the lamp black process. The facilities used today are, however, not comparable with those of olden times. Ancient carbon black ovens, smoking chimneys, and settlement chambers have been replaced by modern manufacturing units with high sophisticated filtering systems.

The lamp black apparatus as shown in Figure 5.5 consists of a cast iron pan on which the liquid feedstock is applied. The pan is surrounded by a fireproof flue hood lined with refractory bricks. The stream of the incoming air, the gap between the pan and the hood, and a vacuum present in the system, help regulate the conditions for the pyrolysis and allow the manufacturer to fine-tune the properties of the resulting carbon black. Radiant heat from the hood brings the raw materials to vaporization, partially to combustion and thereby to conversion into carbon black [4].

The process gases containing carbon black are passed through a filter after the cooling stage. The carbon black, which is collected by this means is processed further as described for the furnace black process.

Carbon black powders manufactured using the lamp black process typically show broad primary particle size distributions, ranging from approximately 60 to over 200 nm. They are mostly used in special applications [4].

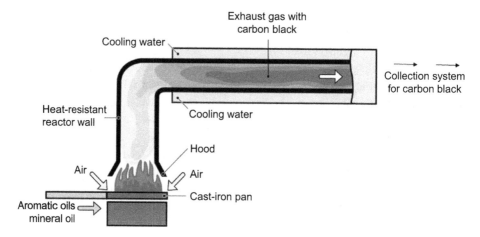

Fig. 5.5: Scheme of lamp black production process (source: MagentaGreen, Wikimedia Commons).

5.2.5 Thermal black process

The thermal black process is a noncontinuous or cyclic process. It does not belong to the thermal oxidative processes because it works without the inflow of air. Natural gas is the most common feedstock, but higher hydrocarbon oils can also be used. Tandem mode plants were developed for the production of thermal black, in order to achieve high efficiency. They consist of two reactors operating alternately in cycles of 5 to 8 min. One of the reactors is heated up with natural gas or an oil-air mixture whereas the other one is fed with pure feedstock, which undergoes thermal decomposition. The carbon black formation occurs in the absence of oxygen and at decreasing temperatures. Properties of the resulting carbon black powders are considerably different from those achieved with thermal oxidative processes.

The formation of thermal blacks usually happens relatively slowly, with the result that coarse particles in the diameter range from 300 to 500 nm are obtained. The use of natural gas as feedstock diluted with inert gases enables the production of thermal black with particle sizes in the range of 100 to 500 nm. However, this special process management is no longer used because other formation routes to achieve such qualities of carbon black are more efficient.

Mechanical rubber goods with high filler contents are one of the main application fields for thermal blacks. However, the importance of thermal blacks for this application is decreasing nowadays because other products such as clays, milled coals, and cokes are used more and more as cheaper substitutes.

5.2.6 Acetylene black process

The acetylene black process was developed at the beginning of the twentieth century. Mixtures of acetylene with light hydrocarbons are used as raw materials for this production route. The decomposition of acetylene is, in contrast to other hydrocarbons, an exothermic process, providing energy for the carbon black formation.

The first manufacturing method for acetylene black was a discontinuous explosion process. Carbon black produced in this way was mainly used for pigment purposes. Modern processes work continuously, whereby acetylene or acetylene-containing gases are fed into a preheated, cylindrical reactor with a ceramic inner liner. The pyrolysis of the acetylene starts after an ignition and is maintained by the decomposition heat, which is released by the exothermic reaction. The formed carbon black in a typical yield of more than 95% is collected in settling chambers and cyclones [4].

Shapes of primary particles of acetylene black are different compared to those of other carbon blacks. There is increased order in the c direction of the crystalline regions. Folded sheets of carbon layers are the main structural units. The application of acetylene blacks is limited to special uses, e.g., in dry cells, because of their relatively high price [4].

5.2.7 Other production processes

Plasma-based processes for the manufacture of carbon black are possible but have not yet been used on an industrial scale [4, 17, 18]. Vaporized hydrocarbons can be converted almost quantitatively into high purity carbon and hydrogen in a plasma arc at temperatures of about 1600 °C by means of electricity. This method can in principle be used to produce carbon blacks with small particle sizes. However, it has not been possible so far to develop a plasma-based commercial process that is economical. The Huels electric arc process for the production of acetylene and ethylene by plasma reactions delivers larger quantities of carbon black as a by-product, which cannot be separated easily and used for pigments or other applications.

The price of raw materials and fuels for the production is among decisive technical factors for the profitability of a business with carbon black. The carbon black industry is, thus, highly dependent on the situation in the petrochemical industry. Consequently, several approaches have been made to replace existing raw materials with new ones. In that light, ways to obtain carbon black directly from coal or to isolate it from old tires have been investigated [4, 19]. These and other attempts have not been successful with regard to commercial importance. Another approach, the use of clay, milled coal, and coke, has found limited interest as a substitute for very coarse carbon blacks, primarily thermal blacks and some furnace blacks.

There is growing use of precipitated silica powders in tires and mechanical rubber goods, mostly in combination with organosilicon coupling agents. The original concept for this development included the search for new non-oil-based fillers. It has led, in the meantime, to new and improved rubber properties.

5.2.8 Oxidative aftertreatment

The application properties of carbon black powders are strongly influenced by functional groups on the surface of the particles. Oxygen-containing functional groups are particularly important here. High amounts of surface oxides lead to a decrease in the vulcanization rate and an improvement in the flow characteristics of inks, as well as to increase of gloss in coatings. The color tone of carbon blacks is shifted under the influence of such functional groups from brownish to bluish, and jetness can be increased.

Oxidative aftertreatment is the typical method used to amend the color properties of carbon blacks. Depending on the oxidizing agent and the reaction conditions selected, different types and amounts of oxygen-containing groups are formed on the surface of the particles (e.g., carbonyl, ether, ketone, peroxide, phenol, lactone, hemiacetal, and anhydride groups).

Limited oxidation of carbon black surfaces can be done with air in a temperature range of 350 to 700 °C. Higher amounts of surface oxides together with better control of the process can be better achieved by using nitric acid, mixtures of nitrogen dioxide and air, ozone, or sodium hypochlorite solutions as oxidizing agents [20–23]. Generally, all strongly oxidizing agents can be used for the treatment, either as a gas or in solution. The oxidation reactions are typically carried out at elevated temperatures.

Oxidative treated carbon black powders contain up to 15 wt% oxygen and are strongly hydrophilic. The latter is the reason why some of them form colloidal solutions in water spontaneously. Better wettability and dispersion behavior in coatings and polar printing inks are achieved by surface oxidation. The consumption of binders can be reduced in this way [4].

The oxidative aftertreatment of carbon black with nitrogen dioxide and air can be carried out on an industrial scale in a fluidized bed reactor [24, 25]. The equipment for the treatment consists typically of a preheating vessel, in which the carbon black is fluidized and heated, a reaction vessel to carry out the surface oxidation, and a desorption vessel in which adsorbed nitric oxide is removed. Temperatures used for the oxidative reaction are in the range of 200 to 300 °C. The reaction time depends on the desired degree of oxidation. The nitrogen dioxide acts primarily as a catalyst and the oxygen in the air as the oxidizing agent [4].

Surface oxidation can also be carried out during pelletizing of carbon black. Nitric acid is used as a pelletizing agent in this case. Oxidation is carried out here while the wet pellets are dried at elevated temperature [26].

5.2.9 Pelletizing of carbon black

For pelletizing of carbon black, two widely used methods are employed: dry pelletizing and wet pelletizing.

The older method is dry pelletizing. Dry loose carbon black is agitated in this process in the presence of seed pellets for several hours until a dry pelletized product is formed. The agitation is usually carried out in a continuous flow process by rolling the loose carbon black in a number of large-volume rotating, pelletizing drums. These drums are supported on trunnions and rotate on their horizontal axis. The loose carbon black is continuously fed into one end of the rotating drum, while the dry pellets continuously flow from the other end thereof.

Wet pelletizing is a method that developed later. Loose carbon black is mixed in this process with 30% to 50% water and then agitated for a few minutes to form wet pellets. The agitation is usually accomplished in a continuous flow pattern by a pin-shaft type agitator rotating within a cylindrical chamber. The cylindrical chamber is stationary, of relatively small volume, and mostly horizontally disposed. Loose carbon black and water are separately fed to the inlet end of the cylindrical chamber or pelletizing box. The wet pellets are continuously discharged from the discharge end into a rotating drying drum. Most of the commercially offered carbon black today, is pelletized using the wet method. The advantage of this method compared to the dry process is that the pelletized product consists of more uniform and durable pellets. These are less dusty and can be shipped and unloaded more easily from bulk storage and shipping equipment [27].

5.3 Pigment properties and uses

Table 5.3 contains select data on carbon black, which can be important for the handling and application of carbon black powders and pellets. Figure 5.6 shows a photograph of a carbon black pigment in powdery form. Figures 5.7 and 5.8 give an impression of the appearance of dry and wet pelletized carbon black products.

Tab. 5.3: Select data on carbon black.

Physical state	Solid: powder or pellet
Lower limit for explosion	50 g/m^3 (carbon black in air)
Minimum ignition temperature:	
VDI 2263 (Germany), BAM Furnace	> 932 °F (> 500 °C)
Godbert-Greenwald Furnace	> 600 °F (> 315 °C)
Minimum ignition energy	> 10 J

Tab. 5.3 (continued)

Burn velocity:	
VDI 2263, ED Directive 84/449	> 45 s; not classifiable as "Highly Flammable" or "Easily Ignitable"
Flammability classification (OSHA)	Combustible solid

Solubility:	
Water	Insoluble
Organic solvents	Insoluble
Specific surface area	10–1000 m²/g
Particle size	10–500 nm
Density	1.8–2.1 g/cm³

Fig. 5.6: Photograph of a carbon black pigment in powdery form (channel black).

Fig. 5.7: Photograph of a dry pelletized carbon black pigment (furnace black).

The specific surface area of industrially manufactured carbon black varies widely from small values of 8 m²/g for coarse thermal blacks up to 1000 m²/g for the finest pigment grades. Carbon blacks used as reinforcing fillers in tire treads have specific surface areas between 80 and 150 m²/g. Carbon black powders with specific surface areas of more than 150 m²/g are mostly porous with pore diameters of less than 1 nm.

Fig. 5.8: Photograph of a wet pelletized carbon black pigment (channel black).

Carbon black species with a large surface area can have particles with an inner pore surface area that exceeds the outer (geometrical) one. The adsorption capacity of carbon blacks for water, organic solvents, and binders is remarkable because of the large specific surface area. The capacity increases with growing specific surface area of the powders. The adsorption occurs chemically and physically at the particle surfaces. It is very important for the wettability and the dispersibility of the carbon black particles in different application systems and consequently also for the decision on the use of a carbon black type as a filler in rubber or as pigment. It is noticeable and relevant for the application that carbon black species with a large specific surface area adsorb up to 20 wt% water when stored in an atmosphere of humid air [4].

Particle sizes of carbon black powders are, in correspondence with the high specific surface areas, very small. Many carbon black grades consist mainly of nanoparticles. Typical particle size distributions for channel and gas blacks comprise 10 to 30 nm, for furnace blacks 10 to 110 nm, for lamp blacks 60 to 200 nm, for thermal blacks 100 to 500 nm, and for acetylene blacks 30 to 50 nm.

Carbon black is electroconductive. Its conductivity is, however, inferior in comparison with graphite. The conductivity of carbon black depends on the structure, the exact composition, and the specific surface area. The electrical conductivity in paint or a plastic material is determined by the distance between conductive neighboring particles in the system. If the distance between a large number of particles is very close and the concentration of particles in the application medium is high enough to establish a conduction path, high electrical conductivity can be achieved. Studies have shown that a negligible quantity of oxygen combined with a high amount of the graphitic phase and of alkali metal compounds leads to high conductivity of carbon black in polar and nonpolar solvents. The conductivity of dispersions increases with a higher structure of the carbon black grade, with larger aggregates, and with a broad particle size distribution [28]. Special qualities of carbon black with suitable conductivity were developed to equip application systems with antistatic or electrically conductive characteristics.

Carbon black grades having high conductivity and high adsorption capacity for electrolyte solutions find application in dry cell batteries.

Carbon black absorbs visible light up to a rate of 99.8%. This property is the reason for the broad use of carbon blacks as black pigments. Not all carbon black pigments are pure black. There are also types showing a bluish or brownish color tone. The character of such color variations depends on the structure of the carbon black particles, the light conditions, and the application system in which the carbon black is incorporated. Carbon black also absorbs infrared and ultraviolet light. Some carbon black grades are used, therefore, as UV stabilizers in plastics. The chemical analysis of carbon blacks varies significantly, as shown in Table 5.4.

Tab. 5.4: Limits for the general composition of carbon black pigments according to elemental analysis [4].

Component	Content (wt%)
Carbon	95.0–99.5
Hydrogen	0.2–1.3
Oxygen	0.2–3.5
Nitrogen	0–0.7
Sulfur	0.1–1.0
Residual ash	<1.0

Manufacturing process, raw materials, and aftertreatment are the most relevant factors for the specific composition of a carbon black type. The residual ash comes primarily from the raw materials and from salts. The latter play a role in the manufacturing process for the control of the carbon black structure. There are also salts in the process water that are not completely washed out.

The surface of carbon black powders can contain small amounts of aromatic substances, which are strongly adsorbed at the end of the production process. These can only be removed by extraction with organic solvents. The quantity of extractable material is, for most industrial carbon blacks below the limit defined by the food laws [4].

The oxygen content of carbon blacks is of special importance for their application properties. Oxygen can be bound at the surface as acidic or basic functional groups. Manufacturing process and aftertreatment have a significant influence on the amount of surface oxides and their chemical nature. Furnace and thermal blacks contain only 0.2 to 2 wt% oxygen because they are produced in a reducing atmosphere. The oxygen is bound here, preferably in form of almost pure, basic surface oxides. Gas and channel blacks, on the other hand, which are manufactured in the presence of air, contain up to 3.5 wt% oxygen. Mainly acidic surface oxides and only small amounts of basic oxides are typical, in this case. Oxidative aftertreatment can increase the amount of acidic oxides.

High temperatures destroy the oxygen-containing surface groups of carbon black. The weight loss at 950 °C is a rough indication of the oxygen amount. Another indication is the pH value measured in aqueous carbon black slurry. In general, pH values greater than 7 indicate low oxygen contents and basic surface oxides (furnace blacks), and pH values lesser than 7 are an indication of oxidized carbon blacks with high oxygen content and acidic surface oxides (gas blacks). The surface properties of carbon black particles can further be modified by additional chemical reactions at the oxidic groups (alkylation, halogenation, esterification) in order to adjust the surface to the application medium.

Hydrogen is bound on the carbon black surface in two different ways. A portion of the hydrogen is directly fused to the carbon. Another portion together with oxygen forms surface bound functional groups. Carbonyl, carboxyl, pyrone, phenol, quinone, lactone, and ether groups have been identified by analytical methods as the oxygen-containing groups bound to the surface of the carbon black particles.

Sulfur appears in carbon black in a variety of forms: in its elementary form, as a bound molecule, and in an oxidized state. High sulfur contents can be responsible for a certain acidity of industrial carbon blacks. Nitrogen, when present, is usually incorporated in the graphite lattice.

Carbon blacks do not spontaneously ignite at temperatures below 600 °C. They glow slowly when ignited in air. Ignition sources must be excluded during processing and storage of carbon black powders and pellets.

The initially mentioned structure of carbon black is characterized by a three-dimensional arrangement of primary particles in aggregates. Extensive interlinking or branching is typical for a "high structure", whereas less pronounced interlinking or branching characterizes a "low structure" of carbon black. Direct determination of the structure of carbon black particles is nearly impossible. A widely used and accepted method for the structural characterization is the absorption of dibutyl phthalate (DBP) measured in ml/g. A specified quantity of carbon black is placed in a Brabender kneader, and DBP is added dropwise. The amount of added DBP is measured, which is needed to reach a predetermined level of torque generated by the kneading machine as it is added. The basis of the measurement is the condition that all interstices known as void volumes in the carbon black aggregates are filled with DBP. For the evaluation of the data it is considered that the surfaces of all carbon black particles are wetted. The modified surface situation is put in relation to the change of the torque of the kneading machine during the DBP addition. Based on this measurement, the added DBP amount allows conclusions on the degree of aggregation of individual carbon blacks. The general rule is: the greater the DBP absorption, the higher the carbon black structure.

A true picture of the primary particles and of the aggregates is obtained by high resolution electron microscopy. The combination of this method with X-ray analytical techniques shows that the primary particles consist of concentrically arranged, graphite-like crystallites. The graphitic layers are often twisted into each other, which leads to a non-ordered state. One single primary particle can consist of up to 1500 of these crystallites.

Investigations with a scanning tunnel microscope have further given an indication that the primary particles consist of superimposed, scale-like layers of graphite [4].

An overview of the most important carbon black applications is shown in Table 5.5. The rubber industry is the main consumer of carbon black with about 90% of the product quantity, using it as reinforcing filler in tires, tubes, conveyor belts, cables, rubber profiles, and other mechanical rubber goods. Carbon black types produced with the furnace black process are mainly used in rubber applications [4].

Tab. 5.5: Applications for carbon black [4].

Area	Application
Rubber	Reinforcing filler in tires and mechanical rubber components
Printing inks	Pigmentation, rheology
Coatings	Full black and tinting application
Plastics	Black and gray pigmentation, tinting, UV protection, conductivity, conductor coatings
Fibers	Pigmentation
Paper	Black and gray pigmentation, conductivity, decorative and photoprotective papers
Construction	Cement and concrete pigmentation, conductivity
Power	Carbon brushes, electrodes, battery cells
Metal reduction, compounds	Metal smelting, friction compound
Metal carbide	Reduction compound, carbon source
Fire protection	Reduction of mineral porosity
Insulation	Graphite furnace, polystyrene and polyurethane foam

Based on their reinforcing capability, carbon blacks are classified as active, semiactive, or inactive blacks. Active blacks have a high reinforcing capability and are used in tire treads. They consist of very fine powders with particle sizes in the range of 15 to 30 nm. Semiactive blacks have a lower reinforcing capability. With particle sizes from 40 to 60 nm, they are used in the tire carcass and in technical rubber components, from screen and door seals to floor mats. Inactive blacks have particle sizes of more than 60 nm and only a limited reinforcing capability. They need high filling levels for use in reinforcing applications.

There are also special carbon black types used in tires. The so-called "adhesion blacks" find application in the improvement of radial steel belt adhesion. Conductive blacks are applied to achieve a certain electrical conductivity at the surface of the tires. The filling level to achieve a sufficient conductivity is comparatively high.

Pigment blacks are quantitatively less important than rubber blacks. With regard to quality, however, pigment blacks are highly sophisticated powdered materials. They are used for a variety of applications, particularly in printing inks, paints, coatings, plastics, fibers, and paper. Depending on the application they are called printing blacks, coating blacks, or plastic blacks.

Carbon black pigments have a number of advantages compared to other black pigments (inorganic and organic) and black organic dyes. Hiding power, color stability, solvent resistance, acid and alkali resistance as well as thermal stability are excellent properties that are not achieved with other blacks.

Pigment manufacturers and users have established a widely accepted international classification system for carbon black pigments. Four groups of products are differentiated:

– high color (HC)
– medium color (MC)
– regular color (RC)
– low color (LC)

The process used for the manufacture is expressed by a third character: (F) for furnace black and (C) for channel or gas black. Oxidative aftertreatment, if used, is indicated by the suffix (o) for "oxidized" (Table 5.6) [4].

Tab. 5.6: Pigment black classification [4].

Designation		Particle size range (nm)
Gas blacks	Furnace blacks	
HCC	HCF	10–15
MCC	MCF	16–24
RCC	RCF	25–35
	LCF	> 36
Gas blacks oxidized		
HCC (o)		10–17
MCC (o)		18–24
RCC (o)		> 25

Jetness and tinting strength are the most important coloristic properties of carbon black pigments. The jetness corresponds to the achievable intensity of blackness. Measurement of the residual refraction (<0.5%) is the most common method for the determination of the blackness. The carbon black sample to be investigated is mixed with linseed oil and measured with a spectral photometer to obtain a M_y value. The finer the carbon black particles, the higher the measured M_y values are, and the higher the value of the blackness is. A modified method is the determination of a M_y factor in an alkyd/melamine resin system (PA 1540). The determined jetness is aligned in this case with group standards and indicated as the relative blackness M_y for optimum reproducibility and consistency [4].

The coloring ability of a carbon black pigment as measured against a white pigment, typically titanium dioxide or zinc oxide, is called tinting strength. The measured value for the tinting strength corresponds to the particle size and the structure of the carbon black investigated. The tinting strength grows with decreasing size of the carbon black particles. It is, therefore, an indirect reference to the surface area and the particle size.

Pigment blacks for printing inks have to fulfill several special requirements in the printing ink itself and in the printed product. Important properties for the pigment in this application are good wettability and dispersibility. Requirements for pigmented printing inks are high pigment concentration, optimum viscosity, good flow characteristics, storage stability, and economy. The printed products have to exhibit good properties in regard to jetness, hiding power, gloss, and rub resistance. A blue hue of the black is desired for most of the printed products. This can best be achieved by the use of carbon black types with finer particle sizes.

Carbon black pigments are used in printing and all other applications in three respects: achievement of pure black color shades (use of carbon black alone), darkening of any color (mixtures of colored pigments with carbon black), or generation of gray effects (mixtures of white pigments with carbon black). Carbon black pigments in printing inks can act as color-giving (inking) and/or rheological components. Inking is represented by properties such as jetness, undertone, and gloss, whereas rheology includes parameters such as viscosity, flow properties, and tack. The inking component is closely related to the particle size of the carbon black pigment. The rheological component depends on the particle size, surface area, structure, and surface chemistry of the carbon black particles. The high surface areas of fine particulate pigment blacks generate strong thickening effects. Increasing particle sizes corresponding with a decreasing surface area exhibit reduced thickening effects and lower viscosity. Consequently, decreasing particle sizes are responsible for high viscosity of the printing ink.

The structure of carbon black pigments is a decisive factor for its use in the preparation of printing inks. Properties like dispersion and flow behavior, viscosity, color density, gloss, rub resistance, and electrical conductivity depend on the structure of the carbon black particles. High structure pigment blacks are characterized by high yield values, high viscosity, and gloss reduction in the application system. Low structure pigment blacks, on the other hand, provide good flow properties, low viscosity, and increased gloss for the application medium.

The electrical conductivity of carbon black pigments is a function of structure. Better conductivity is the result of a higher structure. Sophisticated high structure pigment blacks are used for the production of electrically conductive printing inks. These inks have a relatively high viscosity because of the high filling degree and the special nature of the conductive pigments. Good electrical conductivity on the surface of a printed article is only achieved when a high pigment concentration in the printing ink is used and a high ink film thickness is generated by the printing process.

The use of carbon black pigments for UV curing printing is limited because the black powders strongly absorb the incident light. This applies to the visible and ultraviolet spectral range. The general rule is: the blacker the printing ink, the greater the delay in curing. There are, however, special carbon black types that are suitable for this application to a certain degree.

The surface to be printed, which is commonly called the substrate, is an important factor for the choice of the carbon black pigment. Low structure pigment blacks are the most suitable choice to achieve a high quality of jetness and gloss on coated papers. High structure pigment blacks, on the other hand, generate higher levels of jetness on uncoated papers.

Various inks containing carbon black are used mainly to fulfill the requirements of different printing processes: offset inks, letterpress inks, newspaper printing inks, gravure printing inks for magazines, and decoration and packaging printing inks. Offset inks and letterpress inks are adapted to the surface roughness and the absorbance of the substrate to be printed. Inks with a high yield value are used in these applications to avoid stronger penetration into the uncoated paper and to achieve a greater degree of jetness. Carbon black pigments with particle sizes from 20 to 60 nm are suitable in most cases.

Newspaper printing inks are produced with medium-to-high structure pigment blacks. Conventional newspaper inks have, however, the disadvantage of low rub resistance. The addition of a carbon black type with lower structure can improve the situation. A combination of differently structured pigment blacks is usually the best choice. Newspaper printing inks are manufactured starting from carbon black powders or from granulated pigment material (beads). The use of the latter generates much less dust and, therefore, improves overall handling during automatic ink production. Available beaded versions include oil beads, dry beads, and wet beads. The oil-beaded carbon black form is preferred in many cases because of its favorable dispersion behavior.

Gravure printing for magazines requires very high printing speeds. A low viscosity of the printing ink is, therefore, very important. This can be achieved by using low structure pigment blacks with medium-sized particles, which simultaneously fulfill the requirements for excellent hiding power, jetness, and gloss.

The application for decorative and packaging printing inks typically takes place with flexographic and gravure printing. Printing substrate and ink must be well adjusted to each other. A variety of binder systems is available for the application of these printing inks. Gas and furnace blacks in oxidized and nonoxidized forms are most often used. Packaging inks are also offered in water-based form. High structure pigment blacks produce the best printing results here, due to a high level of jetness. However, adjustments are often necessary, because the high structure of the carbon black affects the flow characteristics of the printing ink in an unfavorable way. This is why in some applications, e.g., for printing on coated paper and cardboard as well as on nonabsorbent surfaces, oxidized pigment blacks with a low structure are used. Thus, better flow properties of the printing ink used can be achieved and higher pigment concentrations are possible [4].

Carbon black pigments are the most frequently used black pigments in the paint and coatings industry. Due to their excellent physical and chemical properties, they are applied in nearly all industrial and automotive coatings systems. Carbon blacks appear to have the best overall performance with respect to requirements of black characteristics (normal black, deep black, black with blue or brown shade), color properties (high tinting strength, hiding power, jetness), resistance behavior (alkali and acid resistance, usability in water-based systems), and weathering properties.

Two technical parameters are of highest importance for the choice of a carbon black pigment in a specific paint or coating application: the average primary particle size (expressed in nanometers or micrometers) and the surface chemistry (expressed by the value of volatile matter in %, which is determined by the thermal treatment of a sample, 7 min at 950 °C).

Black characteristics and color properties are the most decisive factors for the use of a specific pigment in paint or coating system. The average particle size is the most relevant parameter for both factors. The jetness of a coating depends strongly on the average size of the primary particles. As a rule, the smaller the particle size, deeper the jetness. Oxidative aftertreatment can improve the technical application behavior of a carbon black pigment compared with a nonoxidized product. The improvements, which can be achieved by an oxidative aftertreatment comprise dispersion behavior, rheology, jetness, color shade, flocculation stability, gloss, and weather resistance.

The very small particle sizes of carbon black powders, together with their large surface areas, are the reason that these pigments are more difficult to disperse than others. Carbon black pigments can be dispersed with all common grinding units for powders. Sand mills, pearl mills, ball mills, and triple roll mills are examples of the relevant equipment. The wetting properties of the specific carbon black used in paint or a coating are decisive for the pigment–binder ratio.

Pigment black concentrations used to produce a suitable opacity are 2 to 5 wt% calculated on the solid binder. Base coats can contain up to 10 wt%. Electrically conductive coatings with extremely high structured carbon black as the conductive component need pigment concentrations of 30 to 50 wt% to achieve the required conductivity at the surface. An extremely good adjustment of pigment and binder system is an absolute precondition at such high concentrations, because the physical properties of the cured coatings can be changed strongly.

Carbon blacks are also the most important black pigments in the plastics industry. In plastics, they generate not only a variety of coloristic effects, but they also modify the electrical and mechanical properties (filler function) and provide heat and UV resistance to the polymer. Black and gray tintings are among the main applications of carbon black pigments in plastics (polyethylene, polypropylene, polyvinyl chloride, polystyrene, ABS polymers, polyurethane). Carbon black pigments are typically involved in the processing chain in pelletized form.

Full tone tinting is achieved at carbon black concentrations of 0.5 to 2 wt%. The addition of up to 1 wt% carbon black to an unpigmented pure polymer is usually sufficient to generate a strong black appearance. Plastics with a considerable intrinsic color are tinted with pigment blacks of higher tinting strength (concentrations 1 to 2 wt%). The color depth (blackness) of a pigmented polymer depends on the particle size of the carbon black pigment used and on the polymer. Decreasing particle sizes of the black pigments lead to an increasing blackness. Sophisticated HCC and HCF gas and furnace blacks are the solution when optically critical deep black tinting should be achieved. MCC, RCC, and MCF pigments are used when normal black tinting without special brilliance and color depth is required.

Carbon black used for the pigmentation of plastics is typically produced by the furnace black process. Depending on the particle size distribution of the pigments and on the observation conditions (incident light or transmitted light), substantial shifts in hue (bluish or brownish) can be achieved. The bluish-black appearance is commonly preferred for most applications because it also gives the impression of greater color depth.

Dispersing of carbon black pigments requires very intense shear forces, more intense than for other pigments. Optimal dispersion of the previously agglomerated and aggregated particles is, however, a precondition for the development of full coloristic properties for carbon black pigments. Disintegration of agglomerates and aggregates into discrete particles and wetting with the binder system becomes more difficult as the carbon black is finer and as the degrees of agglomeration and aggregation decrease. Tinting strength is negatively influenced by a poor distribution of the particles, especially in gray tints and in the black tinting of opaque plastics. Surface defects in plastic parts and mechanical defects in polymer films and fibers can also be the result of unsatisfactory disintegration of the carbon black pigment.

Pigment blacks are typically incorporated in plastics in two steps to achieve a good distribution of the particles. The process starts with the preparation of a carbon black plastic concentrate (master batch) in a kneader or other suitable equipment. The carbon black content of such a master batch is in the range of 20 to 50 wt%, depending on the binder absorption of the black. Dispersion of the black pigment happens during the preparation of the concentrate. In the second step, the concentrate is diluted with the appropriate polymer amount to obtain the final black content. Plastic concentrates containing carbon black are often produced and purchased by the pigment black manufacturer, not by the plastics industry. Typical forms of master batches are chips or pellets.

The use of carbon black as conductive filler in plastics gives rise to antistatic and conductive phenomena. Sufficient amounts of pigment blacks in plastics can generate antistatic (resistivity 10^6 to 10^9 Ω) or conductive (<10^6 Ω) properties. High conductivities of plastics are achieved when using carbon black fillers with small particle sizes and a high degree of aggregation, together with a sufficient high carbon black concentration. Carbon black types with large amounts of surface oxides and hydrogen-containing surface groups are not suitable for the purpose of conductivity. The pigment should be uniformly distributed in the polymer, but many of the particles must have close contact to each other in order to achieve good electrical conductivity. Bridges of conductivity can be formed between the particles under these conditions, thus promoting the flow of electrons. Processing of the polymer (extrusion, injection molding) often leads to a certain orientation of the carbon black particles, resulting in anisotropic conductivity in the final product. An important factor for the conductivity is also the plastic system itself. Wettability and local concentration by partial crystallization of the polymer have an important influence on its conductivity characteristics. Electrical conductivity and antistatic behavior of plastics are achieved mostly through the use of high structure furnace blacks with relatively fine particles and low contents of volatile components. Concentrations of carbon black in conductive polymers are in the range of 10 to 40 wt%. Antistatic plastics used for floor covering and cable sheathing contain 4 to 15 wt% carbon black [4].

Degradation by UV radiation is a severe problem for many polymers, e.g., for polyethylene. Carbon black pigments can provide long-term stability to such polymers. They absorb UV radiation and act as free radical acceptors. Active intermediate species formed in the degradation process are thus deactivated. Increasing particle sizes and higher carbon black concentrations in the polymer improve the stabilizing effect.

Studies have shown that carbon black could have a potential for renewable energy harvesting and carbon capture. The use of carbon black for renewable energy and other environment saving purposes is expressed in a modern way as "when black turns green" [9]. The production of hydrogen based on the thermal decomposition of methane with carbon black as the catalyst is one approach in this sense. The thermal reaction takes place at about 1000 °C in a movable bed of carbon black particles. The thermal energy is provided by a molten salt mixture consisting of $MgCl_2$ and KCl. An energy flow occurs through a reactor from a hot thermal energy storage tank to a cold thermal energy storage tank and from the cold tank to the hot tank through a higher concentration solar receiver. The reaction takes place at the surface of the carbon black particles that are recycled in a loop until they reach the desired size. This pathway has a minimum energy requirement, much less than the decomposition of water molecules for the generation of hydrogen. An advantage of this process is that carbon black particles of commercial interest are produced simultaneously. A challenge is the design of a high-efficiency reactor for the necessary temperature range that is suitable for a continuous operation [29, 30].

5.4 Toxicology and occupational health

Based on decades of experience in the industrial production and processing of carbon black, under conditions of normal use, there are no known significant harmful effects in humans. This has also been confirmed by numerous epidemiological studies [2].

Carbon black pigments do not exhibit acute toxicity (LD_{50} value rat oral: >5000 mg/kg). They are not irritating to the skin or mucous membranes. Long-term inhalation studies and investigations on animals with the help of the intratracheal instillation have shown that under so-called "lung overload" conditions, chronic inflammation, pulmonary fibrosis, and formation of tumors are possible in rats.

The role of the animal species and of the fine dust, as well as the mechanism of the tumor formation are still not sufficiently understood. The International Agency for Research and Cancer (IARC) made the following recommendation concerning the carcinogenity of carbon black: "Carbon black is possibly carcinogenic to humans (Group 2B)" [31].

An audit of industrial carbon black filtrates revealed an acute fish toxicity LC_{50} and a "no observed effect concentration" (NOEC) of >1000 mg/l. The acute Daphnia toxicity of industrial carbon black filtrates was determined to be >5600 mg/l (NOEC 10,000 mg/l) [2].

Carbon blacks are not explosive under the conditions of daily use. Carbon black/air mixtures can explode only when they are close to a high ignition power, e.g., from a welding torch. In closed silos or poorly ventilated locations, carbon monoxide originating from the carbon black manufacturing process can be present. It is, therefore, recommended that ignition sources are strictly kept away and suitable respiratory protective devices are carried. Storage of carbon black should take place in clean, dry, uncontaminated areas away from exposure to high temperatures, open flame sources, and strong oxidizers. Closed containers are preferred for storage since carbon black adsorbs moisture and chemical vapors.

About three quarters of the carbon black produced is shipped as bulk material, while the remaining quantity is handled in bags. Carbon black for the rubber industry is shipped in containers or tank trucks with a capacity of up to 20 tons. Pigment blacks and smaller amounts of rubber blacks are stored and shipped in paper or plastic bags stacked on palettes.

5.5 Conclusions

Carbon black pigments are manufactured today mainly by modern chemical processes in industrial scale production. They are the most important representatives of black pigments. Carbon black pigments have a number of advantages compared to other inorganic black pigments and black organic colorants. Hiding power, color stability, solvent resistance, acid and alkali resistance as well as thermal stability are excellent properties that are not achieved from other blacks.

Carbon black pigments are applied in most systems where pigments are relevant, such as printing inks, paints and coatings, plastics, and cosmetics. They are produced by several industrial processes. Furnace blacks, channel blacks, and gas blacks have the highest importance among the various carbon blacks. Particle size, particle size distribution, surface quality, and structure determine the coloristic and technical application properties of the individual pigments. Oxidative aftertreatment is used in many cases to modify the surface of the pigments in connection with the stability and compatibility in the application system.

Particle management, sophisticated aftertreatment, pelletizing, and provision of pigment preparations are suitable ways for the improvement of the pigments and the optimization of the dosage form.

Study questions

5.1 What are the most important inorganic black pigments?
5.2 What are the structural motifs of carbon black?
5.3 Which raw materials are used for the manufacture of carbon black pigments?
5.4 What is the main reaction for the carbon black formation in all production processes?
5.5 What are the main advantages of carbon black pigments compared to other black pigments?
5.6 What is decisive for color and technical application properties of carbon black pigments?

Bibliography

[1] Ferch H. Pigmentruße, Vincentz Verlag, Hannover, 1995.
[2] Mathias J, in Kittel – Lehrbuch der Lacke und Beschichtungen, vol. 5, Pigmente, Füllstoffe Und Farbmetrik, 2nd edn. Spille J, ed. S. Hirzel Verlag, Stuttgart, Leipzig, 2003, 214.
[3] Pfaff G, in Winnacker–Küchler: Chemische Technik, Prozesse und Produkte, vol. 7, Industrieprodukte, 5th edn. Dittmeyer R, Keim W, Kreysa G, Oberholz A, eds. Wiley-VCH Verlag, Weinheim, 2004, 358.
[4] Stroh P, in Industrial Inorganic Pigments, 3rd edn. Buxbaum G, Pfaff G, eds. Wiley-VCH Verlag, Weinheim, 2005, 163.
[5] Pfaff G, in Encyclopedia of Color, Dyes, Pigments, vol. 1. Pfaff G, ed. Walter de Gruyter GmbH, Berlin, Boston, 2022, 43.
[6] Boehm HP. Farbe + Lack. 1973,79,419.
[7] Heidenreich RD, Hess WM, Ban LL. J. Appl. Crystallogr. 1968,1,1.
[8] Burgess KA, Scott CE, Hess WM. Rubber Chem. Technol. 1971,44,230.
[9] Khodabakhshi S, Fulvio PF, Andreoli E. Carbon. 2020,162,604.
[10] Jäger C, Henning T, Schlögl R, Spillecke O. J. Non-Cryst. Solids. 1999,258,161.
[11] Patent DE 24 10 565 (Degussa), 1974.
[12] Patent US 3,010,794 (Cabot Corp.), 1958.
[13] Patent US 3,010,795 (Cabot Corp.), 1958.
[14] Patent DE 15 92 853 (Cities Service Co.), 1967.
[15] Patent DE 895 286 (Degussa), 1951
[16] Patent DE 11 29 459 (Degussa), 1960.
[17] Patent US 3,649,207 (Ashland Oil & Refining Co.), 1969.
[18] Patent US 3,986,836 (Phillips Petroleum Co.), 1974.
[19] Kühner G, Dittrich G. Chem. Ing. Tech. 1972,44,11.
[20] Patent DE 742 664 (Degussa), 1940.
[21] Patent US 2,420,810 (Cabot Corp.), 1941.
[22] Patent GB 895 990 (Degussa), 1958.
[23] Patent US 2,439,442 (Cabot Corp.), 1943.
[24] Patent US 3,383,232 (Cabot Corp.), 1968.
[25] Patent US 3,870,785 (Phillips Petroleum Co.), 1975.
[26] Bode R, Ferch H, Koth D, Schumacher W. Farbe + Lack. 1979,85,7.
[27] Patent US 3,333,038 (Continental Carbon Co.), 1965.

[28] Hauptmann N, Vesel A, Ivanowski V, Klanjšek Gunde M. Dyes Pigm. 2012,95,1.

[29] Boretti A. Int. J. Energy Res. 2021,45,21497.

[30] Boretti A, Castellotto S. Renew. Energy Focus. 2022,40,67.

[31] Kuempel ED, Sorahan T, in Views and Expert Opinions of an IARC/NORA Expert Group Meeting, Lyon, France, 30 June – 2 July 2009. IARC Technical Publication No. 42. Lyon, France: International Agency for Research on Cancer. 2010,42,61.

6 Ceramic colors

6.1 Fundamentals and properties

The decoration of porcelain, earthenware, bone china, and other ceramics often needs the use of ceramic colors, also called stains. These colors contain ceramic pigments (glaze or body pigments) with thermal and chemical stability and are mainly used in glazes and enamels (underglaze, inglaze, onglaze colors). The requirements relating to color and application determine the selection of a specific composition for the ceramic color. A temperature-stable ceramic pigment is the color-giving component of such a color. A ceramic color that is ready for sale, often consists of the ceramic pigment (10% to 50%) and the frit (glaze or body, opacifiers, and other components; together 50% to 90%), but can also consist of a ceramic pigment alone. The ratio of network formers and network converters in the frits depends strongly on the firing conditions. Higher proportions of network formers (hard frits) result in better chemical and mechanical resistance, but require higher firing temperatures.

Colored ceramic surfaces are achieved by dissolution of ceramic colors directly in the glaze or enamel or by mass coloration. Such colors are introduced in the ceramic materials before sintering or with the glaze. The use of ceramic colors is the most effective method to adjust defined color tones reproducibly for a glaze or enamel composition. The chemical and the thermal stability of the applied ceramic color including the ceramic pigment in the glaze or enamel have to be high enough that the color is not essentially affected by formation of a melt during the firing procedure [1–9].

Ceramic colors or pigments are often classified based on their chemical composition (Table 6.1). Most of these colors contain fine crystalline mixed crystals or solid solutions

Tab. 6.1: Ceramic colors and pigments for glazes and mass coloration.

Chemical composition	Examples	Remarks
Non oxides	$Cd(S_{1-x}Se_x)$	
Metals	Au, Ag, Pt, Cu	
Metal oxides	Cu_2O, CuO, NiO, Fe_2O_3, Cr_2O_3, CO_3O_4, MnO, MnO_2, SnO_2, TiO_2, SnO_2	
Complex compositions	Spinel	Intrinsically colored
	Pyrochlore	
	Olivine	
	Garnet	
	Phenacite	
	Periclase	
	Zircon	Colored by addition of colored substances
	Baddeleyite	
	Corundum	
	Rutile	
	Cassiterite	
	Sphene	

https://doi.org/10.1515/9783110743920-006

of metal oxides. Color-giving components are often ions of the transition metals vanadium, chromium, manganese, iron, cobalt, nickel, copper, cerium, neodymium, and praseodymium that absorb in the visible spectral range because of not completely filled d- and f-orbitals. There are host lattices for the coloring components that consist of or contain colorless or whitening oxides of aluminum, calcium, silicon, zirconium, titanium, tin, antimony, zinc, and/or magnesium. The color depends on the respective oxidation state, the valency of the transition metal ions, and their coordination in the lattice. Ceramic color types based on a zirconia or a zircon lattice have gained special importance with the incorporation of vanadium, praseodymium, or iron.

The manufacture of ceramic colors typically starts with intensive dry or wet mixing of the fine-sized raw materials, which are examined at the beginning using analytical methods and firing tests. The subsequent reaction of the mixed raw materials takes place at temperatures of up to 1400 °C in a batch process in a truck chamber kiln or continuously in a tunnel or a rotary kiln. After cooling, the reaction product is wet ground in a ball or an agitator ball mill or dry ground in a steam-jet mill, a countercurrent jet mill, or a spiral mill. The grinding of the obtained product is necessary to adjust the particle size distribution and to achieve the desired color parameters. The quality control is done specifically for each application field, e.g., in stoneware bodies for unglazed tiles, in opaque or transparent glazes, or in other frits. Specific test glazes are available for this purpose [7].

Different products are available for the two main areas of application for ceramic colorants and pigments. For mass coloring, mostly color-intensive, low-cost ceramic colors or pigments in the shades blue, turquoise blue, yellow, pink, and black are used. Examples are cobalt blue (blue), vanadium zircon blue (turquoise), chromium rutile yellow (yellow), iron zircon coral (pink), and iron chromium black (black). For glaze coloring (frits or fast firing glazes), ceramic colors with pure brilliant shades, high color strength, great chemical stability of the glaze and the kiln atmosphere as well as high firing stability are required. Examples are cobalt chromium spinel (blue), praseodymium zircon yellow (yellow), and cobalt nickel iron chromium spinel (black). Not all color shades can be mixed advantageously as they may react with each other during firing to form new uninteresting colored compounds [7].

6.2 Types and applications of ceramic colors

6.2.1 Cadmium sulfide and sulfoselenides

Cadmium sulfide and sulfoselenides are important pigments because of their ability to achieve yellow, orange, and red colors for glazes (Section 4.9). They are basically not suitable for firing to higher temperatures, but they can be stabilized by encapsulating them in a vitreous or crystalline matrix. The cadmium chalcogenide pigments responsible for the color are thus occluded in a colorless matrix using a sintering process where two phases are formed.

The most relevant matrix system is zirconium silicate (zircon) [10–12]. In the first step of the process, at about 900 °C, $ZrSiO_4$ is initially formed starting from SiO_2 and ZrO_2 with the support of mineralizers. Hexagonal crystals of cadmium sulfoselenide are formed by the reaction of CdS and selenium or of $CdCO_3$, sulfur, and selenium. The process proceeds via a liquid vitreous phase of low melting compounds formed under the influence of the mineralizers. Under these conditions, zircon grows around the sulfoselenide crystals. The so-formed stains are relatively expensive due to the described process, and the color range is limited.

6.2.2 Metals

Metals such as gold, silver, platinum, and copper are used as ceramic colors in the form of metallic colloids. The most important color is pink, which is achieved with colloidal gold. Selenium is often used in glasses, whereas other colloids give less interesting colors. Colloidal metallic gold is synthesized by adding tin(II) chloride to an acidic solution of gold chloride. The color of the settled particles ranges from pink to violet, depending on the ratio of tin and gold (Cassio's purple). In order to produce stable colors at high temperatures, the settling of purple is performed in a slip of kaolin or clay to avoid coagulation. The metallic gold particles are separated by the clay particles under these conditions. Adding silver chloride shifts the color towards reddish, whereas the addition of cobalt oxide leads to a change towards violet.

6.2.3 Metal oxides

Metal oxides in their use for ceramic colors or pigments are typically dissolved in the vitreous matrix. They give glazes a colored transparent appearance, which goes back on the formation of metal ions in the system. The following colors can be achieved by the addition of suitable metal oxides:

α-iron(III) oxide Fe_2O_3
- yellowish pink (iron coordination VI), stable color at low temperatures
- red-brown (iron coordination IV)

Chromium(III) oxide Cr_2O_3
- green, synthesis in the absence of zinc oxide to avoid the formation of brown $ZnCr_2O_4$ spinel, stable at low temperatures, limited importance due to easy reaction with silicate matrix

Copper(II) oxide CuO
–   blue and green possible (copper coordination in most cases VI)

Cobalt(II,III) oxide / cobalt(II) oxide Co_3O_4 / CoO
–   blue (cobalt coordination IV), Co_3O_4 decomposes to CoO and O_2 at about 900 °C, a coordination change to VI at higher temperatures in boric and phosphate glasses and a purple color appears

Manganese(IV) oxide / manganese(III) oxide MnO_2 / Mn_2O_3
–   brown

Nickel(II) oxide/nickel(III) oxide NiO/Ni_2O_3
–   yellowish purple, color changes due to the effect of retro-polarization of alkaline oxides

There are several disadvantages when using these metal oxides as colorants in ceramics and enamels. MnO_2 and Co_3O_4, for instance, decompose with the liberation of oxygen during the firing process with possible defects in the glaze. Most of the oxides have a significant sensitivity to the necessary high-temperature conditions. Some of the oxides form different oxidation states. Oxidation/reduction and temperature conditions have a high effect on the redox balance and, therefore, on the color tone formed.

Metal oxides also have a high sensitivity to the chemical composition of the glaze where they are incorporated. Copper, for instance, usually gives a green color, but it can lead to blue in glazes with high alkaline or alkaline earth content.

6.2.4 Mixed metal oxides and silicates

Several mixed metal oxides belong to the complex compositions used for coloring of ceramic and enamels (Section 4.4). They are manufactured by solid-state reactions at high temperatures starting from the single oxides or salts, commonly in the presence of mineralizers such as alkali chlorides, fluorides, borates, or carbonates [13–15]. Typical formation temperatures are in the range from 800 to 1400 °C. The pigments remain unaltered in the glaze during thermal treatment, even when used in a finely dispersed form.

Cobalt alumina blue (cobalt blue, cobalt aluminate, $CoAl_2O_4$) is a mixed oxide of cobalt and aluminum, although the composition may include lithium, magnesium, titanium, or zinc. Spinel structured pigments based on cobalt blue are used as underglaze colorants but also for coloring glass, pottery, and vitreous enamels. Incorporation of chromium in the spinel structure, possibly along with magnesium, silicon, titanium, zinc, or strontium leads to blue-green pigments that are also used for coloring glazes. Chromium alumina pink is a suitable ceramic pigment for glazes with high zinc and alumina content.

Chromium rutile yellow is suitable for enamels and for mass coloring (unglazed tiles), but can be dissolved in glazes under certain circumstances. The addition of titanium dioxide has a stabilizing effect in most cases. Pigments in which chromium is replaced by manganese exhibit a brown coloration. In the yellow color range, tin vanadium yellow is also used. Pure orange shades can be produced using lead antimonate. The pigment can be applied up to temperatures of 1050 °C and is used in enamels and glazes. Additional stabilization of lead antimonate is achieved by adding zirconium silicate and tin dioxide.

In the brown color range, chromium iron spinel, chromium iron zinc spinel, or chromium iron zinc nickel spinel are mainly used. All these compositions show high stability in many glazes and are also suitable in certain enamels. Gray shades are obtained by using tin antimony gray. Important applications for this pigment are to be found in tile glazes. Blue-gray and neutral gray shades with a silky appearance are possible when using tin antimony gray. For black colorations, several spinel-based compositions are available: chromium iron spinel (mass colored tiles), chromium iron cobalt nickel spinel (tile and sanitary glazes), and chromium iron cobalt nickel manganese spinel (transparent glazes).

Several silicate-based compounds that are intrinsically colored also belong to the complex compositions used for ceramics and enamels. The most important representatives have crystal structures belonging to the pyrochlore, olivine, garnet, phenacite, or periclase type. Victoria green, based on $CaO/SiO_2/Cr_2O_3$ is a garnet silicate, which is mainly used for coloring ceramic glazes. Cobalt silica blue is based on the olivine structure. Its composition may include aluminum or calcium as a modifier. Addition of zinc produces a lighter blue shade, while phosphorous in the structure changes the shade from deep blue to a violet-blue color tone. Nickel silicate green crystallizes in the olivine structure, as well. It is mainly used for coloring clay bodies.

6.2.5 Zirconia-based compositions

Zirconia and zircon-based ceramic colorants or compositions are the most important groups of ceramic colors, with more than 50% of world production. At room temperature, the stable crystalline phase of ZrO_2 is monoclinic. Above 1200 °C, zirconia is transformed into the tetragonal phase accompanied by a lattice contraction. Further heating to 2370 °C leads to the transformation into the cubic phase, and a lattice expansion takes place. Zirconia has a melting point of 2600 °C, and during cooling down, the reverse crystallographic changes occur [11, 12].

Incorporation of vanadium pentoxide into the zirconia lattice leads to yellow pigments. Intimate mixtures of zirconia (80% to 99% by weight) and ammonium metavanadate (1% to 20% by weight) are typically used for the production. The mixtures are calcined at temperatures of 1400 °C. Small amounts of indium oxide added before calcination lead to more intense yellow colors. Other colorless metal oxides, when added together with vanadium pentoxide to modify the zirconia, also contribute to color shades ranging from greenish yellow to orange-yellow. Trivalent ions with suitable ionic radius can be introduced, minimizing or eliminating V^{3+} ions. Zirconia accepts only V^{5+} ions when gallium, indium, or yttrium is present. The formation of V^{3+} ions is suppressed in this case, and pure orange-yellow shade can be achieved.

Vanadium zirconia-based ceramic colorants are stable in most types of glazes and ceramic bodies for calcination temperatures up to 1350 °C. These ceramic colors are also suitable for blending with all other types of ceramic pigments.

6.2.6 Zircon-based compositions

Zirconia and silica can be calcined in the presence of vanadium to produce zirconium silicate (zircon), which has a blue color. Praseodymium acts in a similar way to generate bright yellow zirconium silicate. It is also possible to introduce iron in zirconium silicate to achieve a coral color. The already-described encapsulation of cadmium sulfoselenide leads to orange and red [5, 11].

The demands for high-temperature applications are fulfilled in a very effective manner by the zircon-based ceramic colorants or compositions vanadium zircon blue, praseodymium zircon yellow, and iron zircon coral. The three colorants are stable up to temperatures of 1350 °C in all types of glazes. Cross mixing allows a wide range of high-temperature stable colors.

Vanadium zircon blue is seen as the best known ceramic colorant. Its production proceeds by calcining a mixture of zirconia, silica, a vanadium compound, and an alkali metal compound, mostly a lithium, sodium, or potassium halide. The calcination step is done in the range of 550 to 1200 °C. Vanadium zircon blue consists basically of zirconium silicate. The presence of alkali metal halides, mainly of sodium fluoride, in the reaction mixture promotes the color formation. The color depends on a number of physical as well as chemical factors, such as the proportions of vanadium pentoxide and sodium fluoride, the purity of the raw materials used, the particle sizes of the components, the distribution of the reactants and their initial degree of compaction, and the atmosphere used for the high-temperature reaction. The color strength of the resulting vanadium zircon blue depends on the amount of vanadium in the zircon lattice. The incorporation of the vanadium depends on the mineralizer component used in the reaction mixture. Mineralizers such as alkali or alkaline earth metal salts promote zircon formation.

Praseodymium zircon yellow was developed in the 1950s due to the demand for a clean, bright yellow for high-temperature applications. Zircon yellows were produced in the beginning from mixtures of rare earth compounds containing some praseodymium.

Later, the pigments were manufactured using sodium chloride and sodium molybdate as mineralizers. The addition of cerium oxide leads to orange shades, whereas lead compounds as mineralizers instead of alkali metal halides produce more intense colors. Stronger colors can also be achieved by the addition of ammonium nitrate to the alkali metal halide. Praseodymium zircon yellow is brighter and cleaner than other high-temperature-stable yellow colorants. The temperature stability is better than for the yellow pigments cadmium sulfide or lead antimonate.

Iron zircon coral is produced by calcining zirconia and silica together with an iron compound and a suitable mineralizer at temperatures in the range of 1000 to 1100 °C. Glaze stable oranges and reds can be produced by the combination of the thermally stable zircon with the less thermally resistant cadmium sulfide or sulfoselenides.

It is necessary to work with very pure chemicals for the production of zircon-based colorants. An expensive grade of ZrO_2 is usually required for the production of ceramic colorants with highest chroma. The high price of zirconia goes back to the processing of this material, where the separation from mineral zircon sand is necessary to obtain a suitable starting quality. Therefore, a new process has been developed in order to manufacture zircon-based colorants directly from zircon sand, which is comparatively cheap and available in suitable qualities. Equimolar proportions of ZrO_2 and SiO_2 are used to form $ZrSiO_4$, leading to intense colorants. In a special step, zircon is calcined with an alkali compound to form an alkali zirconate silicate. Acid treatment of this compound leads to decomposition and formation of a zirconia–silica mixture. After calcination, a zircon-based ceramic colorant is formed. This process can be used for the production of all important zircon colorants.

6.3 Toxicology and occupational health

Information on the toxicology of ceramic colors or pigments can be found in several cases in the related chapters for the single pigments. This concerns the following compounds and compositions used for ceramic colors: cadmium sulfide and sulfoselenides (Section 4.9.4), α-iron(III) oxide (Section 4.2.6), chromium(III) oxide (Section 4.3.4), and mixed metal oxides (Section 4.4.4).

Metals used for ceramic colors in the form of colloids (gold, silver, platinum, copper) do not exhibit acute or chronic toxicity. They can be handled safely because they are harmless when the usual precautionary measures are followed. While transporting and storing, the requirements of the regulations for chemicals, safety, and environment have to be respected.

When dealing with copper(II) oxide, cobalt(II/III) oxide, cobalt(II) oxide, manganese(IV) oxide, manganese(III) oxide, nickel(III) oxide, and nickel(II) oxide, the necessary safety measures are to be followed. All these oxides are more or less critical, in regard to toxicity and occupational health, when they are inhaled or pass into solution. Different health-damaging effects are associated with these oxides. They are regarded as water-polluting and must not be released into the environment.

Zirconia and zircon-based compositions used for ceramic colors are considered as noncritical in regard to toxicity, occupational health, and environment. On the other hand, they may contain one or more heavy metals, which may cause toxic or carcinogenic effects. These heavy metals, however, act in complex inorganic pigments like completely different compounds. They are closely incorporated in the ZrO_2 or $ZrSiO_4$ lattice where they are insoluble and not bioavailable. The common regulations for the use of chemicals, especially of pigments, must be respected.

6.4 Conclusions

Ceramic colors are pigments (ceramic pigments) or contain pigments with high temperature and chemical stability. They are mainly used in glazes and enamels for the coloring of porcelain, earthenware, bone china, and other ceramics. The intended application determines the selection of the ceramic color including the ceramic pigment. This pigment with its high thermal stability is the color-giving component of a ceramic color. Glaze or body, opacifiers, and other components are additional parts of ceramic color composition.

Colored ceramic surfaces can be realized by dissolution of metal oxides in the glaze or enamel themselves or by the use of ceramic colors. These colors are introduced in the ceramic materials before sintering or with the glaze. The use of ceramic colors is the most effective way to achieve reproducibly defined color tones for a glaze or enamel composition. The thermal and chemical stability of the ceramic color in the glaze or in the enamel have to be high enough that the color is not essentially affected when a melt is formed during the firing process.

Technically relevant pigments, which are components of ceramic colors include cadmium sulfide and sulfoselenides; metals such as gold, silver, platinum and copper; diverse metal oxides; mixed metal oxides and silicates as well as zirconia and zircon-based compositions.

Study questions

6.1   What are ceramic colors and where are they used?
6.2   How can colored ceramic surfaces be achieved?
6.3   Which pigments are technically important as components of ceramic colors?

Bibliography

[1]   Mettke P. Keram. Z. 1984,36,538.
[2]   Eppler RA, in Ullmann's Encyclopedia of Industrial Chemistry, vol. A 5, 5th edn. VCH Verlagsgesellschaft, Weinheim, 1986, 545.
[3]   Eppler RA. Ceram. Bull. 1987,66,1600.
[4]   Bell BT. Rev. Prog. Color. 1978,9,48.
[5]   Bell BT. JSDC. 1993,109,101.
[6]   Strohbauch-Homberg G, Oel H-J. Ber. DKG. 1995,72,180.

[7]   Etzrodt G, in Lehrbuch der Lacke und Beschichtungen, vol. 5, 2nd edn. Kittel H, ed. S. Hirzel Verlag, Stuttgart, Leipzig, 2003, 101.

[8]   Monros G. Encyclopedia of Color Science and Technology. Springer Science+Business Media, New York, 2013. DOI: 10.1007/978-3-642-27851-8_181-3#.

[9]   Pfaff G, in Encyclopedia of Color, Dyes, Pigments, vol. 1, Pfaff G, ed. Walter de Gruyter GmbH, Berlin, Boston, 2022, 95.

[10]   Batchelor RW. Trans. Br. Ceram. Soc. 1974,73,297.

[11]   Bayer G, Wiedemann H-G. Chem. Unserer. Zeit. 1981,15,88.

[12]   Kleinschmitt P. Chem. Unserer. Zeit. 1986,20,182.

[13]   Brussaard H, in Industrial Inorganic Pigments. 3rd edn. Buxbaum G, Pfaff G, eds. Wiley-VCH Verlag, Weinheim, 2005, 116.

[14]   White J, in High Performance Pigments. 2nd edn. Faulkner EB, Schwartz RJ, eds. Wiley-VCH, 2009, 41.

[15]   Maloney J, in High Performance Pigments. 2nd edn. Faulkner EB, Schwartz RJ, eds. Wiley-VCH, 2009, 53.

7 Transparent pigments

7.1 Fundamentals and properties

Pigments show transparent behavior in application systems when the size of their particles is in the nanometer range below 100 nm. They become more transparent as their particle sizes gets smaller. Transparent inorganic pigments are typically divided into colored and colorless. Colorless pigment types such as transparent titanium dioxide and transparent zinc oxide do not absorb light in the visible spectral range. They can also be regarded as functional nanomaterials with special properties. Colored pigment types such as transparent iron oxides and cobalt blue are mainly characterized by their color properties. In some cases, however, they also have functional properties [1, 2].

Transparency and hiding power are very important pigment properties. They depend mainly on the light scattering behavior of the particles in the application medium. Scattering is related to the size of the particles in the system in which they are embedded. In the case of colorless transparent pigments, scattering additionally depends on the difference between the refractive indices of the pigment and the binder. Increasing sizes lead to stronger scattering by the pigment particles. Maximum transparency of a medium with a given pigment is achieved when the pigment is dispersed in the optimal manner. The destruction of aggregates and agglomerates during the incorporation of the pigment in the application system is important for the effect to be achieved. The smaller the particles are the larger is their surface area and more the effort usually necessary to destroy aggregates and agglomerates. Typical values for the specific surface area of transparent pigments exceed 100 m^2/g. Organic transparent pigments have nearly no technical importance.

The objective to be achieved for all pigments, and also for transparent pigments, is an optimal dispersion with the highest possible number of primary particles in the application medium. The dependency of absorption, scattering, transparency, and opaqueness (hiding power) of a colored pigment on the particle size are shown in Figure 7.1 and can be summarized as follows:

- very small particles (30 to 100 nm): high transparency, very low scattering, high absorption
- particles in the range of 100 to 800 nm: increasing scattering up to a maximum in the range of 300 to 700 nm, medium absorption, high opaqueness
- particle in the range of 800 to 2000 nm: decreasing scattering, low absorption, medium transparency, medium opaqueness

Aggregated and agglomerated units act comparably to larger particles, resulting in an appearance of a pigment that has lost its transparent character.

Technically important transparent pigments include iron oxides (α-FeOOH, α-Fe$_2$O$_3$), cobalt blue, iron blue, titanium dioxide (micronized TiO$_2$), and zinc oxide.

https://doi.org/10.1515/9783110743920-007

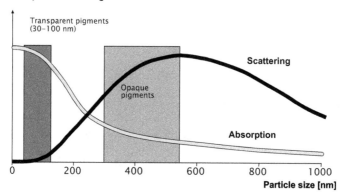

Absorption / Scattering

Transparent pigments
(30–100 nm)

Opaque
pigments

Scattering

Absorption

0   200   400   600   800   1 000

Particle size [nm]

Fig. 7.1: Absorption and scattering of colored pigments as a function of their particle size [1].

Transparent pigments are mostly applied in decorative paints (effect lacquers, e.g., frost effects), printing inks, plastics, and wood protection varnishes, where the high UV absorption and lightfastness is used for the protection of the matrix or the subsurface. The manufacturing processes correspond to those of conventional inorganic pigments with the exception that strong particle growth must be avoided.

7.2 Transparent iron oxides

Table 7.1 shows the relationship between the particle size of α-Fe$_2$O$_3$ pigments and their properties. Primary particle sizes of transparent iron oxides, α-FeOOH and α-Fe$_2$O$_3$, are from 1 to 50 nm in the typical range of nanoparticles. These two pigments are applied in the color range of yellow to red. Effect coatings (together with effect pigments), pure

Tab. 7.1: Selected properties of α-Fe$_2$O$_3$ pigments [1].

Property	Particle size (µm)					
	0.001	0.01	0.1	1	10	100
	◄——————►		◄——————►		◄——————►	
Pigment	Transparent iron oxide red		Opaque iron oxide red		Micaceous iron oxide	
Shade	Yellowish shade		Yellowish red to violet		Metallic brown to black	
Hiding power	Transparent		Very high		Low	
Specific surface	——————————————————————————————————►					
	high					low
Dispersibility	——————————————————————————————————►					
	difficult					easy

yellow and red shades (in some cases, together with titanium dioxide pigments), and wood protection coatings are the main applications of transparent iron oxides. Baking finishes (for transparent α-FeOOH only up to 180 °C), water-based binders, acrylic-isocyanate systems, acid curing systems, amine curing systems, and air drying binders are suitable application systems for both transparent pigment types. For the coloring of transparent plastics, only transparent α-Fe$_2$O$_3$ can be used due to the required high heat stability. Dispersing of the transparent iron oxide pigments is extremely difficult. This is why they are preferably applied in preparation form.

The manufacturing process has a considerable influence on the size and shape of the primary particles. These are mostly needle-shaped, when manufactured by a precipitation process (α-FeOOH and α-Fe$_2$O$_3$), or spherical, when prepared by the reaction of iron pentacarbonyl with oxygen (only α-Fe$_2$O$_3$).

Transparent iron oxide yellow is produced by the precipitation of extremely fine iron(II) hydroxide or iron(II) carbonate crystallites from very diluted aqueous iron(II) salt solutions.

$$FeSO_4 + 2\ NaOH \rightarrow Fe(OH)_2 + Na_2SO_4 \qquad (7.1)$$

Subsequent oxidation, mostly with air, leads to the formation of nanosized α-FeOOH particles in the suspension.

$$2\ Fe(OH)_2 + 1/2\ O_2 \rightarrow 2\ \alpha - FeOOH + H_2O \qquad (7.2)$$

The quality of the pigment is dependent upon concentrations of the solutions, temperature, oxidation time, pH value, and maturing time of the precipitated particles in the suspension. The pigment is washed by decanting to remove remaining salt, filtered, dried, and ground. At the end of the process, needle-shaped particles with lengths of 50 to 150 nm and thicknesses of 2 to 5 nm are formed [3].

One of the possible methods to produce transparent iron oxide red is the thermal treatment of transparent iron oxide yellow at temperatures of 300 to 500 °C with the release of water. A dried, crushed filter cake of α-FeOOH is preferably used for this reaction.

$$2\ \alpha - FeOOH \rightarrow \alpha - Fe_2O_3 + H_2O \qquad (7.3)$$

The pigment thus obtained is subsequently ground to destroy agglomerates and aggregates and to adjust the particle size distribution.

Transparent iron oxide red pigments with a Fe$_2$O$_3$ content of about 85% can be directly manufactured by the precipitation of iron(II) hydroxide or iron(II) carbonate from iron(II) salt solutions and oxidation with air at a temperature of 30 °C in the presence of MgCl$_2$, CaCl$_2$, or AlCl$_3$ [4].

Semitransparent α-Fe$_2$O$_3$ pigments of very high chemical purity are producible by the combustion of iron pentacarbonyl in excess of air at temperatures of 580 to 800 °C. α-Fe$_2$O$_3$ and carbon dioxide are formed [1].

$$2 \ Fe(CO)_5 + 6\,1/2\ O_2 \rightarrow \alpha - Fe_2O_3 + 10\ CO_2 \tag{7.4}$$

The color of the resulting pigments ranges from orange to red and the particle size from 10 to 20 nm. The particles are X-ray amorphous and isometric in shape [5, 6]. The dispersion behavior of transparent α-Fe$_2$O$_3$ derived from the carbonyl process is favorable in comparison with the product obtained from the precipitation route. The transparency, however, is lower for the pigment from the carbonyl process due to the larger particle larger sizes [1].

The fastness properties of transparent iron oxides are comparable with those of hiding iron oxide pigments. They have, however, much higher color strength and a significantly higher color purity. Selected properties of typical transparent iron oxide pigments are summarized in Table 7.2.

Tab. 7.2: Selected properties of transparent iron oxides [1].

Transparent iron oxide	Yellow, needles	Red, needles	Red, isometric
Chemical composition	α-FeOOH	α-Fe$_2$O$_3$	Fe$_2$O$_3$
Structure	Goethite	Hematite	X-ray amorphous
Color Index	Pigment Yellow 42	Pigment Red 101	Pigment Red 101
Manufacturing process	Precipitation	Precipitation	Iron pentacarbonyl
Density, g/cm^3	3.9	4.5	4.9
Specific surface (BET), m^2/g	80–100	90–110	20–40
pH value	3–10	3–9	5–7
Stability to heat, °C	<180	300	300

The high UV absorption of transparent iron oxide pigments is used in a favorable way for the coloring of plastic bottles and food packaging films. It is also applied effectively in wood protection coatings.

7.3 Transparent cobalt blue

Transparent cobalt blue pigments consist of very small and thin flakes of mainly hexagonal shape. The pigments with the composition CoAl$_2$O$_4$ crystallize like conventional cobalt blue in the spinel lattice. The primary particles of transparent CoAl$_2$O$_4$ pigments have diameters in the range of 20 to 100 nm, thicknesses of about 5 nm, and specific surface areas of about 100 m^2/g. Transparent cobalt blue is characterized by very good chemical stability and excellent light and weather fastness. These properties are the reason why transparent cobalt blue pigments easily surpass organic blues of comparable transparency. Transparent CoAl$_2$O$_4$ pigments are, however, only niche products, because the low color strength limits their use in relevant application systems.

The manufacture of transparent cobalt blue takes place by precipitation of cobalt hydroxide and aluminum hydroxide from diluted solutions followed by filtration, washing and drying. The mixture of both hydroxides in the correct stoichiometric ratio is then annealed to temperatures of about 1000 °C.

$$2\ Al(OH)_3 + Co(OH)_2 \rightarrow CoAl_2O_4 + 4\ H_2O \tag{7.5}$$

$CoAl_2O_4$ is formed by a solid-state reaction and water is liberated. Finally, the pigment is ground to destroy aggregates and agglomerates and to achieve the required particle size distribution [7].

Basically, transparent nanosized cobalt blue can also be synthesized by other wet chemical routes, e.g., by the sol–gel process [8]. One of these is the polymeric route, where polyacrylic acid is used as a chelating and gelling agent. Evaporation of the solvents used, drying, and thermal treatment at higher temperatures lead to the pigment with its spinel structure. Cobalt can be substituted partially by nickel and aluminum by chromium to obtain compositions of modified blue or of green character, e.g., green-colored $Co_{0.5}Ni_{0.5}AlCrO_4$ [9]. The synthesis of other modified nanosized spinel compositions, e.g., of blue-colored $Co_{1-x}Ca_xAl_2O_4$, is possible using the polyacrylamide gel method [10].

7.4 Transparent iron blue

Transparent nanosized iron blue pigments (transparent iron blue) are niche products like transparent cobalt blue. They have special weaknesses in regard to thermal and alkali stability. Iron blue pigments with particle sizes below 20 nm are produced as described in Section 4.8.2. The only difference from the procedure described there is the use of solutions of higher dilution. At the end of the synthesis, specific grinding and deagglomeration steps are used.

7.5 Transparent titanium dioxide

Titanium dioxide with primary particle sizes of 5 to 30 nm is characterized by transparent properties in the visible spectral range. Transparent TiO_2 exists in both rutile and anatase forms. The specific surface areas of transparent titanium dioxide pigments correspond to their small sizes and are in the range of 80 to 200 m^2/g.

Primary particles of this small size and below do not or only slightly scatter visible light, which is the reason for their transparency. Micronized TiO_2 of this dimension has, however, a strong UV absorption ability. The main application for transparent TiO_2 pigments is based on this property. As UV absorbers, they are used for the protection of organic materials such as plastics and coatings. The protection function extends also to human skin where it is used in a variety of sunscreen products. The large surface areas of TiO_2 nanopigments, however, are accompanied by a relatively high reactivity. This must be taken into account when using these pigments in products that are in contact with the human skin.

Transparent TiO_2 pigments can also be used for the achievement of so-called frost effects and of hue shifts in coatings pigmented with colored pigments [11]. Frost effects are achieved by the combined application of transparent rutile pigments and metal effect pigments in base coats of metallic coatings [12]. A certain subdued color travel effect can be seen by the human eye if the observation angle on a coated surface is changed. The explanation of this effect goes back to a different interaction of the wavelengths of the visible spectrum with the TiO_2 nanoparticles. The red and green parts of the spectrum are scattered only a little whereas the blue parts are scattered strongly at the TiO_2 particles and leave the coated surface at a flat angle.

Hue shifts in coatings pigmented with colored pigments and modified by the addition of a transparent TiO_2 pigment are not dependent on the observation angle [13]. The TiO_2 particles present in the coating are responsible for an observable hue shift, e.g., from red to magenta.

Nanosized TiO_2 powders are also used for catalysis (DENOX catalysts), gas purification (by adsorption at the large surface), and heat stabilization of silicone rubber.

Various processes can be used for the synthesis of transparent titanium dioxide pigments. Some of them lead almost exclusively to rutile. Starting materials and reaction parameters have an important influence on the quality of the final pigments. Untreated transparent TiO_2 types are very active in regard to photochemical processes. Regarding this, they are similar to TiO_2 white pigments. They are also often coated with a surface treatment consisting mostly of various combinations of inorganic oxides (e.g., SiO_2, Al_2O_3, and ZrO_2) [1]. There is no need to say that all starting compounds must be of highest purity.

Transparent titanium dioxide pigments can be produced by one of the following wet chemical routes:

$$TiOSO_4 + H_2O \rightarrow TiO_2 + H_2SO_4 \quad \text{(mainly anatase)} \tag{7.6}$$

$$TiOCl_2 + H_2O \rightarrow TiO_2 + 2\ HCl \quad \text{(mainly rutile)} \tag{7.7}$$

$$TiCl_4 + 4\ NaOH \rightarrow TiO_2 + 4\ NaCl + 2\ H_2O \quad \text{(mainly rutile)} \tag{7.8}$$

$$Na_2TiO_3 + 2\ HCl \rightarrow TiO_2 + 2\ NaCl + H_2O \quad \text{(mainly rutile)} \tag{7.9}$$

$$Ti(OC_3H_7)_4 + 2\ H_2O \rightarrow TiO_2 + 4\ C_3H_7OH \quad \text{(amorphous } TiO_2) \tag{7.10}$$

Each of the processes consists of the steps precipitation, neutralization, filtration, washing, drying, and micronizing. Surface treatment is applied, if required, prior to the drying step.

It is also possible to produce nanosized TiO_2 by a gas-phase process. The most suitable approach is the combustion of titanium tetrachloride in a mixture of hydrogen and oxygen at temperatures below 700 °C (flame pyrolysis) [14].

$$TiCl_4 + 2\,H_2 + O_2 \rightarrow TiO_2 + 4\,HCl \quad (anatase/rutile) \tag{7.11}$$

7.6 Transparent zinc oxide

Zinc oxide with primary particle sizes in the lower nanometer range is characterized by transparent properties in the visible spectral range, comparable to the related titanium dioxide. Like TiO_2, it has distinct UV absorption properties. Transparent zinc oxide powders are, therefore, mainly applied for sun protection purposes and have to be regarded as functional pigments.

One of the industrially used manufacturing processes is the reaction of vaporized high purity zinc metal with oxygen in a plasma chamber (combustion) [1].

$$2\,Zn + O_2 \rightarrow 2\,ZnO \tag{7.12}$$

Transparent zinc oxide pigments with a primary particle size distribution from 20 to 30 nm are formed.

Another process is the hydrolysis of diluted zinc organic compounds [15].

$$Zn(OR)_2 + H_2O \rightarrow ZnO + 2\,ROH \tag{7.13}$$

R, in the equation stands for an organic rest, mainly for an alkyl group, e.g., methyl or ethyl. The primary particles obtained by this route have a size in the range of about 15 nm. The process is comparatively expensive and time-consuming.

Transparent zinc oxide pigments can also be produced via sol–gel routes or precipitation reactions in the presence of protective colloids. The function of the colloids is the limitation of particle growth [16]. Purified zinc sulfate and zinc chloride solutions are typically used as starting compounds. Hydrothermal techniques seem to have a potential for the synthesis of specific nanosized ZnO powders with high chemical and morphological homogeneities [17].

7.7 Toxicology and occupational health

Transparent pigments belong to nanomaterials. Therefore, special risk management for the manufacture and application of the pigments is recommended. From the occupational health and safety perspective, operations with nanoscale dust especially requires attention. Measures must be taken, therefore, to minimize exposure to nanomaterials for affected employees. Such measures are: the substitution of free flowing pigment powders by pastes and granulates, the replacement of spraying processes by painting and dipping methods, and the use of closed technical and extraction systems. The implementation of organizational measures (minimization of exposure times, limited access) and the use of personal protective equipment should accompany these measures.

Information on the toxicology of the transparent pigments can be found in the related chapters for the single pigments.

7.8 Conclusions

Transparent pigments are characterized by very small particles with sizes in the range below 100 nm and large specific surface areas. Most of the technically relevant pigments consist of particles that are even smaller than 30 nm. They are classified as nanomaterials. Nanosized primary particles do not or only slightly scatter visible light, which is the reason for their transparency. Pigmentation with these pigments leads to a transparent appearance of the application systems. There are colored and colorless transparent pigments. If colored pigments such as α-FeOOH, α-Fe$_2$O$_3$, and CoAl$_2$O$_4$ are used, the application medium receives color but it remains transparent. Colorless transparent pigments are nearly invisible in the application system. Only in combination with effect pigments or conventional colored pigments, are effects with a special appearance possible, e.g., frost effects.

There are two industrially relevant colorless transparent pigments, TiO$_2$ and ZnO. Both pigments find their main applications as functional powders. They are characterized by strong UV absorption and are, therefore, applied as UV absorbers. As such, they are used for the protection of organic materials such as plastics and coatings. In addition, they are used in a broad variety of sunscreen products where they protect human skin.

The manufacture of transparent pigments takes place mostly using wet chemical or gas-phase reactions. Particular attention should be paid to avoiding aggregates and agglomerates. Their presence in the application medium would influence the effect of transparent pigments negatively, in a way seen in nontransparent inorganic pigments.

? Study questions

7.1   When do pigments show transparent behavior in application systems?
7.2   Which are the most important types of colorless and colored transparent pigments?
7.3   What are the properties of transparent TiO$_2$ and ZnO pigments and where are they applied?
7.4   How are transparent pigments produced?

Bibliography

[1] Etzrodt G, in Industrial Inorganic Pigments, 3rd edn. Buxbaum G, Pfaff G, eds. Wiley-VCH Verlag, Weinheim, 2005, 261.

[2] Pfaff G, in Encyclopedia of Color, Dyes, Pigments, vol. 3., Pfaff G, ed. Walter de Gruyter GmbH, Berlin, Boston, 2022, 1195.

[3] Etzrodt G, in Kittel – Lehrbuch der Lacke und Beschichtungen, vol. 5, Pigmente, Füllstoffe und Farbmetrik, 2nd edn. Spille J, ed. S. Hirzel Verlag, Stuttgart, Leipzig, 2003, 86.

[4] Patent DE 25 08 932 (Bayer), 1975.

[5] Patent DE 22 10 279 (BASF), 1972.

[6] Patent DE 23 44 196 (BASF), 1973.

[7] Patent DE 28 40 870 (BASF), 1978.

[8] El Jabbar Y, Lakhlifi H, El Ouatib R, Er-Rakho L, Guillemet-Fritsch S, Durand B. J. Non-Cryst. Solids. 2020,542,120115.

[9] El Jabbar Y, Lakhlifi H, El Ouatib R, Er-Rakho L, Guillemet-Fritsch S, Durand B. Solid State Commun. 2021,334(335),1.

[10] Zhou N, Li Y, Zhang Y, Shu Y, Nian S, Cao W, Wu Z. Dyes Pigm. 2018,148,25.

[11] Winkler J. Titandioxid, Vincentz Verlag, Hannover, 2003, 108.

[12] Patent US 4,753,829 (BASF), 1986.

[13] Patent DE 43 23 372 (Sachtleben Chemie), 1993.

[14] Patent EP 0 609 533 (Degussa) 1994.

[15] Patent DE. 199 07 704 (Bayer), 2000.

[16] Patent JP. 07 232 919 (Nippon Shokubai Co.), 1994.

[17] Rosowska J, Kaszewski J, Witkowski B, Wachnicki L, Kuryliszyn-Kudelska I, Godlewski M. Opt. Mater. 2020,109,110089.

8 Effect pigments

8.1 Fundamentals and properties

Effect pigments are subdivided into metal effect pigments (metallic effect pigments) and special effect pigments, including pearl luster pigments (pearlescent pigments, nacreous pigments) and interference pigments. They consist of μm-sized thin platelets that show strong lustrous effects when oriented in parallel alignment in application systems.

Effect pigments are colorants that provide additional color effects, such as angular color dependence (luster, iridescence, color travel) or texture, when applied in an application system [1]. The term "luster pigments" is also often used for effect pigments because almost all effect pigments provide lustrous effects in their applications. Luster pigments predominantly consist of platelet-like particles that readily align with a parallel orientation to the surface to which they are applied. A characteristic luster is generated, arising from the reflection of incident light from the smooth surface of the pigment platelets.

In earlier times, the terms "metal effect pigments" and "pearl luster pigments" were used for the classification of effect pigments. The term "metal effect pigments" completely describes the effects of this pigment class. The term "pearl luster pigments", on the other hand, does not comprehensively express the effects of the large variety existing today of luster pigments without the metallic character of metal effect pigments. Therefore, the term "special effect pigments" has become more commonly used for this category.

Metal effect pigments are luster pigments consisting of metallic flakes. After the parallel orientation in their application medium, they show a metal-like luster by reflection of light at the surface of the platelet-like metal particles. Special effect pigments, on the other hand, are luster pigments consisting of transparent flakes with high refractive index. They can also be oriented in a parallel way to show a characteristic pearl luster generated by multiple reflections. Pearl luster pigments showing interference colors are also referred to as nacreous or interference pigments [1]. Interference pigments are defined as effect pigments whose color generating mechanism is based completely or predominantly on the phenomenon of interference [1].

There are several effect pigments that also consist of transparent platelets and follow the optical phenomenon of multiple reflection by light interaction but do not show the characteristic pearl luster. Other interference pigments are based on nontransparent platelets, which lead to different coloristic effects in the application system. All these varieties have made classification more complex. The term "special effect pigments" was, therefore, established, which comprises all platelet-like effect pigments that do not belong to the category of metal effect pigments. Special effect pigments are, therefore, pearl luster pigments as well as transparent and nontransparent interference pigments, which lead in the application medium to pearl lustrous or nonpearl lustrous effects, in combination with interference phenomena depending on the composition and structure of the respective pigment [1].

Special effect pigments are almost exclusively inorganic pigments, distinguishing themselves by high luster, brilliance, and iridescent color phenomena based on optically thin films. Their visual appearance originates in reflection and deflection of

https://doi.org/10.1515/9783110743920-008

light at thin single and multiple layers. Effects of this kind are also found in nature, for example in pearls and clamshells. There are also many fascinating examples in the world of birds, fish, gemstones, minerals, and insects. Studies on the understanding of the optical principles of the natural pearl luster show that the brilliant colors can be attributed to structured biopolymers and layered structures, which are formed by bio-mineralization [2–6].

Figure 8.1 shows a comparison of optical principles for the interaction of light with absorption (colored and black) and white pigments as well as with metal effect and special effect pigments (pearl luster pigments).

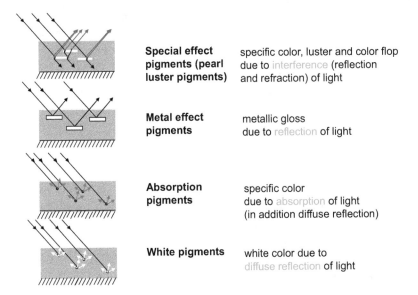

Special effect pigments (pearl luster pigments)	specific color, luster and color flop due to interference (reflection and refraction) of light
Metal effect pigments	metallic gloss due to reflection of light
Absorption pigments	specific color due to absorption of light (in addition diffuse reflection)
White pigments	white color due to diffuse reflection of light

Fig. 8.1: Optical interaction of visible light with particles of different pigment types in an application system, e.g., in a coating.

White, colored, and black pigments have typical particle sizes of 0.1 to 1 μm. These sizes are partly in the range of the wavelengths of the visible light. Effect pigments, on the other hand, have particle sizes predominantly in the range of 5 to 100 μm. The sizes are in some cases even above these values. With regard to their average platelet diameter, effect pigments are, therefore, significantly larger than the wavelengths of visible light. However, with some minor exceptions the thickness of the platelets shows values below 1 μm, which is of the order of the thickness of optical interference layers. Thus, effect pigments mostly have high aspect ratios (ratio of diameter to thickness). Values for the aspect ratio of up to 200 are possible.

The mostly irregularly formed particles of white pigments interact with visible light mainly by diffuse reflection of the incident light in all directions (scattering) without absorption. The particles of colored pigments reflect part of the light diffusely and absorb another wavelength-sensitive part of the light (selective absorption). The particles of black pigments absorb all wavelengths of the spectral range that are visible to the human eye (complete absorption).

Metal effect pigments are nontransparent for light and reflect the entire incident light in one direction. The pigment particles act similar to small mirrors and lead when parallelly orientated in the application system to a reflecting metal luster (metallic effect). Special effect pigments including pearl luster pigments and transparent interference pigments directly reflect only part of the light when striking the even platelet surface. Another part of the light enters the transparent particles and is there partly reflected at the interfaces on the inside of the pigments or at the bottom side of the platelets. The remaining portion of the light leaves the platelets on the underside and can interact with pigment particles located underneath to create further reflections (multiple reflection). The reflected light leads to interference phenomena due to optical superimposition, and combines with multiple reflections to give the appearance of a luster coming from the depths, which is often comparable with pearl luster. The difference in the refractive indices between the highly refractive materials comprising special effect pigments (typically 1.8 to 2.9) and the application medium (typically 1.5 to 1.6) plays a decisive role in these optical processes.

8.2 Metal effect pigments

Metal effect pigments consist of platelet-like metallic particles, so-called metal flakes. They are supplied in the form of powders, pastes, pellets, suspensions, or color concentrates [7, 8]. Most of the metal effect pigments consist of aluminum (so-called silver bronzes), copper and copper/zinc alloys (so-called gold bronzes).

Besides the metal effect pigments, there are also metallic pigments and powders that are used for functional purposes in coatings, e.g., for anticorrosion, IR reflectivity, heat resistance, and electrical conductivity. Metals used here include zinc, stainless steel, and silver.

Aluminum pigments with their silver color are used in their pure form, but they may also be combined with other pigments or with dyes. A typical combination is that of aluminum pigments with special effect pigments. The color of gold bronze pigments depends on the composition of the copper/zinc alloy and can range from copper (0 wt% Zn) to pale gold (approx. 10 wt% Zn) and rich pale gold (approx. 20 wt% Zn) to rich gold (approx. 30 wt% Zn). The pigments provide the appearance of silver and gold, respectively, with the aura of quality and prestige that these metals imply.

Oxidized gold bronze pigments are also available, which achieve deeper color shades by a controlled oxidizing process during manufacturing [7–9].

The origin of metallic pigments goes back to the ancient art of gold beating. Early civilizations, for example the Egyptians, used metallic gold, treating it mechanically in order to form thin sheets, which were applied to overlay wood, bone, or other materials with the precious metal. This technique spread to other regions, and the demand for thin gold sheets increased. It became necessary to produce thinner and thinner foils. In a further step, very thin gold foils were crushed to generate gold powders consisting of small platelets. Such gold powders were used for ornamental artwork and printing inks. The extremely high cost of gold was most probably the reason for an interest in alternative materials. Thus, gold bronze came into use, which was neither metallic gold nor bronze, but it showed golden effects. Gold bronze as the oldest substitute for metallic gold, with its components copper and zinc, is of course not a bronze but a brass alloy. In later times silver and tin were combined to form silver bronze powders. However, the discovery of aluminum smelting led to the replacement of silver/tin alloys by aluminum and thus to the development of the largest group of metal pigments.

The manufacture of gold flake pigments started in Germany and can be dated back to around 1820. A breakthrough in the process was obviously the development of the stamping process by Bessemer in the middle of the nineteenth century [10]. Main parts of the equipment for the process were steel hammers, which fell on steel anvils, thus forming the desired metal into a flake form comparable with that of modern metallic pigments. From that time on, it was possible to replace the very expensive gold and silver bronze powders by cheaper metals. A further decisive improvement was the development of the electrolytic processes for the production of aluminum in the 1880s and 1890s. The metal now available was quickly introduced into the manufacture of metallic flakes. The aluminum powder used as the raw material for flakes came initially from the production of thin foils or from the waste of the aluminum production.

From about 1910, it was possible to produce aluminum powder and granules specifically by industrial-scale production. In addition, a grinding process was invented in which aluminum powders and granules could be deformed to flakes using large ball mills [11]. This process was not yet explosion-proof because the production of the aluminum flakes took place in dry conditions. Important steps for the development and improvement of metal flakes, which were more and more used as effect pigments, were the introduction of the Hametag dry grinding process for gold bronze pigments and of the Hall wet grinding process for aluminum pigments [12, 13]. Only with the introduction by Hall of mineral spirit as a grinding medium was it possible to produce aluminum flakes safely and in large quantities. The processes were refined over the course of the years, associated with an improvement in the pigment qualities. Typical stages of development for aluminum pigments were the conventional "cornflake type", the lenticular "silverdollar type", and the "VMP type" (vacuum metallized pigments). The latter was prepared for the first time in the 1970s and is also called the "PVD type" (physical vapor deposition) [14, 15]. Zinc flake pigments are also produced for a long time, but are of minor importance.

The development and manufacture of colored aluminum pigments is based on the idea of expanding the effects achievable with metallic pigments, especially to combine the advantages of metal effect pigments with color. Concepts used for this purpose include the formation or deposition of absorption and interference layers on the surface of aluminum flakes or the inclusion of colored pigment particles in an additionally coated layer on the flakes, which consists mostly of silica. Pigments based on metal oxide-coated aluminum flakes take an intermediate position between metal effect pigments and special effect pigments. They offer strong reflectivity, together with color effects, from interference and absorption.

Pigments with strong angle-dependent color effects, based on the Fabry-Perot structure, which have aluminum as a central layer, are possible to produce using the PVD technology. Diffractive pigments consisting of two dielectric layers and a central nontransparent aluminum layer can also be produced by the PVD process. A structured polymer film template is used to achieve the diffractive properties of these pigments. Pigments with the Fabry-Perot structure as well as diffractive pigments are usually considered special effect pigments [1, 8].

8.2.1 Optical properties

The metallic effect of metal effect pigments has its origin in the reflection of light at the planar surfaces of the flakes, which is overlaid by the scattering of light at the edges and at irregularities on the metal surface. The metallic effect can be optimized by increasing the portion of directed reflection at the surface of the platelets and by the simultaneous reduction of scattering at the edges and at irregularities.

The optical impression depends primarily on the following factors [16]:
- type of metal (aluminum, copper, copper/zinc alloy)
- wetting behavior of the pigment (leafing or non-leafing)
- surface smoothness of the pigment particles
- particle size and particle size distribution
- aspect ratio of the platelets (ratio of diameter to thickness)
- nature of the interface (air or application medium)

Metal effect pigments, with the exception of vacuum-metallized pigments, are subjected to strong mechanical forces during their manufacture. These forces lead to surface defects. The absence of surface irregularities means a large degree of directed reflection. A description of the visual impression of the metallic effect is not easy without information on the orientation of the particles. Visual impression and measurement of the effect consist of the combination of characteristic single effects [16]:
- color shade
- brightness (whiteness)
- brilliance
- hiding power (tinting strength)
- optical roughness (sparkle effect)

Brightness describes how bright an effect pigment is recognized by an observer. Brilliance characterizes the reflectivity of a pigment and its "metallic" character. Pigments are brilliant if they reflect the incident light to a large degree and if only little undirected light scattering or absorption takes place. Metal effect pigments are in general more brilliant as the pigment surfaces are more perfect or as the particle size distribution is narrower. Very small and irregularly shaped pigment particles lead to a shift of the color shade to gray. Coarse metal effect pigments with average particle sizes above 25 µm are known as sparkle types. The human eye may in this case perceive the mirror planes of single particles during the incidence of light.

Three characteristics are of special importance with regard to the gloss of a metallic coating, especially an automotive coating [16]:
- flop effect (travel effect, two-tone-effect, brightness flop)
- distinctiveness of image (DOI)
- hue saturation

The flop effect describes the phenomenon where the physically experienced brightness of metallic effect coatings depends very much on the viewing angle. It varies from extremely bright at or close to the gloss angle to rather dark at inclined angles. The flop behavior of metallic coatings is determined in practice by using the brightness contrast of coated surfaces during observation under different viewing angles.

The DOI value is determined by a gloss measurement and describes the topcoat gloss close to the reflection angle. The determination is done either subjectively in a glow box or by measurement. The measured values are relative numbers from 0 to 100. A higher DOI value corresponds to a sharper appearance of the image of high contrast objects reflected in the coated surface. The DOI value of an effect coating is influenced by the particle size and the particle size distribution of the pigment used, the binder compatibility of the pigment, and the general spreading properties of the clear coat.

The hue saturation is defined as the ability of a metal effect pigment to cover or influence a color shade generated by a colored pigment. There is a direct relation to its hiding power. Generally, the hue saturation is higher as the metal effect pigment is finer and as its particle size distribution is broader [16].

The particle orientation within a dry pigmented film is an important factor in the quality of the final coating. The orientation depends to a large extent on the dispersion and wetting of the metal flakes, the formulation, the pigment concentration, and the method of application. The optimum situation is that all the flakes are oriented in parallel in the coating and reflect light in a parallel manner. The result in such a case is a maximum of brightness, brilliance, and flop. Poor orientation results in an irregular reflection, causing "salt and pepper effects", "cloudiness" and a poor metallic effect and flop. A disorientation of the flakes can be caused by overly wet films (or slow solvents) or by overly dry films (or fast solvents) [7].

An important factor for the quality of the final metallic effect is also shrinkage during the solvent evaporation in the drying stage. During drying, the flakes find their final parallel orientation in the system. There is little or no shrinkage in high solid systems and even more extremely in powder coatings, which explains why the pigment orientation is poorer here than in low solid or medium high solid coatings [7].

Optical measurements for the characterization of the metallic effect of coatings containing metal effect pigments are typically carried out with goniophotometers or goniospectrophotometers in order to take the flop effect into account [16–19].

8.2.2 Production of metal effect pigments

8.2.2.1 Aluminum pigments

Aluminum pigments are manufactured starting from aluminum granulate (grit). This granulate is produced from aluminum ingots, which are mostly melted in induction furnaces at temperatures above 700 °C and thus above the melting point of aluminum. The molten aluminum is then sprayed (atomized) under high air pressure through a nozzle. The atomized metal cools down immediately after this process step and is now available in passivated form as isometric granulate This granulate, commonly referred to as aluminum grit, is the raw material for the following grinding step, which leads to effect pigments with platelet-shaped particles called flakes [16].

Grinding under formation of flakes is necessary because the spherical grit is much too coarse for utilization in thin coating or printing films. In addition, the grit does not have the required properties of an effect pigment with respect to brilliance and hiding power.

The aluminum grid is now subjected to a grinding procedure in ball mills in the presence of mineral spirits and a lubricant (Hall process) [13]. C18 carboxylic acids, such as oleic acid or stearic acid, are usually added as lubricants. The grit is not only deformed to flakes during this process step but also crushed to smaller particles. The resulting highly reactive surfaces of the formed flakes are immediately oxidized, and a hydrophilic aluminum oxide layer is formed. The lubricant is adsorbed by the oxidized surface of the aluminum flakes. The lubricant layer at the surface of the particles is important for the prevention of uncontrolled reactions (cold-welding, formation of agglomerates). The process heat formed during grinding is dissipated via cooling systems. The concentration of the aluminum pigments in the mineral spirits during the production, especially during grinding and sieving, is typically below 30% (safety reasons).

After the grinding process, the pigment slurry is sieved, classified (agglomerates and coarse particles are separated), and pressed on a filter press to remove excess solvent. The press cake, which consists of around 80% aluminum and 20% mineral spirits is blended with organic solvents to form an aluminum paste, typically containing 65% pigment and 35% solvent [16]. The average size and size distribution as well as the shape of the aluminum particles in the paste depend strongly on the manufacturing conditions. As a result, a distinction of aluminum pigments produced by the Hall process is made between the conventional "cornflake type" and the more sophisticated, lenticular "silverdollar type" (Figures 8.2 and 8.3). The latter shows a more perfect shape and a smoother surface. Consequently, aluminum pigments of the "silverdollar type" provide more attractive metallic effects in coatings and other applications.

Fig. 8.2: Scanning electron micrograph of an aluminum pigment: "cornflake type" (source: Carl Schlenk AG).

Fig. 8.3: Scanning electron micrograph of an aluminum pigment: "silverdollar type" (source: Carl Schlenk AG).

The main steps of the production of flaky aluminum pigments are summarized in Figure 8.4. A scanning electron micrograph of aluminum grit as it used for the manufacture of aluminum platelets is shown in Figure 8.5.

There are applications where aluminum powder is requested, e.g., in powder coatings, or where remaining mineral spirits would not be compatible with the application medium, e.g., in waterborne coatings, waterborne inks, and masterbatches. The mineral spirits of the press cake is removed in such cases in vacuum driers and substituted by some other kind of solvent, water, plasticizer, mineral oil, or other liquids [7].

Aluminum pigments, as well as other metal effect pigments, are also supplied in the form of pellets or granulates. These preparations are dust-free and easy to handle, with a certain percentage of resin. Pellets and granulates offer the end user a broad variety of formulation possibilities, besides other advantages. They are mainly used in printing inks and masterbatches [7]. Special types of aluminum pigments are coated with inorganic (mostly silica) or organic (polymer) materials to improve their chemical and thermal stability in waterborne coatings, powder coatings, or masterbatches. Such aluminum types

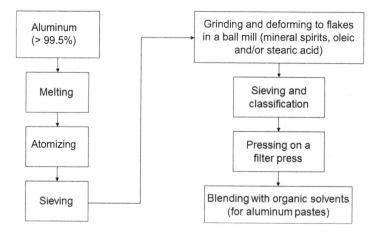

Fig. 8.4: Process steps for the manufacture of aluminum effect pigments.

Fig. 8.5: Scanning electron micrograph of aluminum grit (source: Carl Schlenk AG).

are used in all relevant pigment applications, but especially in automotive coatings and other outdoor uses.

The production of aluminum pigments prepared by vacuum metallization (physical vapor deposition) leads to very thin flakes with a very unique mirror effect (Figure 8.6). They differ considerably from aluminum pigments of the "cornflake type" or the "silver-dollar type". The production starts from highly purified aluminum, which is vaporized in high vacuum and deposited in the form of a very thin film on a carrier material.

The actual manufacturing starts with the coating of a conventional foil (mostly polyethylene polypropylene) with a so-called release coat. Materials for the release coat are coatings or binders, comparable with those used in printing applications, e.g., acrylates, cellulose systems, or vinyl resins. The foil coated with the release layer is now vacuum-deposited with an extremely thin aluminum layer. The vaporized aluminum at this point in the high vacuum is directly deposited on the release coat. The

Fig. 8.6: Scanning electron micrograph of an aluminum pigment: "VMP type" (source: Carl Schlenk AG).

evaporation step plays a key role for the pigment quality achieved at the end of the process. The decisive pigment properties, such as particle thickness and surface quality, are derived from the evaporation parameters.

In the further course of the process, the foil coated with the release layer and the aluminum layer runs through a solvent bath (typically ketones, esters, and alcohols) where the release coat is dissolved and the aluminum layer is removed in the form of coarse particles. The resulting dispersion, consisting of aluminum fragments, residues of the release coat, and solvents, is then concentrated and washed. The filter cake is dispersed now in the desired solvent and comminuted by means of a high speed stirrer to the final average particle size of 10 to 12 µm [16].

8.2.2.2 Gold bronze pigments

Starting materials for the production of gold bronze pigments consisting of copper or copper/zinc alloys are electrolytically obtained copper and zinc. Copper and zinc are alloyed with the addition of some aluminum as reducing agent. The copper-zinc ratio determines the color of the alloy. The copper content is usually above 70%.

In the first production step, irregularly shaped, spattered metal grit is generated by atomization of the metal melt followed by cooling of the atomized material. Grinding of the metal grit to platelets takes place with steel balls in the presence of stearic acid as lubricant. The latter serves primarily to prevent the buildup of cold-welding of pigment particles, comparable with the situation during the production of aluminum pigments. A scanning electron micrograph of a gold bronze pigment with its irregularly shaped particles is shown in Figure 8.7.

Unlike aluminum flakes, gold bronze pigments are manufactured by dry grinding (Hametag process) [12]. Dry gold bronze pigments do not tend to dust explosions in the presence of atmospheric oxygen and can, therefore, be treated, under appropriate precautions, in the dry state in ball mills. The pigment powders formed are universally usable in nearly all solvents and binders. The pigment particles also have a platelet-like shape. Their density is, however, three times higher than for aluminum flakes.

Fig. 8.7: Scanning electron micrograph of a gold bronze pigment (source: Carl Schlenk AG).

The adjustment of particle size and particle size distribution is done using several grinding steps and different grinding conditions. Design of the mill, grinding speed, size of the grinding balls, grinding time, and the lubricant play an important role for the quality of the final pigment. Classification is done with cyclones. The pigment is fractionated in an air stream using gravitational and centrifugal forces. The properties of gold bronze pigments in coatings can be improved by a subsequent additional treatment of the surface with lubricants. Thus, the best possible metallic gloss for a pigmented coating can be achieved [16].

The range of copper and brass colored natural shades can be expanded by the formation of an oxide layer on the surface of the metal particles. Atmospheric oxygen reacts with the metal under defined conditions (temperature, time) to form thin oxide layers on the metal platelets. Interference and reflection in combination lead to interesting color nuances. Common color shades are lemon yellow, fine gold, and flame red.

8.2.2.3 Zinc pigments

The production of flaky zinc powders (zinc pigments) is analogous to the manufacture of aluminum flakes. Zinc grit is obtained by spraying molten zinc metal under high air pressure through a nozzle. In the following step, the grit is converted to flakes by wet grinding in mineral spirits or by dry grinding and deformation. Stearic acid is mostly used as lubricant for the production of zinc flakes.

8.2.2.4 Colored aluminum pigments

There are several technical concepts for colored aluminum pigments:
– controlled wet chemical oxidation of aluminum flakes, formation of an aluminum oxide/hydroxide layer on the surface of the flakes; champagne effect based on interference [20]
– fixation of colored pigment particles on the surface of aluminum flakes by embedding in a coating matrix preferably consisting of silica; the silica layer is typically formed using a sol-gel process and has a thickness of about 200 nm; effects depend

on the color and amount of the colored pigment and the quality of the aluminum flakes; color is based on absorption [21–25]
- coating of iron oxide layers on aluminum flakes by fluidizing them at 450 °C in an atmosphere of nitrogen in a fluidized bed reactor; the reactants $Fe(CO)_5$ and O_2 are streamed in the reactor together with the agitated Al flakes; α-Fe_2O_3 is formed directly on the surface of the Al flakes:

$$2\,Fe(CO)_5 + 6.5\,O_2 + Al \rightarrow \alpha\text{-}Fe_2O_3/Al + 10\,CO_2 \tag{8.1}$$

Golden, orange, and red effect pigments with high color brilliance are possible; color is based on the interaction of absorption and interference; angle-dependent color effects are possible by introducing a layer with a low refractive index (preferably silica) between the Al flake and the iron oxide layer (optically variable pigments, pigments with strong color travel); the pigments are often considered to be special effect pigments [1, 26, 27]
- coating of iron oxide layers on aluminum flakes by wet chemical procedures starting with the deposition of a passivation layer (preferably silica) on the Al flakes followed by the precipitation of iron oxide hydroxide; the conversion to iron oxide takes place by thermal treatment; an example for the final pigment structure is $Fe_2O_3/SiO_2/Al/SiO_2/Fe_2O_3$; golden, orange, and red effect pigments with high color brilliance are possible; color is based on interaction of absorption and interference; the pigments are often considered to be special effect pigments [28–30]
- PVD technology for the manufacture of pigments with Fabry-Perot structure that show strong angle-dependent color effects; the resulting pigments are based on a metal-dielectric multilayer structure, the thicknesses of the single layers determine the color travel and the color intensity; the pigments are produced in a series of special coating installations (roll coaters) that are situated in vacuum chambers; a release layer is first deposited on a moving transfer foil before the PVD process starts; typically five layers are successively deposited on the roll coaters to generate the desired color effect of the final pigment: layer of a semi-transparent absorber metal (e.g., chromium), dielectric layer (e.g., MgF_2), non-transparent metal layer (e.g., aluminum), dielectric layer (e.g., MgF_2), layer of a semitransparent absorber metal (e.g., chromium); a typical pigment arrangement after removal of the multilayer system from the belt by dissolving of the release layer consists, therefore, of $Cr/MgF_2/Al/MgF_2/Cr$, the resulting pieces of the deposited material are collected, broken down into particles and fractionated; the pigments are typically considered to be special effect pigments [1, 31, 32]
- PVD technology for diffractive pigments, which create special rainbow-like color effects by the bending of light at surfaces with a regularly repeating structure; production of the pigments by vacuum deposition of thin films onto a structured polymer film template; typical examples of the layer arrangement on the polymer film are $MgF_2/Al/MgF_2$ or $SiO_2/Al/SiO_2$; a release layer is used for the removal of the deposited layer structure; the resulting pieces of the deposited material are

collected, broken down into particles and fractionated; typical particle thicknesses of less than 1 µm are achieved; the periodic repetition of the structure is around 1 µm with a depth of several nanometers; the pigments are typically considered to be special effect pigments [1, 33, 34].

8.2.3 Pigment properties and uses

The wetting behavior in different liquid systems is of special interest for the application of metal effect pigments. These are basically divided into leafing and non-leafing types (Figure 8.8).

Leafing pigments - the particles flood to the surface of the application system, e.g., an ink

Ink layer
Substrate

LEAFING NON-LEAFING

Paste containing a metal effect pigment and an aromatic solvent (e.g., toluene)

Non-leafing pigments - the particles are uniformly distributed in the applications system, e.g., an ink

Substrate

Fig. 8.8: Leafing behavior of metal effect pigments: leafing and non-leafing types [8].

Leafing pigments move toward the surface of a wet, not yet dried film (coating, ink) and orient themselves parallel to the surface. They thereby form a dense metallic layer on top of the film with high reflectivity and significant barrier properties for the protection of the film. Leafing pigments can be used to create a distinct metallic chrome-like effect ("chrome effect") in decorative coatings and printing inks. They are also valuable for manifold applications in functional coatings, e.g., for roof coatings, tank coatings, or anticorrosive coatings.

Non-leafing pigments show favorable wetting behavior in most application systems. They orient themselves predominantly parallel to the substrate on the bottom of the wet film. The dried films are rub-resistant and exhibit a beneficial adhesion behavior for the subsequent clear coats. Coatings pigmented with non-leafing metallic pigments can easily be tinted with colored pigments to achieve polychromatic metallic effects.

The particle shape of metal effect pigments depends primarily on the process parameters used for their production. The question of whether the irregularly shaped "cornflake type" or the lenticular "silverdollar type" of an aluminum pigment is formed is already pre-decided with the selection of the grit quality. The thickness of the particles can vary from 0.1 to 1 µm, depending on the treatment. Particles of vacuum metallized pigments, however, are with a thickness of about 0.03 µm – extremely thin. Their surface is very smooth and highly reflective.

Average particle size (diameter of the metal flakes) and particle size distribution also depend on manufacturing conditions. The size of the particles may differ from a few micrometers up to 100 µm, depending on the optical effect required.

The flaky shape of the metal effect pigments should be considered during dispersion steps and application. High shear forces may injure the pigments and the desired effects, and should, therefore, be avoided.

Main applications of metal effect pigments are paints and coatings, printing inks, graphic arts, and plastics.

Automotive coatings (OEM and refinish coatings, interior and exterior vehicle parts), general industrial coatings, can coatings, coil coatings, powder coatings, anti-corrosion coatings, roof coatings, marine paints, and conductive coatings are areas of application for metallic pigments. The selection of a specific pigment depends very much on the final application. There is a broad variety of high performance pigments for automotive coatings, waterborne coatings, and powder coatings available, besides the conventional leafing and non-leafing types.

Metal effect pigments are used in all common printing inks and in graphic arts. Typical areas of application are offset, gravure, flexographic, silk screen, and textile printing, as also paper, textile, and leather coating. Pigments used in these application segments are supplied in form of powders, pastes, granulates, suspensions ("VMP types"), color concentrates, or ready-for-press printing inks. Particle sizes used depend on the printing process, and range from 2 to 6 µm for offset printing, 6 to 14 µm for flexographic printing, 10 to 14 µm for rotogravure printing, 10 to 25 µm for screen-printing, to 12 to 40 µm for textile printing. The selection of solvents in pigment pastes and suspensions depends on the printing technique as well. Solvents commonly used are mineral oils (offset inks); alcohols, esters, and aromatic hydrocarbons (rotogravure and flexographic inks); water, alcohols, and glycols (waterborne inks); and reactive thinners (UV inks) [7]. Important end products for printing inks containing metal effect pigments are packages for cigarettes, food and non-food articles, illustrations, labels, gift wrap papers, wallpapers, and tissues.

Aluminum as well as gold bronze pigments are applied in plastics mainly for the tinting of masterbatches, which are used for injection molding, blow molding, extrusion, calendaring, and other processing operations for thermoplastics. Thermosetting materials, like polyester putties, polyester and epoxy floorings, and art objects, make up another application segment for metal effect pigments. The pigments are here preferably supplied in the form of powders (only gold bronze pigments), pastes, or pellets. There are specifically coated types of gold bronze pigments, which are in use for high temperature processes (heat-resistant gold bronze pigments for plastics). Special chemically resistant aluminum and gold bronze pigments are used in PVC.

Increasing demands to reduce the amount of organic solvents (VOCs) in coatings and printing inks have led to the development of waterborne and high solid systems as well as of solvent-free powder coatings. Metal effect pigments are applied in all these application systems. Another way is the use of radiation curing systems (UV radiation). Reactive thinners are used here instead of organic solvents. These thinners are incorporated in the dry film without the emission of solvents.

Aluminum reacts in aqueous systems, especially in alkaline or acidic media, under formation of hydrogen gas. The use of conventional aluminum pigments in waterborne coatings would be a safety risk and would destroy the metallic effect. The way to overcome this problem is the use of stabilized aluminum pigments. High-performance aluminum pigments are equipped with an inhibiting treatment, which protects the aluminum against the attack by water. Stabilization can be achieved for example with layers of phosphor organic compounds, chrome treatments, or by organic or inorganic encapsulation [35, 36].

Metal effect pigments in powder coatings can basically be applied using coextrusion, dry blending, or bonding. However, coextrusion is not recommended for the application of metallic pigments because the high shear forces during extrusion could mechanically destroy the flaky particles and thereby reduce the metallic appearance. The advantage of dry blending is the gentle handling of the metal flakes and a maximum of metallic effect. There is, however, the disadvantage that the metal platelets seem to be differently charged, compared to the resin particles and, therefore, tend to separate in the overspray. In the bonding process, the metallic flakes are thermally and mechanically bonded to the surface of the resin powder, avoiding a separation in the overspray and thereby a problem-free reuse of the material. There are also optical advantages in bonded metallic coatings due to the very homogeneous distribution of the metal effect pigments [7]. Applications of metallic effect pigments in powder coatings are found in steel furniture, tools, transportation, household equipment, and architectural elements.

Uncoated aluminum and gold bronze pigments are also used in anhydrous cosmetics and personal care formulations. The recommended pH value for the formulations is between 6 and 8. Examples of the use of metallic pigments in cosmetic products are lipsticks, cosmetic pencils, eye shadows, mascara, and nail polishes. Pigments for cosmetics are manufactured using the purest metals [37].

Zinc powders have been used for a long time for corrosion protection. The common type of application is zinc dust. Flake-shaped zinc pigments offer an alternative to zinc dust and can generate a comparable anticorrosive behavior with significantly less zinc in the dry coating [38].

8.2.4 Toxicology and occupational health

After long-term experience in production and application, aluminum, gold bronze, and zinc pigments are regarded as nontoxic. However, the inhalation of copper vapor may cause so-called "metal vapor fever". Dust containing copper may have a similar effect if it is present in the ambient atmosphere in higher concentrations. Chronic poisoning initiated by aluminum, gold bronze, and zinc pigments is not known [39].

Metal effect pigments in the form of powders and preparations can be safely handled, stored, and transported under specified conditions. The risk potential is different for various pigments and dosage forms. Assessment and handling, therefore, requires careful consideration of the safety data sheet of each individual product.

Hazards caused by aluminum and zinc pigments are flammability, formation of explosive dust-air mixtures, and the formation of flammable hydrogen when coming into contact with acids or bases (also with water over a longer period of time). A hazard of gold bronze pigments is flammability under certain conditions. Furthermore, pure copper is classified as marine pollutant and falls under transport regulations for dangerous goods [40].

The properties of metal effect pigments are strongly affected by the specific surface area of the particles. Fine powders have a much higher tendency to react than coarser types because of the larger surface area. Very fine aluminum powder can easily be stirred up and an air metal dust cloud is formed. When such a cloud comes into contact with an ignition source, an explosion can occur. The raising of aluminum or zinc dust must be strictly avoided. Precautions must be taken to avoid sources of ignition (sparks, hot surfaces, electrostatic discharges). Inert atmosphere (nitrogen) has to be used where appropriate [40].

Aluminum and zinc powders may become electrostatically charged by friction during production and application, and sparks can arise. The processing equipment must, therefore, be grounded. This is done for containers and other equipment by the use of grounding cables during handling and storage. The original metal containers have to be used only when transferring and filling aluminum or zinc pigments.

The use of pigment preparations reduces the risk of dust formation and explosion very efficiently. The type of solvent used, however, can cause new hazards, especially in case of irritating or flammable solvents [40].

Metal powder fires are extinguished with dry sand or special fire-fighting resources. A sufficient amount of dry sand should always be placed in close proximity to the working environment. Long-handled shovels must be available to immediately form a boundary around the fire and cover it. Stirring up the burning metal powder must be strictly avoided, because an explosion is already possible with a small dust cloud [40].

8.3 Special effect pigments

Special effect pigments are natural or synthetic pigments, characterized by high luster, brilliance, and iridescent colors, known from optically thin films [1–4, 41–43]. The visual impression has its origin in the reflection and refraction of light at thin single and multiple layers. Nature provides many fascinating lustrous and colorful examples based on these optical phenomena: pearls, clamshells, birds, fishes, gemstones, minerals, and insects (Figure 8.9). Fundamental investigations on natural pearls, for example, have shown that the visual appearance has its origin in layered and structured biopolymers, formed by biomineralization.

Pearl luster pigments as typical representatives of special effect pigments simulate the luster of natural pearls or shells of mollusks. These consist of alternating transparent layers with differing refractive indices. The layers consist of $CaCO_3$ in the crystal

Fig. 8.9: Examples of luster and interference effects in nature.

modification aragonite (high refractive index) and the complex protein conchiolin (low refractive index). Figure 8.10 shows the structure of the shell of a mollusk [5, 6].

The difference in the refractive indices, arising equally on the interface between air and an oil film or an oil film and water, is a precondition for the known iridescent color phenomena of these systems. Thin highly refractive platelets of pearl luster pigments align themselves parallel to each other in their application media such as coatings, paints, printing inks, or plastics. Interference effects are generated when the distances of the various layers or the thicknesses of the platelets have appropriate values.

Pearl luster pigments are either transparent, semitransparent, or light-absorbing platelet-shaped crystals or layer systems. They can consist of single crystals as in the case of $Pb(OH)_2 \cdot 2\ PbCO_3$ and $BiOCl$ or possess a monolayer or a multilayer structure in which the layers have different refractive indices and light absorption properties (Figure 8.11).

The use of pearls and nacreous shells for decorative purposes goes back to antiquity. The history of pearl pigments started around 1650 in France where the rosary maker, Jaquin, isolated a silky lustrous suspension from fish scales (pearl essence) and applied this to small beads to create artificial pearls [44]. Some 250 years later, pearl essence material could be isolated from fish scales in the form of guanine/hypoxanthine platelets. Experiments were carried out to develop synthetic pearl colors as

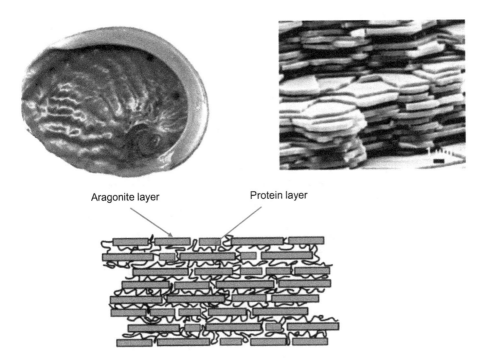

Aragonite layer   Protein layer

Fig. 8.10: Structure of the shell of a mollusk [6].

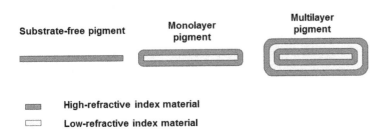

Substrate-free pigment   Monolayer pigment   Multilayer pigment

High-refractive index material

Low-refractive index material

Fig. 8.11: Basic structural principles of transparent effect pigments (pearl luster pigments, interference pigments), left: platelet-like particles consisting of an optically homogeneous material (substrate-free); middle and right: platelet-shaped particles, consisting of layer-substrate structures (monolayer, multilayer).

organic or inorganic, transparent, highly refractive formulations. Beginning in 1920, hydroxides, halides, phosphates, carbonates, and arsenates of zinc, calcium, barium, mercury, bismuth, lead, and other metals were prepared in the form of thin platelet-shaped crystals. Only natural fish silver, basic lead carbonate, and bismuth oxychloride have achieved substantial importance as pear luster pigments [1].

The demand for pearl effects at the beginning came mainly from the growing coatings and plastics industries. Both wanted to improve the acceptance and popularity of their products through a more attractive design. There was also interest from artists and designers to create new visual effects similar to those known from nature. The breakthrough for pearl luster pigments was the development of pigments consisting of mica platelets, coated with metal oxides in the 1960s. Thus, the most successful type of special effect pigment was born.

Pearl luster pigments, and even more, the entire class of special effect pigments, are used to obtain pearl, iridescent (rainbow), or metallic effects, and in transparent color formulations to obtain brilliance or two-tone color, luster flops, and color travel effects. The most important applications are automotive and industrial coatings, printing inks, plastics, and cosmetic formulations.

Table 8.1 presents an overview of inorganic effect pigments. These can be classified with respect to their composition as metal flakes, oxide-coated metal flakes, oxide-coated mica platelets, oxide-coated silica, alumina and borosilicate flakes, platelet-like single crystals, and comminuted PVD films. Monocrystalline natural fish silver (natural pearl essence), platelet-shaped Cu-phthalocyanine, and special polymerized cholesteric liquid crystal compositions (LCP-pigments) are the only organic effect pigments used in practice. Their application, however, is limited to very specific products [1, 4].

Tab. 8.1: Overview of inorganic effect pigments.

Pigment type	Examples
Metallic platelets	Al, Zn/Cu, Cu, Ni, Au, Ag, Fe (steel), C (graphite)
Oxide-coated metallic platelets	Surface oxidized Cu-, Zn/Cu-platelets, Fe_2O_3 coated Al-platelets
Coated mica platelets[1]	Nonabsorbing coating:
	TiO_2 (rutile), TiO_2 (anatase), ZrO_2, SnO_2, SiO_2
	Selectively absorbing coating:
	FeOOH, Fe_2O_3, Cr_2O_3, TiO_{2-x}, TiO_xN_y, $CrPO_4$, $K[Fe^{III}Fe^{II}(CN)_6]$,
	organic colorants
	Totally absorbing coating:
	Fe_3O_4, TiO, TiN, $FeTiO_3$, C, Ag, Au, Fe, Mo, Cr, W
Platelet-like single crystals	BiOCl, $Pb(OH)_2 \cdot 2\ PbCO_3$, $\alpha\text{-}Fe_2O_3$,
	$\alpha\text{-}Fe_2O_3 \cdot n\ SiO_2$, $Al_xFe_{2-x}O_3$, $Mn_yFe_{2-y}O_3$,
	$Al_xMn_yFe_{2-x-y}O_3$, Fe_3O_4
Comminuted thin PVD films	Al, $Cr/MgF_2/Al/MgF_2/Cr$

[1] instead of mica, other platelets such as silica, alumina, or borosilicate can be used

8.3.1 Optical principles of special effect pigments

The basic optical principles of pearl luster (interference) pigments are shown in Figure 8.12 for a simplified case of nearly normal incidence without multiple reflection and absorption. A part of the original light beam L is reflected (L_1) and a part is transmitted (i.e., refracted) (L_2) at the interface P_1 between two materials

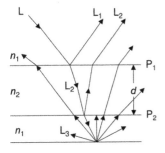

Fig. 8.12: Simplified scheme showing nearly normal incidence of a light beam (L) from an optical medium with refractive index n_1 through a thin solid film of thickness d with refractive index n_2. L_1 and L_2 are regular reflections from the phase boundaries P_1 and P_2. L_3 represents diffuse-scattered reflections of the light from the ground [42].

with refractive indices, n_1 and n_2. The intensity ratios of the partial beams depend on n_1 and n_2. In a multilayer arrangement, as found in natural pearl or pearl luster and iridescent materials, each interference produces partial reflection. After penetration through several layers, virtually complete reflection is obtained in the case that the materials are sufficiently transparent.

Pearl effect pigments have the potential to simulate natural pearl effects. The simplest case is a platelet-shaped particle with two phase boundaries P_1 and P_2 at the upper and lower surfaces of the particles. Such particles are thin and transparent and are characterized by a refractive index that is higher than that of its surroundings. The properties of thin platelets with a thickness in the range of up to several hundred nanometers can be understood, based on the physical laws of thin optical films.

Multiple reflection of light on a thin solid film with a high refractive index causes interference effects in the reflected light and in the complementary transmitted light. For the simple case of nearly perpendicular incidence, the intensity of the reflectance (I) depends on the refractive indices (n_1, n_2), the layer thickness (d), and the wavelength (λ):

$$I = \frac{A^2 + B^2 + 2AB \cos\Theta}{1 + A^2 B^2 + 2AB \cos\Theta} \tag{8.2}$$

$$A = \frac{n_1 - n_2}{n_2 + n_1}, \quad B = \frac{n_2 - n_1}{n_2 + n_1}, \quad \Theta = 4\pi\frac{n_2 d}{\lambda} \tag{8.3}$$

With given values for n_1 and n_2, the maximum and minimum intensities of the reflected light can be calculated, which are visible as interference colors. The calculated data agree well with experimental results. Values for the refractive index of materials relevant for pearl luster pigments are shown in Table 8.2.

Pearl luster and interference pigments are synthesized with a layer thickness d calculated to produce the desired interference colors (iridescence). Most of the pigments consist of at least three layers of two materials with different refractive indices. Thin flakes (thickness approximately 500 nm) of a material with a low refractive index (mica, silica, alumina, borosilicate glass) are coated with a highly refractive metal oxide (TiO$_2$, Fe$_2$O$_3$, layer thicknesses of 50 to 250 nm). Each pigment particle is, therefore, a small thin film system with four interfaces. The spectral characteristics of more complex multilayer

Tab. 8.2: Refractive indices of materials relevant for pearl luster and interference pigments.

Material	Refractive index
Vacuum/air	1.0
Water	1.33
Plastics, paints, printing inks	1.4–1.7
Mica (muscovite)	1.55–1.62
Mica (phlogopite)	1.53–1.62
$CaCO_3$ (aragonite)	1.68
Natural pearl (guanine, hypoxanthine)	1.85
$Pb(OH)_2 \cdot 2\ PbCO_3$	2.0
BiOCl	2.15
TiO_2 (anatase)	2.5
TiO_2 (rutile)	2.7
α-Fe_2O_3 (hematite)	2.9
SiO_2 (amorphous thin layer)	1.46
Al_2O_3 (corundum)	1.77
Borosilicate glass	1.47

pigments containing additional thin layers of low or highly refractive materials can also be calculated on the basis of appropriate optical parameters.

The color effects of special effect pigments depend on the viewing angle. The platelets of the pigments split white light into two complementary colors depending on their thickness. The reflected (interference) color dominates under regular (maximum) reflection. This applies when the object is observed at the angle of regular reflection. The transmitted color dominates at other viewing angles under diffuse viewing conditions when there is a nonabsorbing (white) or reflecting background. Variation of the viewing angle leads at one point to a sharp peak reflectance (luster) and the color changes between two complementary colors. The complex interaction of luster and color is measured goniophotometrically in reflection and at different angles. A special effect pigment is characterized by a minimum of three $L^*a^*b^*$ data sets (CIELAB-system), measured under different conditions (e.g., 0°/45° black background, 22.5°/22.5° black background, 0°/45° white background). These data specify such an effect pigment with respect to its hiding power, luster, and hue [1, 45–48].

The transmitted light is absorbed from a black background or in a blend of the effect pigment with carbon black, and the reflected interference color is seen as the mass tone (i.e., overall color) of the material. In blends of special effect pigments with absorbing colorants, the particle size of the latter should be very small to reduce scattering to a minimum. The pearlescent effect or the iridescent reflection is otherwise quenched by the hiding pigments. This also has to be taken into consideration for blends with strongly reflecting metal effect pigments. Blends of transparent effect pigments with different interference colors obey an additive color mixing law. For example, appropriate blends of blue and yellow interference pigments generate a green interference color.

8.3.2 Substrate-free pigments

There are only three substrate-free pearl luster pigments of historical and commercial importance: natural fish silver (natural pearl essence), basic lead carbonate, and bismuth oxychloride. Further substrate-free effect pigments are micaceous iron oxide (industrially relevant), titanium dioxide (industrially irrelevant due to the absence of an adequate production process), and colored organic flakes (small industrial importance due to limited optical and stability properties) [1].

8.3.2.1 Natural fish silver

Natural fish silver is an organic effect pigment, but is mentioned here because it was the first pearl luster pigment. It is isolated as a silky lustrous suspension from fish scales (herring, carp, sardine). The pigment particles in the suspension are platelet-shaped with a very high aspect ratio. They consist of 75% to 97% guanine and 3% to 25% hypoxanthine. The manufacture starts from an aqueous suspension of white fish scales, which is treated with organic solvents in a complicated washing and phase-transfer process to remove proteins and irregular guanine crystals. More than four tons of fish are necessary to produce one kilogram of fish silver. The pigment particles tend to agglomerate, which is the reason that they are only handled as dispersions. They have a refractive index of about 1.85 and a relative low density of 1.6 g/cm^3, and they exhibit high but soft luster. Natural fish silver is almost exclusively used in expensive cosmetic formulations [1–4].

8.3.2.2 Basic lead carbonate

Basic lead carbonate was the first commercially successful synthetic pearl luster pigment. It consists of hexagonal lead salt crystals, especially basic lead carbonate $Pb(OH)_2 \cdot 2\,PbCO_3$. Basic lead carbonate is precipitated from aqueous lead acetate solutions with carbon dioxide under carefully controlled conditions.

$$3\,Pb(CH_3COO)_2 + 2\,CO_2 + 4\,H_2O \rightarrow Pb(OH)_2 \cdot 2\,PbCO_3 + 6\,CH_3COOH \qquad (8.4)$$

The resulting monocrystalline platelets have an aspect ratio greater than 200. They are less than 0.05 μm thick and show hexagonal dimensions of about 20 μm. With their refractive index of 2.0 and an even surface, they exhibit a very strong luster. An increase of the platelet thickness leads to interference pigments [1]. The crystals are fragile and are, therefore, handled in dispersions. Their high density of 6.14 g/cm^3 is responsible for an unfavorable fast settling behavior. Further disadvantages are the low chemical stability and the toxicological properties of basic lead carbonate. The risks, when working with lead compounds have especially led to a significant reduction in the use of basic lead carbonate for pigment applications [1].

8.3.2.3 Bismuth oxychloride

Bismuth oxychloride pigments are synthesized by the hydrolysis of very acidic bismuth salt solutions in the presence of chloride (pH value < 1.0).

$$Bi(NO_3)_3 + HCl + H_2O \rightarrow BiOCl + 3\,HNO_3 \qquad (8.5)$$

The quality of the BiOCl crystals formed can be varied by careful adjustment of bismuth concentration, temperature, pH value, pressure, reactor geometry, and addition of surfactants. Crystals with a tetragonal bipyramidal structure are usually formed. These can be flattened to platelets with a high aspect ratio by modified reaction conditions. Products with an aspect ratio of 10 to 15 exhibit low luster and offer very good skin feeling. They are used as fillers in a variety of cosmetic products. Crystals with higher aspect ratios show strong luster. They are mainly used in cosmetics for nail polishes. Besides use in cosmetics, the pigments are also applied in interior coatings, printed surfaces, buttons, and jewelry [1, 49, 50].

Disadvantages of BIOCl pigments are low light stability and fast settling caused by a high density of 7.73 g/cm^3. Light stability can be improved to some extent by doping with transition metals and UV absorbers in the application system. The low light stability of BiOCl is characterized by graying of the pigmented system during light exposure. High luster bismuth oxychloride pigments are mostly sold in the form of preparations. Special pastes are used with pigment contents of 50% to 80%, adjusted to specific paint and printing formulations. These pastes can be equipped with suitable UV stabilizers [1].

8.3.2.4 Micaceous iron oxide

Micaceous iron oxide (MIO) consists of pure or doped hematite (α-Fe$_2$O$_3$). There are large natural deposits of MIO. Micaceous iron oxide is also formed by hydrothermal synthesis in alkaline media. The dull dark color of the so-obtained products is as unappealing as that of the natural product. The incorporation of substantial amounts of dopants leads to increased aspect ratios of up to 100, resulting in a much increased luster. The color can be influenced by the composition, and more attractive reddish-brown tones are achieved, which can be used for decorative purposes [27, 34, 42].

The most important dopants for synthetic MIO are Al$_2$O$_3$, SiO$_2$, and Mn$_2$O$_3$. These dopants can enforce the formation of thinner platelets, often crystallizing in the spinel structure. Iron(II) sulfate is the most suitable starting material for the synthesis of micaceous iron oxide pigments. Iron(II) is oxidized with air to iron(III) in an agitated vessel. The so-formed iron(III) is precipitated as α-FeOOH by alkali hydroxide addition.

$$2\,FeSO_4 + 4\,NaOH + O_2 \rightarrow 2\,\alpha\text{-FeOOH} + 2\,Na_2SO_4 + 2\,H_2O \qquad (8.6)$$

The α-FeOOH in the suspension is converted in an autoclave in the presence of aluminum oxide and potassium permanganate to platelet-shaped hematite-containing Al and Mn. The autoclave is equipped with a stirrer and heating apparatus. The size of the platelets is controlled by the reaction parameters, such as temperature, pressure, reaction time, concentration of alkali, and α-FeOOH content. The type and quantity of the doping agents are very important for the quality of the growing crystalline platelets [51, 52].

$$2\,FeOOH + (Al_2O_3,\ NaOH,\ KMnO_4) \longrightarrow Al_xMn_yFe_{2-x-y}O_3 + H_2O \qquad (8.7)$$

The α-FeOOH suspension, together with the doping agents, is heated up for several hours to temperatures above 170 °C, typically to 250 to 300 °C. In the second reaction step, the pH-value is further increased in order to achieve platelets in the required final size and shape. Platelet-shaped doped and undoped iron oxide, which has a refractive index of about 2.9 and a density of 4.6 to 4.8 g/cm^3, is preferably used as effect pigment in paints and coatings [51, 52].

8.3.2.5 Titanium dioxide flakes

Titanium dioxide flakes can be manufactured by breaking down continuous films of TiO$_2$. Such films can be obtained using a web coating process where TiOCl$_2$ is thermally hydrolyzed on the surface of the web [53]. Substrate-free TiO$_2$ flakes can also be achieved from TiO$_2$-mica pigments by dissolving the substrate in strong acids or hydroxides [1, 34]. The resulting titanium dioxide flakes do not consist of single crystals. They are polycrystalline and quite porous. Their main disadvantage is insufficient mechanical stability for most applications, especially for technical uses where stress is exerted. TiO$_2$ flakes show typical interference colors that are dependent on the respective thickness of the platelets [4, 53]. Further variations of color and luster of these effect pigments can be achieved by coating the platelets with different metal oxides in single and multilayers [54, 55].

8.3.3 Pigments based on the layer-substrate principle

Pigments based on the layer-substrate principle represent the dominant class of special effect pigments. The variation and combination possibilities of layer and substrate materials are very broad (Figure 8.13). Special effect pigments, based on natural mica as substrate material, are the most important representatives of layer-substrate pigments. Other substrates, such as alumina, silica, borosilicate glass, and synthetic mica are used in form of platelets as alternatives to natural mica.

8.3.3.1 Metal oxide mica pigments

8.3.3.1.1 Fundamentals

Metal oxide mica pigments consist of platelets of natural mica coated with thin films of transparent or semitransparent metal oxides. The mica platelets act as templates

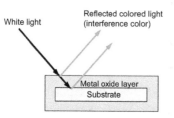

Substrate materials (low-refractive):
mica (natural, synthetic), SiO_2, Al_2O_3, borosilicate glass

Layer materials (high-refractive):
TiO_2 (anatase, rutile), Fe_2O_3, Fe_3O_4, Cr_2O_3 and other metal oxides

Layer thickness: 40 – 250 nm

Color:
Interference color and luster
depend on substrate type and
geometry, number, type and thickness
of the metal oxide layers

Fig. 8.13: Structure of layer–substrate pigments.

for the synthesis and as a mechanical support for the deposited thin metal oxide layers. Most pearl luster and interference pigments are based on natural muscovite mica; some of them on synthetic fluorophlogopite mica. Micas belong to the class of layered aluminosilicates. Pure mica samples are optically transparent. Muscovite occurs worldwide in large deposits, but only a few of these are suitable for pigment production. Natural mica is classified as biologically inert and approved for use as filler and colorant [1–4].

The selection and preprocessing of the mica is very important for the final quality and appearance of the resulting special effect pigments. Rough mica blocks are ground and classified into different particle size ranges at the beginning of the process. Particle size distribution and aspect ratio of the thus obtained mica platelets have a large influence on the properties of the final pigments. The thickness of the mica platelets is typically in the range from 300 to 600 nm. The platelet diameters can be adjusted by the choice of the grinding and classification parameters. Typical ranges are <15, 5 to 25, 10 to 50, or 30 to 110 μm. Light incident on the final pigment particles is regularly reflected from the metal oxide and the mica planes and scattered from the edges. Pigments based on larger mica platelets, therefore, have high brilliance and low hiding power. Pigments on smaller platelets, on the other hand, have accordingly lower brilliance and higher hiding power.

There are several deposits at various locations around the world for natural muscovite mica, with the basic composition $KAl_2[AlSi_3O_{10}](OH)_2$ of appropriate quality and only slight impurities. Synthetic muscovite mica is not accessible via technical synthesis because the special formation conditions at high temperatures and pressures are only difficult to realize. Natural fluorophlogopite mica of the purity required for effect pigments is practically not available. Synthetic fluorophlogopite mica, on the other hand, can be produced by a high temperature crystallization process, starting from Al_2O_3, MgO, SiO_2, and K_2SiF_6. It has the basic composition $KMg_2[AlSi_3O_{10}]F_2$. The necessary synthesis temperatures are considerably above 1200 °C. The mica blocks formed are ground and classified, comparably with natural mica [56].

Disadvantages of synthetic fluorophlogopite (also called artificial mica) compared to natural muscovite are poorer cleavability, often resulting in higher platelet thicknesses, and hydrogen fluoride as byproduct during the synthesis. However, as a synthetic material, it has a more controllable composition. In particular, the absence of iron and other metal impurities results in a very bright, pure body color of the synthetic mica. This is particularly noticeable in white application systems pigmented with mica-based effect pigments, especially at high pigment volume concentrations or when viewing effect coatings at a flat angle.

A mica platelet coated on all sides with a metal oxide has three layers, with two different refractive indices (n_1 and n_2) and four phase boundaries (P_1–P_4): P_1/TiO$_2$/P_2/mica/P_3/TiO$_2$/P_4 (Figure 8.14). Interference of light is generated by combined reflections at pairs of phase boundaries, some of which are equal: $P_1P_2 = P_3P_4$, $P_1P_3 = P_2P_4$, P_1P_4, and P_2P_3. The thicknesses of the mica platelets correspond to a statistical distribution. As a consequence, interference effects involving the phase boundaries between the mica substrate and the oxide coating add together to generate a white background reflectance. For this reason, the interference color of a large number of particles depends only on the thickness of the metal oxide layer on both sides of the mica.

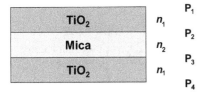

P_1

n_1

P_2

n_2

P_3

n_1

P_4

Fig. 8.14: Schematic structure of a titanium dioxide mica pigment with four existing interfaces (P_1–P_4).

The most important mica-based effect pigments consist of TiO$_2$-mica (TiO$_2$ in the form of anatase and rutile) and α-Fe$_2$O$_3$-mica (Figure 8.15). TiO$_2$-mica pigments represent the transparent pearlescent and interference types. α-Fe$_2$O$_3$-mica pigments combine interference color with the absorption color of α-iron(III) oxide and cover the range of bronze, copper, red, red-violet, and red-green color shades. The so-called combination pigments are brilliant, mass tone colored types consisting mostly of TiO$_2$ and another metal oxide, with one color (interference color same as mass tone) or two colors (interference and mass tone different) on the mica. Multilayer structures with more than three layers of different refractive indices extend the series of mica-based effect pigments by very intensive shades and show, in part, strong angle-dependent color effects. The color of all single pigments depends on the specific composition and on the viewing angle.

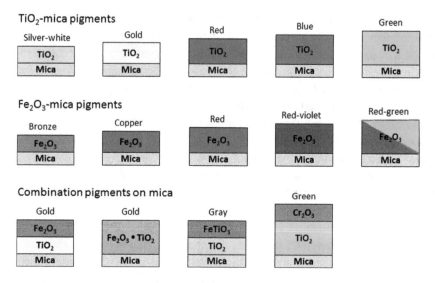

Fig. 8.15: Schematic illustration of different metal oxide mica pigments.

8.3.3.1.2 Production of metal oxide mica pigments

Two different processes are used for the production of TiO_2-mica pigments. They are referred to as homogeneous hydrolysis ("sulfate process") and titration ("chloride process"). Both processes take place in aqueous suspension.

Homogeneous hydrolysis at 100 °C:

$$TiOSO_4 + mica + n\,H_2O \rightarrow TiO_2(hydrate)/mica + H_2SO_4 \tag{8.8}$$

Titration:

$$TiOCl_2 + mica + H_2O + 2\,NaOH \rightarrow TiO_2(hydrate)/mica + 2\,NaCl \tag{8.9}$$

In both processes, the resulting freshly coated mica particles are separated from the suspension by filtration, washed with water, dried, and calcined in air at temperatures of about 800 °C.

$$TiO_2(hydrate)/mica \rightarrow TiO_2/mica + H_2O \tag{8.10}$$

The titration process is preferred for interference pigments with thick TiO_2 layers because it is easier to control.

When TiO_2 is precipitated onto mica platelets under reaction conditions unfavorable for side precipitation, e.g., pH value >1.5 and temperatures in the range of 60 to 90 °C, only the anatase modification is formed. Even after annealing at 1000 °C, no rutile phase is detected in the TiO_2 layers. On the other hand, titanium dioxide hydrate precipitated in an aqueous solution without the presence of substrate particles, can be completely converted into rutile at about 700 °C [57].

However, the formation of rutile on the mica platelets is also desirable because of the higher refractive index and a better light stability compared to anatase. The higher refractive index leads to stronger reflectivity and pearlescent effects. Consequently, processes have been developed to create rutile layers on mica platelets. Tin dioxide hydrate is initially precipitated as a thin layer onto the substrate, and then the TiO_2 hydrate layer is deposited using the common process. $SnCl_4$ is used in most cases as the precursor for the SnO_2 hydrate precoating.

$$SnCl_4 + mica + 4\,NaOH + H_2O \rightarrow SnO_2(hydrate)/mica + 4\,NaCl \qquad (8.11)$$

$$SnO_2(hydrate)/mica + TiOCl_2 + 2\,NaOH$$
$$\rightarrow TiO_2(hydrate)/SnO_2(hydrate)/mica + 2\,NaCl \qquad (8.12)$$

SnO_2 formed during the calcination acts as a template for TiO_2. It crystallizes only in the rutile structure and from there forces the formation of the rutile structure for the titanium dioxide.

$$TiO_2(hydrate)/SnO_2(hydrate)/mica \rightarrow TiO_2(rutile)/SnO_2/mica \qquad (8.13)$$

Figure 8.16 (left) shows the SEM picture of a TiO_2-mica pigment. The shape and size of the pigment particles are clearly visible. Figure 8.16 (right) shows a cross section of an individual particle. The layer of TiO_2 formed during coating and calcination is easily recognizable on both sides of the mica platelet. The TiO_2 layer, which has a sintered granular structure, covers the mica platelets completely. Its thickness can be adjusted with highest precision in the nanometer range during the manufacturing process. Exact control of the layer thickness is decisive for the reproducible production of special effect pigments based on the layer-substrate principle.

The production of multilayer pigments needs the precipitation of an additional material with low refractive index. Typically, SiO_2 layers are deposited between TiO_2 layers. An example of a multilayer composition based on mica is the layer structure, $TiO_2/SiO_2/TiO_2/mica/TiO_2/SiO_2/TiO_2$ [58].

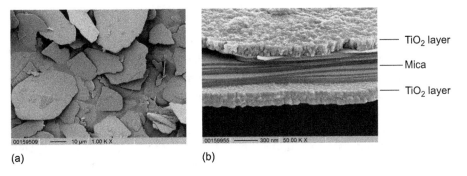

TiO₂ layer

Mica

TiO₂ layer

(a) (b)

Fig. 8.16: Scanning electron micrographs of a titanium dioxide mica pigment; (a) overview image; (b) cross section through one particle; layer structure: TiO_2-mica-TiO_2 (source: Merck KGaA).

The desired interference color determinates the thickness of the TiO_2 layer on the mica. For the manufacture of a silver-white pigment, 40 to 60 nm of TiO_2 is necessary and for a blue interference color about 120 nm. The sequence of interference colors obtained in practice with increasing TiO_2 layer thickness is in accordance with physical color calculations in the color space (Figure 8.17). The color values in the diagram are calculated using the equations for the optics of thin films and are expressed by the $L^*a^*b^*$-values according to Hunter [1, 2] (Figure 1.12). The calculation of the TiO_2 layer thickness necessary to obtain a specific color is done by superpositioning of the a- and b-curves. As an example, to obtain a pure interference blue, a layer thickness should be selected that lies beneath the maximum value for $-b$, and where the a-value goes through the zero point. The purest blue can, therefore, be obtained with a TiO_2 layer thickness of about 125 nm. At 110 nm, the combination of the high value for $+a$ and the high value for $-b$ generates a violet shade.

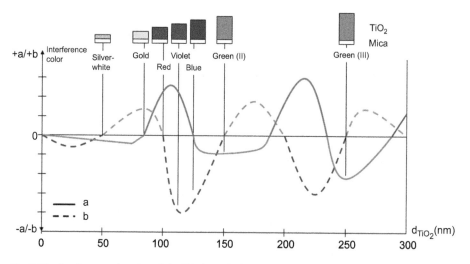

Fig. 8.17: a/b-values as a function of the TiO_2 layer thickness for TiO_2-mica pigments.

Interference pigments show a distinct color flop (color changing effect, flip-flop effect). The color varies with the angle of observation. The interference color is visible from a certain angle, the glancing angle. The color that is seen in a specific application of TiO_2 interference pigments depends strongly on the background. Due to the transparency of the pigments, the background color, e.g., white, black, or colored, plays an important role for the entire color appearance. TiO_2-mica pigments as well as other special effect pigments are often used in combination with other pigments, e.g., with metal effect pigments, white, black and colored inorganic pigments, and also organic pigments. Color and effects achieved by the use of special effect pigments in the different application systems have, therefore, a very complex origin.

Iron(III) oxide is suitable for the coating of mica platelets, similar to titanium dioxide. It combines a high refractive index (metallic luster) with sufficient hiding power and excellent weather resistance. Fe_2O_3-mica pigments are manufactured by precipitation of goethite from $FeCl_3$ or $FeSO_4$ solutions onto mica platelets in aqueous suspension followed by filtration, washing, drying, and calcination of the resulting coated particles at 700 to 900 °C:

$$FeCl_3 + mica + 2H_2O \rightarrow \alpha\text{-FeOOH}/mica + 3HCl \qquad (8.14)$$

$$2\alpha\text{-FeOOH}/mica \rightarrow \alpha\text{-Fe}_2O_3/mica + H_2O \qquad (8.15)$$

Hematite layers are formed on the mica platelets during calcination and water is removed. The thickness of the layers depends on the amount of α-FeOOH deposited during the coating process.

The α-Fe_2O_3 nano-crystallites formed at the beginning of the annealing procedure have diameters of 20 to 40 nm and show a preferred orientation, which has its origin in a template effect arising from the crystallographic structure of the muscovite mica. The initially generated granular layer structure is sintered during calcination to form a less porous, crystalline, and coherent phase [59].

Fe_2O_3-mica pigments have typical thicknesses of the iron oxide layer of 50 to 150 nm and show brilliant, intense colors. Absorption and interference colors are produced simultaneously and vary with the layer thickness. The red shades are intense because interference and absorption phenomena enhance each other. A green-red flop visible from different viewing angles is possible at a thickness of the α-Fe_2O_3 layer, which corresponds to green interference.

Combination pigments containing TiO_2 and other colorants together on mica platelets have extended the class of special effect pigments. Iron(III) oxide is the most important metal oxide for combination with titanium dioxide for such pigments. Brilliant golden pigments with combined layers can be synthesized using two different routes. The pigment structures formed differ considerably from each other. In the first route, a thin layer of Fe_2O_3 is deposited onto the surface of a TiO_2-mica pigment. The generated interference color is the result of both metal oxide layers. The mass tone of such a pigment is determined by the properties of the Fe_2O_3 layer. Interesting gold pigments, e.g., with reddish-golden color, are possible. In the second route, coprecipitation of hydrated titanium dioxide and iron oxide on mica particles followed by calcination leads to greenish-golden pigments. The mass tone in this case is further modified by the highly refractive yellowish iron titanate phase Fe_2TiO_5 (pseudobrookite), which is formed under these conditions [1–4].

Other inorganic colorants used instead of iron oxide for combination pigments are Cr_2O_3 (green), iron blue (blue), Fe_3O_4 (black), and $FeTiO_3$ (silver-gray). In the case of black materials, the interference color is seen as the mass tone. The optical situation is similar to blends with black pigments in a color formulation where the transmitted part of the light is absorbed. Additional coating of a TiO_2-mica pigment with an organic colorant for a mass tone or two-tone effect is executed by precipitation and deposition on the surface of the mica pigment in aqueous suspension, assisted by complexing agents or surfactants. Another method is the fixation of the colorant as a mechanically stable layer by using proprietary additives.

Mica platelets can be coated with a variety of other compounds to achieve attractive color effects. Solid-state reactions extend the possibilities for the synthesis of mica-based pigments. Further options are offered by calcination of the materials in the presence of inert (N_2, Ar) or reactive gases (NH_3, H_2, hydrocarbons). The formation of some colored phases that are not producible by working in air is possible under these conditions [1, 26, 60–63]. Table 8.3 contains examples of mica-based effect pigments with special color properties, especially for the gray, silver-gray, and black range.

Tab. 8.3: Examples of mica-based effect pigments with special coloristic properties.

Pigment composition	Preparation	Color properties
$TiO_{2-x}/TiO_2/mica$ [60, 61]	$TiO_2/mica + H_2$ (Ti, Si) T > 900 °C (solid-state reaction)	Gray, blue-gray
$TiO_xN_y/TiO_2/mica$ [26]	$TiO_2/mica + NH_3$ T > 900 °C (solid-state reaction)	Gray, blue-gray
$FeTiO_3/TiO_2/mica$ [61]	$Fe_2O_3/mica + H_2$ T > 600 °C (solid-state reaction)	Gray
$Fe_3O_4/mica$ [26, 62]	$Fe_2O_3/mica + H_2$ T > 400 °C (solid-state reaction)	Black
	$Fe^{2+} + O_2 + mica$ (wet chemical process)	Black
	$Fe(CO)_5 + O_2 + mica$ (CVD-process)	Black

Simple blending of transparent absorption pigments with pearl luster pigments is one way to generate new color effects. Another way is the production of pearl luster pigments coated with a layer of a transparent absorption colorant to realize more pronounced brilliant colors with a sharper color flop. An additional advantage of such pigments is the avoidance of dispersion problems associated with transparent absorption pigments due to their high surface area and small particle size [1, 2].

Mica-based and other special effect pigments appropriate for outdoor applications must meet the highest standards for color fastness and weather resistance. These pigments are additionally coated with thin layers of transparent and colorless inorganic and/or organic compounds (surface treatment). These layers increase the light resistance by reducing the photoactivity of the titanium dioxide surface. In addition, the interaction between the pigment and the binder is optimized. This also applies to iron oxide or chromium oxide pigment types. Pigments developed with a surface treatment are preferably used in automotive coatings and in industrial outdoor applications. A criterion for their use in these applications is mostly passing of the "Florida test". In this case, test paint systems containing the surface treated effect pigment are subjected to the sunny and humid climate of Southern Florida for at least two years and must not show any change in color or visibly deteriorate in any other way [1].

8.3.3.1.3 Functional metal oxide mica pigments

Initially, metal oxide mica pigments were developed purely for their special coloristic properties. They may, however, also be the basis for functional pigments. These consist typically of metal oxide layers on platelets of natural mica. The used metal oxides possess electrically conductive, magnetic, infrared reflecting or other physical properties, and give rise to the functional behavior of the pigment. Table 8.4 contains information on technically relevant functional pigments based on mica.

Light-colored conductive pigments are applied as conductive component in coatings with permanent antistatic or electrostatic-dissipative (ESD) properties. They are also used as an additive for conductive fillers for the electrostatic spraying of paints. The conductive metal oxide layer on the mica platelets consists in most cases of tin dioxide containing antimony with the general formula $(Sn, Sb)O_2$. Such compositions show semiconductive-to-conductive properties. Antimony is incorporated into the lattice of tin dioxide on lattice sites of tin. The production of the pigments starts from mica suspensions, soluble tin, and antimony salts. The manufacturing process is comparable with that used for the coating of mica platelets with TiO_2 or Fe_2O_3 layers. The conductivity of the pigmented system depends strongly on the pigment volume concentration [1, 64–67].

Tab. 8.4: Examples for functional metal oxide mica pigments.

Pigment composition	Property	Application
$(Sn, Sb)O_2$/mica [64–67]	Electrical conductivity	Antistatic flooring, electrostatic painting
Fe_3O_4/mica [1]	Magnetism	Magnetic surfaces
TiO_2/mica [68, 69]	Solar heat reflection	Green houses, transparent roof lights
TiO_2/mica,	Laser marking of polymers	Laser marking of plastics
$(Sn, Sb)O_2$/mica		
[69, 70]		

Magnetic pigments based on mica coated with magnetite can be manufactured either by the reduction of Fe_2O_3-containing mica pigments or by precipitation and oxidation, starting from iron(II) salt solutions. The magnetite layer formed on the mica platelets gives these pigments their black absorption color; however, at certain thicknesses, they appear colored at the glancing angle. The magnetic properties are less favorable, compared with the characteristics of conventional magnetic pigments, such as γ-Fe_2O_3, Fe_3O_4, Co-containing iron oxides, metallic iron, or CrO_2 [1].

Solar heat reflecting pigments based on mica consist almost completely of transparent platelet-shaped particles that reflect a certain part of the infrared radiation. They are used for architectural and horticultural purposes, where it is important to have plenty of light but little heat. The pigments absorb only a small amount of light and substantially reduce the heating caused by intense sunlight. Metal oxide layers on mica used for solar heat reflecting pigments consist mostly of TiO_2 and SiO_2, with precisely tuned thicknesses [1, 68].

Pigments for the laser marking of polymers may also be based on the layer-substrate principle that is used in special effect pigments. In this case, mica platelets are coated with specific layers of either TiO_2 or $(Sn, Sb)O_2$. The laser technology offers several advantages, compared with the sharp lettering achieved by printing. Writing speed is very high and the marking is immediately permanent and indelible. Furthermore, it is an ink-free marking process. The platelet-shaped pigment particles absorb the incidental laser light (suitable lasers are CO_2-laser with 10.6 µm, Nd:YAG-laser with 1064 or 532 nm, and excimer laser with 193 to 351 nm), and they themselves become a temperature sink. The temperature of the pigment particles increases rapidly, and the polymer material surrounding the pigment is carbonized. The polymer turns black, specifically at the laser-targeted point. Carbonizing can take place together with the formation of fine gas bubbles; thus producing light gray-to-whitish markings. The quality of the laser marking depends strongly on the pigment volume concentration, the polymer material, and the laser parameters used [69, 70].

8.3.3.2 Effect pigments based on silica flakes

Silica flakes (SiO_2 flakes) can be used instead of natural or synthetic mica as a substrate for special effect pigments. The flakes are produced by a web coating process. SiO_2 flakes offer several advantages over the use of mica. Their thickness can be controlled during preparation so that at the end a pigment with a true optical three-layer system is obtained, the interference color of those systems being stronger than that for the conventional mica pigments, where the effect of the mica is levelled by a broad thickness distribution. As a synthetic substrate material, silica does not have iron impurities that cause the slightly yellow mass tone of some natural mica qualities. Finally, SiO_2 has a lower refractive index (1.46) than mica (about 1.58) and, therefore, leads to stronger interference effects [1, 71–74].

Figure 8.18 shows a cross sectional diagram of a metal oxide-coated mica platelet and a metal oxide-coated silica flake. Characteristics of silica flake pigments are uniform and controllable substrate thickness, smooth and uniform substrate surface, absolutely transparent substrate with no mass tone color, excellent reflection and chroma, and intensive color travel.

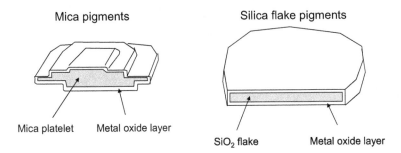

Mica pigments   Silica flake pigments

Mica platelet   Metal oxide layer

SiO$_2$ flake   Metal oxide layer

Fig. 8.18: Schematic view of a section through a mica platelet coated with a highly refractive metal oxide, in comparison to a metal oxide-coated SiO$_2$ flake.

The web coating process for the production of silica flakes uses a polymer web that moves at a controlled speed in an endless loop. The web is wet at a certain point of the loop with an aqueous solution of a silica compound. The wet film generated on the web is dried and a stable coating layer is formed. The thickness of this layer can be selected from 50 to 1000 nm by adjustment of the process parameters. The wet layer is removed from the web and processed further to produce flakes that are fractioned and coated. The transparent flakes can be coated with various metal oxides, such as titanium dioxide and iron oxide, using the procedure known from the mica pigments [75].

The color of pigments based on silica flakes can exhibit extreme angle-dependent effects. Objects painted with them will change appearance with the direction of lighting and the location of the observer. Color changes from gold-silver to green, to green-blue, to dark blue. Strong color travel is seen even under subdued lighting conditions. Decisive for the effect are the uniform thicknesses of the silica substrate and the metal oxide layer. Both layer thicknesses have to be adjusted and controlled precisely in a range of less than 5 nm.

Special effect pigments with improved color strength and very high luster are produced on silica flakes with a specific interference color by deposition of titanium dioxide or iron oxide layers of a thickness required to generate the same interference color. Such pigments show stronger chroma than comparable mica-based types.

The scanning electron micrographs in Figure 8.19 provide information about particle size and shape as well as the layer thicknesses of a TiO$_2$ silica flake pigment. It shows clearly that both the SiO$_2$ flake and the TiO$_2$ layer thickness are precisely controlled.

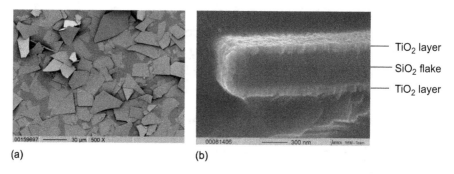

TiO$_2$ layer
SiO$_2$ flake
TiO$_2$ layer

(a)   (b)

Fig. 8.19: Scanning electron micrographs of a silica flake pigment coated with titanium dioxide; (a) overview image; (b) cross section through one particle; layer structure: TiO$_2$-SiO$_2$ flake-TiO$_2$ (source: Merck KGaA).

8.3.3.3 Effect pigments based on alumina flakes

Thin hexagonal monocrystalline alumina flakes (Al$_2$O$_3$ flakes) suitable as substrate material for special effect pigments, which show an extreme sparkle in coating applications, can be produced by a specific crystallization process. The flakes are produced starting from an aqueous aluminum salt solution. Small amounts of dopants are added to control the subsequent crystal growth. An aluminum-containing precipitate is formed, which is filtered off, dried, and heated up to temperatures above 1000 °C. After washing of the cooled material, very thin flakes are obtained, which consist of corundum and show a high aspect ratio, a very narrow thickness distribution, and very smooth surfaces. The thickness can be controlled by the reaction parameters. The resulting flakes can act as substrate particles for metal oxide layers in the process, which is also used for mica and silica flakes. Special effect pigments of very high luster are obtained [1, 71–74, 76].

Pigments based on alumina flakes produced in this way possess a narrow particle size distribution of about 5 to 30 µm as well as a high aspect ratio. The striking sparkle effect has its origin in the optimized thickness of all layers, including that of the Al$_2$O$_3$ substrate. The effect is additionally intensified by directed illumination, e.g., by sunlight. The sparkle appearance in coating applications is controlled by the quantity of the added pigment. The effect, which is already visible at concentrations of 0.1% in the paint system, can be intensified by an increase of the pigment concentration of up to 2%. The single light spots arising from the pigment structure and orientation seem to spring back and forth when a pigmented surface is tilted.

The scanning electron micrographs in Figure 8.20 show alumina flakes coated with titanium dioxide. The special shape of the particles, resulting from the crystal growth process for the alumina flakes and the fine structure of the TiO$_2$ layer, are clearly visible.

(a) (b)

Fig. 8.20: Scanning electron micrographs of an alumina flake pigment coated with titanium dioxide; (a) overview image; (b) image of the fine-grained structure of the TiO_2 layer (source: Merck KGaA).

8.3.3.4 Effect pigments based on borosilicate flakes

Special effect pigments with neutral body color, high luster, and attractive sparkling properties can be achieved from borosilicate flake substrates (glass flakes) [1, 42, 77, 78]. Such pigments consist mostly of calcium-aluminum borosilicate platelets coated with metal oxide layers, e.g., SiO_2, TiO_2 or Fe_2O_3. Synthetically manufactured high quality flakes are absolutely planar and show a very smooth surface. They are characterized by a relatively uniform thickness of each single particle and, therefore, typically show a homogeneous color of each platelet.

Manufacturing of borosilicate pigments consists of the process steps, raw flake production, grinding and classification of the flakes, and coating of the flakes with highly refractive metal oxides, preferably, titanium dioxide and iron(III) oxide. The synthesis of the borosilicate flakes is carried out, starting from a glass melt. Raw materials of high purity are used to achieve the desired colorlessness and transparency of the resulting flakes, which are produced directly from the melt [79]. They can be used after cooling, grinding, and classification as substrate platelets for effect pigments. Coating of the borosilicate flakes and the subsequent working steps are done analogous to the processes used for mica and other platelet-like substrates.

Special effect pigments, based on borosilicate flakes possess in most cases particle diameters below 200 μm and a thickness below 2 μm. Depending on the flake geometry and the coating, the pigments provide a unique and striking multicolored sparkle effect in application systems. This appearance, which particularly is visible in systems pigmented with silver-white types, is called the multicolor effect.

Figure 8.21 shows scanning electron micrographs of borosilicate flakes coated with titanium dioxide. The flakes are perfectly covered with the TiO_2 layer leading to high light reflection and good skin feeling, which is important for the application of the pigments in cosmetic formulations [77].

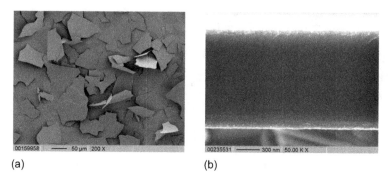

(a)   (b)

Fig. 8.21: Scanning electron micrographs of a borosilicate flake pigment coated with titanium dioxide; (a) overview image; (b) cross section through one particle (source: Merck KGaA).

8.3.3.5 Comparison of different effect pigments based on the layer-substrate principle

Today, a broad number of special effect pigments, based on different substrates, are available for various applications. Natural mica, synthetic mica, silica, alumina, and borosilicate glass are the most widely used substrate materials for this class of pigments. Figure 8.22 shows a scanning electron microscopical comparison of the platelets/flakes of the five relevant substrate materials for special effect pigments together with bismuth oxychloride platelets. Each of the substrates has specific properties, advantages, and disadvantages that are transferred to the resulting pigments when coated with metal oxides. Effects like classic silver-white and interference (natural

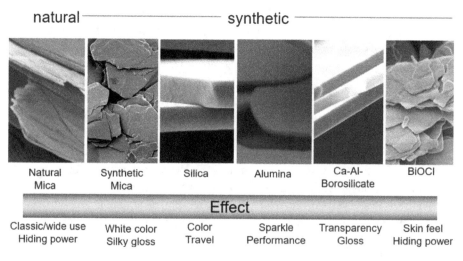

Fig. 8.22: Scanning electron micrographs of different substrate platelets/flakes for special effect pigments, together with an image of BiOCl platelets (source: Merck KGaA) [43].

mica, synthetic mica), color travel (silica flakes), sparkle (alumina flakes), transparency (borosilicate flakes), and special skin feel, combined with hiding power (BiOCl), can be obtained, depending on the substrate material chosen for the production of a specific pigment. The user is always faced with the question of which effect he would like to achieve and which pigment is most suitable for his desired application.

8.3.4 Pigment properties and uses

Special effect pigments are used alone or as part of color formulations where traditional pigments are applied, but where additional color depth, brilliance, iridescence, color travel, and other spectacular effects are required [1–4]. They are characterized amongst others by pearl and interference effects, brilliance, stability, and user-friendly behavior in different application systems. The pigments require transparent or at least translucent binders. Formulations with absorption pigments have to take their transparency and color mixing rules into consideration.

Important for the achievement of desired effects is the parallel alignment of the platelet-shaped pigment particles during their application in coatings, paints, plastics, printing inks, or cosmetic formulations. The pigments can be dispersed homogeneously in most cases, simply by carefully stirring the liquid application system.

Special effect pigments are standard components of solvent- and water-based automotive and industrial coatings. Decorative effects and technical quality are decisive factors for the broad use of the pigments in these application areas. Automotive designers have a wide range of color and composition possibilities at their disposal. Attractive and individual coatings are provided not only by using the various colors of special effect pigments, but also by the different particle sizes available in combination with absorption pigments. Examples for industrial paint applications of special effect pigments are bicycles, airplanes, rail wagons, cans and containers, building exteriors, interior decoration, furniture, artificial pearls, and jewelry [33, 80, 81].

The application of special effect pigments in plastics is widespread. The pigments are used in nearly all kinds of thermoplastic and thermosetting polymers. All relevant techniques for polymer processing are applied for adding color and special effects to plastic parts. The orientation of the pigment platelets parallel to the object surface occurs in almost all applications through flow movements in the corresponding matrix. Usually 0.5% to 2% of an effect pigment is added to color a specific polymer [70].

Special effect pigments are also used in a broad variety of printing applications where they fulfill the highest esthetic requirements. Many products are printed with these pigments especially in cases where articles should reflect a touch of luxury and artistic style. The pigments can be processed with all relevant printing techniques such as offset, screen, gravure, and flexographic printing. Coating techniques such as overprint-varnishing (OPV) and paper coating are likewise used. The pigments can be applied in solvent- and water-based printing formulations as well as in UV inks [82–84].

Cosmetic applications require specific effect pigments approved for use according to cosmetic regulations. Such applications include color cosmetics (e.g., lipsticks, eye shadows, blushers, eye pencils, makeup, mascaras, and nail lacquers), personal care products (e.g., shampoos, body washes, lotions, creams, oral care products, and hair gels) and special effect items (e.g., theatrical and costume makeup and temporary hair color sprays) [49, 50].

Special effect pigments for ceramic and glaze applications are coated with additional layers to achieve stabilization against the aggressiveness of the frits at high temperatures [85].

Some of the special effect pigments, mainly those consisting of mica coated with TiO_2 or Fe_2O_3, are approved food and pharmaceutical colorants. A series of pigments is available, ranging from silver-white and gold up to the iron oxide shades of red and bronze. Such pigments are used for confectionery (e.g., liquorice, fruit gums, jelly beans, and chewing gums), chocolate, biscuits, ice cream, and beverages. They meet international standards for food and drug safety [86].

8.3.5 Toxicology and occupational health

In examinations for acute toxicity, no toxic effects were found for special effect pigments based on mica. The pigments do not show acute toxicity (LD_{50} value rat oral: >15,000 mg/kg). Metal oxide mica pigments demonstrate no irritation or sensitivity to skin or mucous membranes. Also, for comparable products based on other substrate platelets such as silica, alumina, or borosilicate flakes, no irritation to the skin or mucous membranes was found [87].

No hazard to humans has been found through contact with substrate-based effect pigments in the working place. Studies on the genetic toxicity of mica-based and other comparable pigments have not shown mutagenic effects in vitro and in vivo. Ames and microcore tests have not shown a mutation of the inheritance. Chronic toxicity tests with mica-based effect pigments have not resulted in ill effects on rats over a period of two years [87, 88].

Metal oxide mica pigments as well as the corresponding pigments on silica, alumina, and borosilicate are highly stable and inert. In aqueous suspensions, virtually no metal ions can be leached from the surface. As a consequence, these pigments are regarded as nontoxic and harmless with regard to ecological hazards. As inorganic materials that are insoluble in water, the pigments are non-biodegradable [87].

All titanium dioxide and iron oxide mica pigments without a special surface treatment are approved for use in food packaging, e.g., as a colorant in related printing inks and plastics. Some of the pigments are even approved as food and pharmaceutical colorants.

Bismuth oxychloride pigments are considered to be nontoxic and nonhazardous. Extensive and direct contact of the pure powder with the skin should be avoided due to possible skin irritation.

Basic lead carbonate has to be considered like other lead compounds as dangerous for the environment. Lead compounds are classified as very toxic to aquatic organisms and may cause long-term adverse effects in the aquatic environment.

Special effect pigments can be handled under usual hygienic working conditions. The inhalation of pigment dust during production and application should of course be avoided as in the case of other pigments.

8.4 Conclusions

Effect pigments, which are subdivided into metal effect pigments (metallic effect pigments) and special effect pigments, including pearl luster pigments (pearlescent pigments, nacreous pigments) and interference pigments, consist of μm-sized thin platelets that show strong lustrous effects when oriented in a parallel manner in application systems.

Metal effect and special effect pigments are used in all relevant application systems, such as coatings, paints, plastics, artists' paints, cosmetics, printing inks, leather, construction materials, paper, glass, and ceramics. Specific composition, particle size distribution, and surface quality determine the coloristic and application technical properties of the individual pigments. Surface treatments are used for many pigments to improve the quality with respect to stability and compatibility with the application system.

Metal effect pigments consist of thin metallic flakes. These are offered in the form of powders, pastes, pellets, suspensions, or color concentrates. Nearly all metal effect pigments consist of aluminum (so-called silver bronzes), copper and copper/zinc alloys (so-called gold bronzes), and zinc. They are manufactured starting from a metal granulate (grit), which is then ground under formation of flakes. Metal effect pigments are nontransparent for light. Parallel-oriented metal flakes reflect the entire incident light in one direction similar to a mirror, which leads to the effect of metal luster (metallic effect).

Special effect pigments are, in most cases, synthetic pigments characterized by high luster, brilliance and iridescent colors, known from optically thin films. Their visual appearance has its origin in the reflection and refraction of light at thin single and multiple layers. The pigments are either transparent, semitransparent, or light-absorbing platelet-shaped crystals or layer systems. They can consist of single crystals, but also of monolayer or multilayer structures, in which the layers have different refractive indices and light absorption properties.

Pigment, based on the layer-substrate principle, represent the technically most important class of special effect pigments. There are many variations and combination possibilities for the layer and substrate materials used. Special effect pigments based on platelets of natural or synthetic mica, on alumina, silica, or borosilicate flakes are the main representatives of layer-substrate pigments. They are manufactured in most cases by wet chemical processes combined with high temperature processes. The production of substrate particles includes mechanical processes, but also crystal growth, glass formation, and web coating.

Major efforts have been made to develop new sophisticated effect pigments and to improve existing ones. Optimized particle management and surface treatment are key factors for the success of new pigment types.

Study questions

8.1	How are effect pigments defined?
8.2	Into what classes are effect pigments subdivided?
8.3	How are metal effect pigments characterized?
8.4	How are special effect pigments characterized?
8.5	What types of metal effect pigments are practically used to achieve metallic effects?
8.6	What types of aluminum pigments are distinguished?
8.7	What are the main steps for the production of aluminum pigments?
8.8	What are the most important substrate-free special effect pigments?
8.9	What are special effect pigments based on the layer-substrate principle?
8.10	What are the main steps for the production of metal oxide mica pigments?
8.11	What are functional pigments based on the layer-substrate principle?

Bibliography

[1]   Pfaff G, in Special Effect Pigments, 2nd edn. Pfaff G ed. Vincentz Verlag, Hannover, 2008, 16.
[2]   Franz K-D, Emmert R, Nitta K. Kontakte (Darmstadt). 1992,2,3.
[3]   Pfaff G. Chem. Unserer Zeit. 1997,31,6.
[4]   Pfaff G, Reynders P. Chem. Rev. 1999,99,1963.
[5]   Snow MR, Pring A, Self P, Losic D, Shapter J. Amer. Miner. 2004,89,1353.
[6]   Katti KS, Katti DR. Mater. Sci. Eng. C. 2006,26,1317.
[7]   Besold R, in Industrial Inorganic Pigments, 3rd edn. Buxbaum G, Pfaff G ed. Wiley-VCH Verlag, Weinheim, 2005, 252.
[8]   Pfaff G, Bartelt MR, Maile FJ, in Encyclopedia of Color, Dyes, Pigments, vol. 2, Pfaff G ed. Walter de Gruyter GmbH, Berlin, Boston, 2022, 829.
[9]   Wheeler I. Metallic Pigments in Polymers, Rapra Technology Limited, Shawbury, 1999, 7.
[10]   Hoenigsberg O, in Beiträge Zur Geschichte der Technik Und Industrie, Band 2, Matschoss C ed. Springer, Berlin, 1910, 271.
[11]   Patent US 964,024 (Montgomery GS, Hotchkiss CR) 1908.
[12]   Patent US 1,832,868 (Hartstoff Metall AG) 1928.
[13]   Patent US 2,002,891 (Metals Disintegration Co) 1931.
[14]   Patent US 3,949,139 (Avery Corporation) 1972.
[15]   Patent US 4,321,087 (Revlon, Inc.) 1978.
[16]   Wissling P, Kiehl A, in Metallic Effect Pigments, Wissling P ed. Vincentz Verlag, Hannover, 2006, 12.
[17]   Rodrigues ABJ. Die Farbe. 1990,37,65.
[18]   Rösler G. Die Farbe. 1990,37,121.
[19]   Cloppenburg H, Schmittmann D. Farbe + Lack. 1989,95,631.
[20]   Patent DE 195 20 312 (Eckart) 1995.
[21]   Patent US 5,037,475 (Showa) 1988.
[22]   Patent EP 0 737 212 (Silberline) 1994.
[23]   Patent DE 195 01 307 (Eckart) 1995.
[24]   Patent EP 0 755 986 (Toyo Aluminium) 1995.
[25]   Patent EP 0 810 370 (Toyo Aluminium) 1996.
[26]   Ostertag W, Mronga N, Hauser P. Farbe + Lack. 1987,93,973.
[27]   Ostertag W, Mronga N. Macromol. Chem. Macromol. Symp. 1995,100,163.

[28]   Patent EP 1 412 433 (Merck) 2001.
[29]   Pfaff G, Fritsche K. Eur. Coat. J. 2014,11,34.
[30]   Patent WO 2016/165832 (Schlenk) 2016.
[31]   Phillips RW, Bleikolm AF. Appl. Opt. 1996,35,5529.
[32]   Pfaff G, Reynders P. Chem. Rev. 1999,99,1963.
[33]   Argoitia A, Witzmann M. Society of Vacuum Coaters, 45th Annual Techn, Conf. Proc. 2002,539.
[34]   Maile FJ, Pfaff G, Reynders P. Progr. Org. Coat. 2005,54,150.
[35]   Besold R, Reisser W, Roth E. Farbe + Lack. 1991,97,311.
[36]   Müller B. Surf. Coat Int., Part B, Coat. Transact. (JOCCA). 2002,85,111.
[37]   Graziano M, Kaupp G, Krüger P, in Metallic Effect Pigments, Wissling P ed. Vincentz Verlag, Hannover, 2006, 214.
[38]   Fetz A, Wissling P, in Metallic Effect Pigments, Wissling P ed. Vincentz Verlag, Hannover, 2006, 208.
[39]   Hartl-Gunselmann A, Hartmann T, in Metallic Effect Pigments, Wissling P ed. Vincentz Verlag, Hannover, 2006, 223.
[40]   Lutz D, in Metallic Effect Pigments, Wissling P ed. Vincentz Verlag, Hannover, 2006, 224.
[41]   Pfaff G, in High Performance Pigments, Smith HM ed. Wiley-VCH Verlag, Weinheim, 2002, 77.
[42]   Pfaff G, in Industrial Inorganic Pigments, 3rd edn. Buxbaum G, Pfaff G ed. Wiley-VCH Verlag, Weinheim, 2005, 230.
[43]   Pfaff G, in Encyclopedia of Color, Dyes, Pigments, vol. 3, Pfaff G ed. Walter de Gruyter GmbH, Berlin, Boston, 2022, 1081.
[44]   Wiegleb JC. Handbuch der allgemeinen Chemie, Band 2, Friedrich Nicolai, Berlin and Stettin, 1781, 636.
[45]   Hofmeister F, Pieper H. Farbe + Lack. 1989,95,557.
[46]   Pfaff G, Gabel P. Eur. Coat. J. 2005,6,30.
[47]   Gabel P, in Special Effect Pigments, 2nd edn. Pfaff G ed. Vincentz Verlag, Hannover, 2008, 197.
[48]   Gabel P, in Colour Technology of Coatings, Kettler W ed. Vincentz Verlag, Hannover, 2016, 61.
[49]   Thurn-Mueller A, Hollenberg J, Listen J. Kontakte (Darmstadt). 1992,35.
[50]   Pfaff G, Becker M. Household Personal Care Today, Monogr. Suppl. Ser. "Colour Cosmet". 2012, 1, 12.
[51]   Patent EP 0 068 311 (BASF) 1982.
[52]   Ostertag W. Nachr. Chem. Tech. Lab. 1994,42,849.
[53]   Patent EP 0 837 911 (Merck) 1997.
[54]   Patent EP 0 931 112 (Merck) 1997.
[55]   Patent EP 0 944 677 (Merck) 1997.
[56]   Van Valkenburg A, Pike RG. J. Res. Natl. Bur. Stand. 1952,48,360.
[57]   Kumar KNP, Keizer K, Burggraaf A, Okubo T, Nagamoto H. J. Mater. Chem. 1993,3,1151.
[58]   Patent EP 1 025 168 (Merck) 1997.
[59]   Hildenbrand VD, Doyle S, Fuess H, Pfaff G, Reynders P. Thin Solid Films. 1997,304,204.
[60]   Patent EP 0 632 821 (Merck) 1992.
[61]   Hauf C, Kniep R, Pfaff G. J. Mater. Sci. 1999,34,1287.
[62]   Patent EP 0 354 374 (Merck) 1988.
[63]   Patent EP 0 246 523 (Merck) 1986.
[64]   Brückner H-D, Glausch R, Maisch R. Farbe + Lack. 1990,96,411.
[65]   Hocken J, Griebler W-D, Winkler J. Farbe + Lack. 1992,98,19.
[66]   Pfaff G, Reynders P. Chem. Rundschau Jahrbuch. 1993,31.
[67]   Vogt R, Neugebauer E, Pfaff G, Stahlecker O. Eur. Coat. J. 1997,7/8,706.
[68]   Patent EP 0 659 198 (Hyplast, Merck) 1992.
[69]   Pfaff G. Phänomen Farbe. 2003,7/8,21.
[70]   Kieser M, in Special Effect Pigments, 2nd edn. Pfaff G ed. Vincentz Verlag, Hannover, 2008, 137.

[71] Teaney S, Pfaff G, Nitta K. Eur. Coat. J. 1999,4,90.

[72] Pfaff G. Welt der Farben. 2000,1,16.

[73] Sharrock S, Schül N. Eur. Coat. J. 2000,1/2,20.

[74] Pfaff G. Inorg. Mater. 2003,39,123.

[75] Patent EP 0 608 388 (Merck) 1991.

[76] Patent EP 0 763 573 (Merck) 1995.

[77] Rüger R, Oldenburg N, Schulz E, Thurn-Schneller A. Cosmet. Toiletries. 2004,5,133.

[78] Huber A, Pfaff G. Phänomen Farbe. 2005,25,34.

[79] Patent EP 0 289 240 (Glassflake Ltd.) 1987.

[80] Maile FJ, in Special Effect Pigments, 2nd edn. Pfaff G ed. Vincentz Verlag, Hannover, 2008, 92.

[81] Maile FJ, in Encyclopedia of Color, Dyes, Pigments, vol. 1, Pfaff G ed. Walter de Gruyter GmbH, Berlin, Boston, 2022, 161.

[82] Weitzel J, in Special Effect Pigments, 2nd edn. Pfaff G ed. Vincentz Verlag, Hannover, 2008, 171.

[83] Böhm K, Pfaff G, Weitzel J. Farbe + Lack. 1999,105,30.

[84] Pfaff G, Rathschlag T. Coat. Agenda Eur. 2002,150.

[85] Pfaff G, Hechler W, Brabänder C, Keram Z. 2001,53,1002.

[86] Patent EP 1 469 745 (Merck) 2002.

[87] Pfaff G, in Special Effect Pigments, 2nd edn. Pfaff G ed. Vincentz Verlag, Hannover, 2008, 209.

[88] Bernard BK, Osheroff MR, Hofmann A, Mennear JH. J. Toxicol. Environ. Health. 1989,28,415.

9 Functional pigments

Functional inorganic pigments are characterized by special physical properties. Color and appearance do not play a role for this category of pigments. The two main representatives of inorganic functional pigments are magnetic and anticorrosive pigments, which are described in this chapter. There are further functional pigments that are introduced elsewhere. These include electrically conductive, IR-reflective, and laser-marking pigments (Section 8.2.3) as well as UV-absorbing pigments (Section 7.5).

9.1 Magnetic pigments

Magnetic pigments are finely divided powders with distinct magnetic properties. They are used in magnetic information storage systems, e.g., in audio cassettes, videotapes, floppy discs, hard discs, and computer tapes. They are also applied as toner pigments in photocopiers and laser printers and find application in special printings, e.g., in the printing of banknotes. The pigment particles are dispersed for magnetic applications in suitable binder systems and then applied to the storage system. Main representatives of magnetic pigments are γ-Fe_2O_3, Fe_3O_4, Co-containing iron oxides (Co-γ-Fe_2O_3, Co-Fe_3O_4), metallic iron, CrO_2, and $BaFe_{12}O_{19}$ [1, 2].

9.1.1 Fundamentals and general properties

Magnetism of solids is a cooperative phenomenon. Macroscopic magnetization is composed of the contributions of individual components (atoms, ions, quasi free electrons), which the solid consists. For many materials, the single components themselves have a magnetic moment. However, even among materials containing components with such magnetic moments, only a few show macroscopic magnetization. As a rule, the addition of various moments to the total moment is usually zero. Only if the individual contributions are not cancelled out in sum does a macroscopic magnetism appear. There are five types of magnetism: diamagnetism, paramagnetism, ferromagnetism, ferrimagnetism, and antiferromagnetism. Of these, only ferromagnetism and ferrimagnetism are relevant for magnetic pigments.

In the case of ferromagnetism, the magnetic moments of single components are not independent of each other. They spontaneously align themselves in parallel. The interconnection of the magnetic moments does not extend through the entire material, but instead it is limited to small areas, the Weiss domains. The typical extent of these domains is in the range of 10 nm to a few μm. The orientation of the Weiss domains is distributed statistically, so that the entire material appears nonmagnetic. The domains can, however, be oriented in the same direction by an external magnetic field. This orientation remains after removal of the external field, so that a permanent magnetization

https://doi.org/10.1515/9783110743920-009

is achieved. The magnetization can be destroyed by heating the material to temperatures above the Curie temperature. Representatives of ferromagnetic pigments are metallic iron, CrO_2, and $BaFe_{12}O_{19}$.

In the case of ferrimagnetism too, the magnetic moments of single components are not independent of each other. There are, however, two types of magnetic centers. The spin moments of similar centers align themselves in parallel and the spin moments of different spin centers antiparallel. This leads to a partial elimination of the magnetic moments. All other behavior is similar to that of ferromagnetic materials. Representatives of ferrimagnetic pigments are γ-Fe_2O_3, Fe_3O_4, and their Co-containing derivatives.

Essential characteristics of magnetic pigments are saturation magnetization, remanence, and coercive field strength as well as particle shape and size. The coercive field strength determines the area of application. High coercivity is a prerequisite for high storage densities. The coercive field is the magnetic field required to demagnetize the material. The saturation magnetization is a specific constant for the magnetic material. The magnetic properties of a pigment are typically determined by measurement of hysteresis curves on the powder or on magnetic tapes.

The shape of the pigment particles is extremely important for suitable magnetic properties. The length-to-width ratio of magnetic particles is mainly in the range of 5:1 to 20:1. The average particle lengths vary between 0.1 to 1 μm. The values for the specific surface area are in the range of 10 to 60 m^2/g.

9.1.2 Production, special properties, and uses of magnetic pigments

9.1.2.1 Iron oxide pigments

The ferrimagnetic iron oxides γ-Fe_2O_3 (tetragonal superlattice structure) and Fe_3O_4, (spinel structure) have already been used for magnetic information storage since the beginning of magnetic tape technology. Nonstoichiometric mixed-phase pigments (berthollides) of the type $FeO_xFe_2O_3$, with $0 < x < 1$, were also in use in the early days. Nowadays, γ-Fe_2O_3 pigments are mainly used for the production of low bias audio cassettes and in studio, broadcasting, and computer tapes. Fe_3O_4 pigments are also in use, but they have a lower importance for magnetic applications.

Suitable magnetic properties depend strongly on the shape of the pigment particles. Isometric particles are hardly suitable due to weak magnetic parameters. Needle-shaped particles have clear advantages in this respect and are, therefore, preferred for magnetic applications. Needle-shaped γ-Fe_2O_3 pigments with a length-to-width ratio from 5:1 to 20:1 and a crystal length of 0.1 to 1 μm were prepared for the first time in the late 1940s [3].

Anisometric, needle-shaped forms of γ-Fe_2O_3 and Fe_3O_4 do not crystallize on a direct route. They are usually produced starting from α-FeOOH or γ-FeOOH that forms acicular crystals by precipitation under suitable conditions (Figure 9.1) [4–6]. The two oxide hydroxides are converted by thermal dehydration to α-Fe$_2$O$_3$ and further by reduction at 300 to 600 °C to Fe$_3$O$_4$. Hydrogen, carbon monoxide, or organic compounds, e.g., fatty acids, can act as reducing agents. The acicular particle shape is retained during the reduction step. The iron oxide hydroxide particles can be stabilized prior to the conversion with a protective layer against caking during sintering. Suitable compounds for such a layer are silicates, phosphates, chromates, and fatty acids [1, 2]. The resulting Fe$_3$O$_4$ pigments are often stoichiometric and finely divided. Such pigments are not stable to atmospheric oxidation and are, therefore, stabilized by partial oxidation.

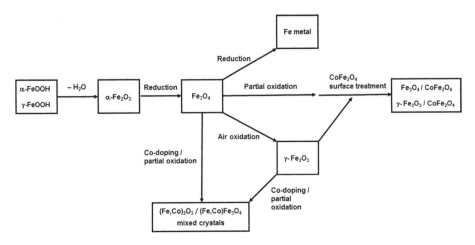

Fig. 9.1: Production of magnetic iron oxide pigments.

Needle-shaped γ-Fe$_2$O$_3$ is obtained by oxidation of Fe$_3$O$_4$ in air at temperatures from 300 to 900 °C. The acicular shape of the Fe$_3$O$_4$ particles is also retained during the thermal treatment in this case.

The hydrothermal route can also be used for the synthesis of γ-Fe$_2$O$_3$ pigments. In the first step, a suspension of precipitated Fe(OH)$_3$ is prepared. The suspension is transferred into an autoclave where γ-Fe$_2$O$_3$ is formed under suitable hydrothermal conditions. The crystal growth is controlled by organic modifiers [7, 8].

Magnetic iron oxide pigments used for recording media are characterized by very different morphological and magnetic properties. The selection of a pigment depends on the concrete field of application and the required magnetic parameters. The largest particles, with a length of about 0.6 µm, are applied in computer tapes. The noise level of the magnetic tapes increases with increasing particle size of the magnetic pigment. Finer pigments are, therefore, used to achieve compact cassettes of high quality.

Table 9.1 shows a few quality requirements for the most important applications of selected magnetic pigments (γ-Fe$_2$O$_3$, Co-γ-Fe$_2$O$_3$, Co-Fe$_3$O$_4$, metallic Fe). For magnetic iron oxide pigments, the saturation magnetization M_s is primarily determined by the Fe^{2+} ion content. The ratio of remanent magnetization to saturation magnetization M_r/M_s for a tape depends mainly on the orientation of the pigment needles with regard to the longitudinal direction of the tape.

Tab. 9.1: Some quality requirements for magnetic iron oxide and metallic iron pigments [2].

Field of application	Pigment type	Approximate particle length, µm	Specific surface area, m^2/g	Coercive field strength H_c, kA/m	Saturation magnetization, M_s, $\mu T \cdot m^3/kg$	M_r/M_s
Computer tapes	γ-Fe_2O_3	0.60	13–17	23–25	86–90	0.80–0.85
Studio radio tapes	γ-Fe_2O_3	0.40	17–20	23–27	85–92	0.80–0.85
IEC I compact cassettes standard (iron oxide operating point)	γ-Fe_2O_3	0.35	20–25	27–30	87–92	0.80–0.90
High grade	Co-γ-Fe_2O_3	0.30	25–37	29–32	92–98	0.80–0.90
IEC II compact cassettes (CrO$_2$ operating point)	Co-γ-Fe_2O_3, Co-Fe_3O_4	0.30	30–40	52–57	94–105	0.85–0.92
IEC IV compact cassettes (metal operating point)	Metallic iron	0.35	35–40	88–95	130–160	0.85–0.90
Digital audio (R-DAT)*	Metallic iron	0.25	50–60	115–127	130–160	0.85–0.90
½" Video	Co-Fe_3O_4	0.26	30–40	52–57	98–105	0.80–0.90
Super-VHS video	Co-γ-Fe_2O_3	0.20	45–50	64–72	94–96	0.80–0.85
8 mm video	Metallic iron	0.25	50–60	115–127	130–160	0.85–0.90
Data storage tape (S-DLT)	Metallic iron	0.06	50–70	190–210	119–130	0.85–0.90

*R-DAT: rotary digital audio tape.

In addition to magnetic and morphological characteristics, other properties such as dispersibility and chemical stability are of importance for the application of magnetic pigments. The adjustment of all relevant pigment properties is achieved by selecting the suitable manufacturing process with the corresponding reaction parameters.

9.1.2.2 Cobalt-containing iron oxide pigments

The cobalt-containing iron oxides $Co-\gamma-Fe_2O_3$ and $Co-Fe_3O_4$ are characterized as magnetic materials with a high coercivity. They are used as an alternative to CrO_2 for the production of videotapes, high bias audiotapes, and high density floppy disks.

Cobalt-containing iron oxide pigments are manufactured in two ways:

–   Cobalt is incorporated into the structure of $\gamma-Fe_2O_3$ and Fe_3O_4 during the synthesis of the two oxides. The final pigments $Co-\gamma-Fe_2O_3$ and $Co-Fe_3O_4$ contain 2 to 5 wt% cobalt homogeneously distributed in the iron oxides. Incorporation of cobalt takes place by coprecipitation of cobalt hydroxide (from soluble cobalt salts) together with the iron oxide hydroxide, which is used as precursor. Alternatively, cobalt hydroxide can be subsequently precipitated onto the particles of the already formed $\alpha-FeOOH$ or $\gamma-FeOOH$ [9].

–   Cobalt is coated on the particles of $\gamma-Fe_2O_3$ or Fe_3O_4 by using a precipitation reaction. The final pigments consist of a core of $\gamma-Fe_2O_3$ or Fe_3O_4 and a coating layer of cobalt ferrite. They contain 1.6 to 4 wt% cobalt. The cobalt ferrite layer is formed during thermal treatment. It has a thickness of 1 to 2 nm and gives the magnetic pigment a higher coercivity, as compared to the pure magnetic iron oxides [10]. The coating layer can be prepared by adsorption of precipitated cobalt hydroxide or by epitaxial precipitation of cobalt ferrite in a strongly alkaline milieu [11, 12].

Cobalt-containing iron oxide pigments have coercive field strengths in the range of 29 to72 kA/m, depending on the application (Table 9.1). Their particle sizes of 0.2 to 0.4 µm are adjusted by the manufacturing process, according to the intended use.

$Co-\gamma-Fe_2O_3$ pigments with coercive field strength of about 70 kA/m and a particle length from 0.15 to 0.2 µm are preferably used for super VHS cassettes. Pigments containing only small amounts of cobalt (0.5 to 1 wt%) and having coercive field strength of about 30 kA/m can be used as alternatives to $\gamma-Fe_2O_3$ for high-quality low-bias audiocassettes.

As in the case of the pure magnetic iron oxides, other properties of pigments, which are relevant to the application must be kept in mind during the manufacturing process. Most important of these are compatibility with the application medium, especially dispersibility and chemical stability.

9.1.2.3 Metallic iron pigments

Metallic iron is characterized by magnetization that is more than three times higher than that of magnetic iron oxides. The values for the coercive field strength of iron can vary in the range of 30 to 210 kA/m, depending on the size and shape of the magnetic particles. With such properties, metallic iron pigments are predestined for use in high density recording media. Earlier problems of instability against oxidation could be solved for the first time in the 1970s [1].

Metallic iron pigments are produced by reduction of acicular crystallizing iron compounds [13]. Starting materials are iron oxide hydroxides or iron oxalates, which are reduced to iron in a hydrogen stream. The resulting iron powders have very high specific surface areas. They are, therefore, pyrophoric and must be stabilized by a passivation process. Passivation can be done by controlled, slow oxidation of the particle surfaces [14].

Pigments for analog music cassettes with a coercivity of about 90 kA/m usually have a particle length of 0.35 μm (Table 9.1). The length-to-width ratio of the needle-shaped particles is about 10:1. Finely divided iron pigments with a particle length of about 0.12 μm and coercive field strength of 130 kA/m are used for 8-mm video and digital audio cassettes, tapes for television broadcasts, and for master videocassettes (mirror master tapes). They also find use in micro floppy disks and backup cassettes.

A comparison with magnetic iron oxides shows that metallic iron pigments have a higher specific surface area of up to 60 m^2/g and a higher saturation magnetization. Their capacity for particle alignment expressed by the ratio M_r/M_s corresponds to that of the iron oxides (Table 9.1).

9.1.2.4 Chromium dioxide pigments

Chromium dioxide is characterized by ferromagnetic behavior and high recording densities, which surpass those of γ-Fe_2O_3. It crystallizes in a tetragonal rutile lattice and under suitable conditions forms small needles. The anisotropic, acicular shape of the particles is an important requirement for magnetic CrO_2 pigments. The morphology of the CrO_2 particles can be modified by doping with antimony, tellurium, or other elements [15, 16]. In addition, the coercive field strength can also be adjusted by suitable dopants. Coercivity values in the range of 30 to 75 kA/m are possible by doping with transition metals, e.g., with Fe^{3+} ions, which modify the magneto-crystalline anisotropy of the CrO_2 [17].

The manufacture of chromium dioxide pigments takes place under hydrothermal conditions starting from intimate mixtures of Cr(III) and Cr(VI) compounds, e.g., Cr_2O_3 and CrO_3, in an autoclave [1, 15, 16, 18]. The CrO_2 formation requires temperatures of about 350 °C and pressure of 300 bar. The process consists of two partial reactions. These allow a separate influence on the nucleation and on the crystallization. Thus, pigments with a very narrow particle size distribution are possible.

$$Cr_2O_3 + 2\,CrO_3 \rightarrow 3\,CrO_2 + CrO_3 \qquad (9.1)$$

$$CrO_3 \rightarrow CrO_2 + {}^1/_2\,O_2 \qquad (9.2)$$

The crystal structure of CrO_2 supports the direct formation of needle-shaped particles. Pure CrO_2 disproportionates slowly in the presence of water. In order to prevent this, the surface of the particles can be converted topotactically under reducing conditions from CrO_2 to β-CrOOH. The β-CrOOH layer acts as a protective layer and also allows the use of CrO_2 pigments in a humid atmosphere [19]. CrO_2 is stable up to temperatures of about 400 °C in the absence of moisture. It decomposes under formation of Cr_2O_3 and oxygen above this temperature.

Chromium dioxide pigments are used exclusively for magnetic recording media. In order to achieve specific magnetic values, it is also used in combination with cobalt-modified iron oxides.

9.1.2.5 Barium ferrite pigments

Barium hexaferrite with the composition $BaFe_{12}O_{19}$, belongs to the ferromagnetic materials and is used among other things for magnetic pigments (barium ferrite pigments). These are especially suitable for the manufacture of nonoriented (floppy disks), longitudinally oriented (conventional tapes), and perpendicularly oriented media [20]. Their use for perpendicular recording systems, where the magnetization is oriented perpendicular to the coating surface, may lead to extremely high data densities, especially on floppy disks. Barium and strontium ferrites are also used for magnetic stripes in checks and identity cards for counterfeit protection.

Hexagonal ferrites exist in a wide range of structures differentiated by stacking arrangements of three basic elements known as M, S, and Y blocks [21]. The M-type structure represented by $BaFe_{12}O_{19}$ is the most important one for magnetic pigments. The magnetic properties of an M-ferrite can be adjusted over a wide range by partial substitution of the Fe^{3+} ions in the structure. Combinations of di- and tetravalent ions such as Co^{2+} and Ti^{4+} are suitable for the replacement of Fe^{3+}. Barium ferrite forms small hexagonal platelets during crystallization. The preferred direction of magnetization is parallel to the c-axis of the crystal structure and is, therefore, perpendicular to the surface of the platelets. The coercive field strength of barium hexaferrite is primarily determined by its magneto-crystalline anisotropy. The particle morphology plays only a minor role in its magnetic properties. Barium ferrite, therefore, has extremely uniform magnetic properties.

Barium hexaferrite pigments have a brown color. Their chemical properties are similar to those of the iron oxides. The diameter of $BaFe_{12}O_{19}$ platelets is adjusted during the crystallization process according to the intended use. Diameters from 40 to 70 nm (nonoriented media) up to 300 to 600 nm (magnetic cards) are required. The thickness of the platelets varies from 15 to 30 nm (nonoriented media) up to 100 to 400 nm (magnetic cards). The values for the specific surface area range from 3 to 60 m^2/g.

Three industrial methods for the manufacturing of barium ferrites are in use. These are known as the ceramic method, the hydrothermal method, and the glass crystallization method. For the ceramic method, mixtures of barium carbonate and iron(III) oxide react at temperatures of 1200 to 1350 °C to form crystalline $BaFe_{12}O_{19}$.

$$BaCO_3 + 6\,Fe_2O_3 \rightarrow BaFe_{12}O_{19} + CO_2 \qquad (9.3)$$

Barium hexaferrite contains agglomerates and has to be ground to achieve a mean particle size of about 1 μm. The ceramic method is only suitable for pigments with high coercivity, which is required for magnetic strips [22].

The hydrothermal method starts with the precipitation of a precursor containing iron(III) and barium, also with dopants if required. The coprecipitate reacts with an excess of sodium hydroxide solution at 250 to 350 °C in an autoclave. The hydrothermal treatment is followed by annealing at temperatures of 750 to 800 °C to obtain the desired magnetic properties for the resulting $BaFe_{12}O_{19}$. In a modification of the process, the hydrothermal synthesis is followed by coating of the hexaferrite particles with cubic ferrites. This process is similar to the cobalt modification of iron oxides and has the intention of improving the magnetic parameters of the material, e.g., the saturation magnetization [23–25].

For the glass crystallization method, the starting materials are dissolved in a borate glass melt at about 1200 °C [26]. The melt is quenched by pouring it onto rotating cold copper wheels. Glass flakes formed in this way are annealed to crystallize the ferrite in the glass matrix. In the final step, the glass matrix is dissolved in acid. The glass matrix may also be produced by a spray drying procedure [27].

Barium hexaferrite pigments are suitable for high density digital recording mainly because of their very small particle sizes and their very narrow switching field distribution. They have a high anhysteretic susceptibility and are, therefore, difficult to overwrite [28]. This property makes media containing barium hexaferrite suitable for the anhysteretic (bias field) duplicating process [29, 30].

An advantage of $BaFe_{12}O_{19}$ in comparison with other magnetic materials used for high density recording is that it is not affected by corrosion [31]. On the other hand, the processing of barium hexaferrite pigments may be problematic. Under certain circumstances, orientating fields can lead to unwanted stacking of the particles with adverse effects on the noise level and the coercive field strength of the magnetic tapes.

9.1.2.6 Toner pigments

Magnetic pigments are applied in photocopiers and laser printers, where they are referred to as toner pigments. Fe_3O_4 pigments with specific properties are used here [32].

Copier and laser printers contain a cylinder (photoconductive drum) equipped with an organic photoconductor coating (OPC). This cylinder is electrostatically charged by a high voltage corona before the actual copying or printing process starts. It is then exposed to light from a laser or projection system (Figure 9.2). The areas of the organic photoconductor coating that are subjected to light lose a varying portion of the charge. The unexposed areas, which correspond to the black dots of the original, retain their charge. The process at this point uses the effect that electrostatically charged objects attract dust. The toner, consisting of a fine, black thermoplastic powder, is sprinkled over the cylinder. Only the charged areas are covered by the black toner dust. Thus, a mirror image of the original is produced on the roller. The black particles are thereafter transferred electrostatically to paper. They are fixed there by heat exposure or pressure to make the image permanent [33].

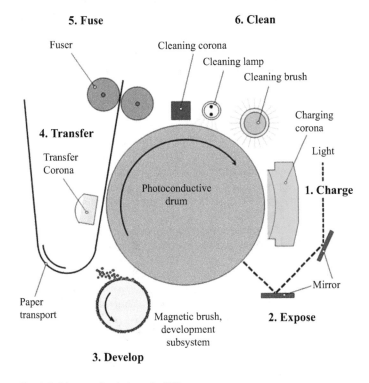

Fig. 9.2: Scheme of a photocopier [32].

The toner powder is transported using a magnetic brush development system. A rotating bar magnet is covered with a magnetizable ferrite powder (component 1, the so-called carrier) when the dual component process is used. The arrangement acts like a magnetic brush. The ferrite cover is dipped into a bath of toner powder (component 2) and transported from there via the cylinder surface, a specifically defined distance away. The powder is attracted only to the areas that are electrostatically charged.

Laser printers and many copiers use the one-component process. Very compact machines can be used for this process. The principle of the magnetic brush is similar to that of the dual component process but the toner in this case is magnetizable. Magnetization of the toner is achieved by adding magnetic Fe_3O_4 particles (30 to 50 wt%) into the dispersion of the thermoplastic resin. Various additives are additionally incorporated to control the electrostatic chargeability, the flow characteristics, and the color. The particle size of such toners is in the range 5 to 15 μm and that of the specific Fe_3O_4 pigments for magnetic toners is in the range 0.2 to 0.5 μm. High-quality toners contain iron oxide particles with a narrow particle size distribution.

Coercivity and saturation magnetization of a toner depend to a certain extent on the application medium used in the copying machine or printer. The values for the coercive field strength of toners are significantly lower than those of pure magnetic pigments. They are adjusted by the production process, particle shape, and particle size. The lowest coercivity is achieved with spherical particles. High-quality toners for laser printers contain such spherically shaped Fe_3O_4 pigments. Most copier toners, however, need a certain higher coercivity and are prepared using octahedral or cubical Fe_3O_4 particles [32].

9.1.3 Toxicology and occupational health

Based on the criterion of composition, most magnetic pigments can be regarded as harmless or, in some cases, as potentially hazardous. From the occupational health and safety perspective, operation with pigment dust requires special attention. The inhalation of the dust is most critical and must, therefore, be avoided by all available means.

Pure γ-Fe_2O_3 and Fe_3O_4 pigments are nontoxic unless they are contaminated with heavy metal compounds. The amount of crystalline silica in the industrially manufactured pigments lies below the detection limit for X-ray diffraction analysis.

Co-γ-Fe_2O_3 and Co-Fe_3O_4 pigments have a substantial amount of cobalt in their composition and must, therefore, be handled like other cobalt oxide powders. Soluble cobalt is problematic for the human body and can cause allergic reactions and other injuries. The low solubility of the pigments in water and organic media is an advantage, in this regard, but dissolved cobalt cannot be completely excluded. The pigments have to be handled, therefore, as water-contaminating. Acute and chronic health risks cannot be ruled out.

Magnetic iron pigments must be handled as flammable substances. Short-term exposure to an ignition source can already be sufficient to ignite fine iron powder. The risk of ignition is higher as the powder is finer. Generally, there is also the danger of a dust explosion. Protection measures must, therefore, be taken to avoid explosion when working with magnetic iron pigments.

CrO_2 is classified as a material with potential for acute and chronic toxicity. It is almost insoluble in water and acts as a weak oxidation agent. Above 70 °C, chromium dioxide can react with organic compounds. Appropriate measures must be taken when using CrO_2 pigments to protect the persons involved.

Barium hexaferrite is classified as a material with acute toxicity. It is almost insoluble in water. Swallowing and inhaling are regarded as critical. Suitable measures must be taken when working with $BaFe_{12}O_{19}$ pigments to protect the health of the persons involved.

9.2 Anticorrosive pigments

Anticorrosive pigments play an important role in the protection of metallic materials against corrosion. Damage arising from the corrosion of metals is enormous and can be up to 5% of the gross domestic product in leading industrial countries [34, 35]. The performance of a protective coating can be significantly improved by the addition of anticorrosive pigments. The corrosion protection of a coating particularly depends on the effective interaction of the anticorrosive pigment used and the binder system chosen. Therefore, the selection of an anticorrosive pigment must be tuned to the binder system used.

9.2.1 Fundamentals and general properties

Corrosion is a process in which a solid, especially a metal, is eaten away by chemical or electrochemical reactions resulting from exposure of the material to weathering, moisture, chemicals, or other agents in the environment in which it is placed. Corrosion can also be described as an electrochemically driven process of energy exchange. The assumption is that the metals, which were originally found in nature in the form of metal compounds in ores, are reconverted to their ores. This also implies that the energy that is released when a metal is reconverted to its ore is the same as for the original metal production. Iron and other corroding metals are not thermodynamically stable. They, therefore, tend to oxidize under the formation of compounds, for example, oxides, sulfides, silicates, in which they have a stable oxidation state. In summary, corrosion can be regarded as the natural tendency of metals to achieve their most stable form [36].

As a consequence, the more energy for the production of a metal that is necessary, the easier the energy is released in the form of corrosion. The corrosion tendency of a metal is consistent with its position in the electrochemical series. Nonprecious metals are highly reactive, corresponding to an easy release of energy. This means that these metals are most susceptible to corrosion.

Metals typically show heterogeneities in their composition. These can particularly be generated by surface contamination or structural inhomogeneities. The contact of such heterogeneous areas of a metal with an electrolyte, e.g., water, and oxygen can lead to corrosion. Different areas of the metal act as anode or cathode, and an active electrochemical cell is formed. Figure 9.3 shows the situation for the corrosion of iron. At the anode, Fe^{2+} ions are formed out of the iron, and electrons are released. The presence of water as an electrolyte enables the electrons to migrate to the cathode where they react with oxygen from the air and with water.

The main electrochemical reactions can be summarized as follows (for example, the corrosion of iron):

$$\text{Anodic reaction:} \qquad Fe \rightarrow Fe^{2+} + 2\,e^- \qquad\qquad (9.4)$$

$$\text{Cathodic reaction:} \quad {}^1/_2\,O_2 + H_2O + 2\,e^- \rightarrow 2\,OH^- \qquad (9.5)$$

$$\text{Summary:} \qquad Fe + {}^1/_2\,O_2 + H_2O \rightarrow Fe(OH)_2 \qquad (9.6)$$

$Fe(OH)_2$ is not stable under these conditions and immediately reacts further with oxygen to form $FeOOH$.

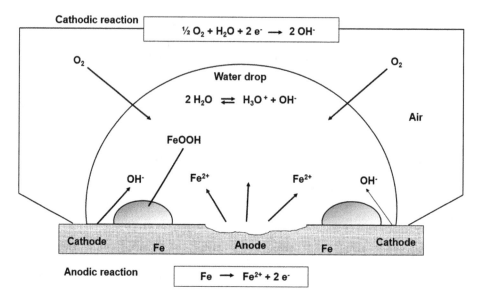

Cathodic reaction

$$\frac{1}{2} O_2 + H_2O + 2\,e^- \longrightarrow 2\,OH^-$$

O_2

Water drop

$$2\,H_2O \rightleftharpoons H_3O^+ + OH^-$$

O_2

Air

FeOOH

OH$^-$ Fe^{2+} Fe^{2+} OH$^-$

Cathode   **Fe**   **Anode**   **Fe**   **Cathode**

Anodic reaction

$$Fe \longrightarrow Fe^{2+} + 2\,e^-$$

Fig. 9.3: Processes during the corrosion of an iron surface.

$$2\,Fe(OH)_2 + {}^1\!/_2\,O_2 \rightarrow 2\,FeOOH + H_2O \tag{9.7}$$

$Fe(OH)_3$ can also be formed in addition to FeOOH in this very complex reaction course. In the further process, Fe_2O_3 is formed by dehydration of FeOOH and $Fe(OH)_3$. Complex iron oxides containing Fe(II) and Fe(III) and in some cases carbonates can be formed, too. The rust already formed (FeOOH) can further accelerate the corrosion process. The hydroxide ions formed at the cathode generate a locally high pH value. Therefore, hydrolysis reactions with components of the binder system are initiated. Finally, the coating film is separated from the metal substrate [37].

One key feature of the described electrochemical corrosion process is the presence of an electrolyte. This electrolyte is the result of rain, snow, and dew. The ions in the electrolyte lead to an ionogenic surface contamination of the metal. The main components of such an electrolyte are various dirt particles and water-soluble salts, e.g., metal chlorides and sulfates. Chlorides as well as sulfates are so-called corrosion stimulators. They have a significant influence on the course of the corrosion reactions [35, 36].

Anticorrosive pigments can affect the performance of protective coatings by:
- prevention of underfilm corrosion
- protection of the metal surface where the paint film is damaged due to mechanical forces
- prevention of undercutting in damaged areas
- improvement of the durability per unit of film thickness
- improvement of the durability of thin films [35]

There are three ways of action for anticorrosive pigments [35, 38]:
- chemical/electrochemical action (active pigments)
- physical action (barrier pigments)
- electrochemical/chemical action (sacrificial pigments)

Active pigments, also referred to as inhibitive pigments interact chemically or electrochemically with the metal surface to reduce the rate of corrosion, either directly or via intermediates. An example of the formation of intermediates is the chemical reaction of the anticorrosive pigment with the binder system.

Some active pigments are able to form a stabilization layer by interaction with the metal on the surface to be protected. This process of stabilization is called passivation. Pigments that prevent corrosion of the metal by the formation of such protective layers on the metal surface are regarded as being active in the anodic areas (anodic passivation). Pigments preventing corrosion due to their high oxidation potential are classified as active in the cathodic areas (cathodic passivation). To summarize, it can be noted that active pigments are able to inhibit one or both of the two electrochemical partial reactions that are responsible for the corrosion of metals. Active pigments can additionally act by the neutralization of corrosive substances such as acids, sulfates, and chlorides. This mechanism results in a constant pH value in the coating over long periods of time.

Barrier pigments act physically in the binder system by building up pigmentary barriers in the anticorrosive coating film. This means they reduce the permeability of the film for agents that promote corrosion. Most barrier pigments are chemically inert; in other words, they are inactive or passive. Pigments with a platelet-like or lamellar particle shape are predestined for the achievement of barrier effects. They can easily be orientated parallel to the surface during the application and build up a staggered arrangement of flat particles in the coating film. Oxygen, water, and any electrolytes have to surmount the particles of the barrier pigments by taking a less direct route through the film to the metal surface.

Sacrificial pigments can be regarded as active pigments with a specific electrochemical and chemical effect mechanism. They consist of metallic particles, which act by cathodic protection when applied on ferrous surfaces. Sacrificial pigments contain a metal with a lower electrochemical potential (lower position in the electrochemical series) than that of the metal to be protected. Under the conditions of corrosion, the sacrificial metallic pigment with its more reactive character compared with the metal to be protected, acts as anode in an electrochemical corrosion cell. The metal itself acts as cathode (cathodic protection). Metallic zinc is the only industrially relevant sacrificial pigment for anticorrosion purposes. It is supplied either as zinc dust or zinc flakes. Sacrificial pigments act not only electrochemically in an anticorrosion coating, but also chemically. The pigments react with components of the atmosphere under formation of insoluble zinc compounds. Such chemically inactive compounds contribute to the protection of the coating by filling cavities.

Finally, there are some pigments that are called film reinforcers, which could be counted among the anticorrosive pigments. The intended application of these pigments is the improvement of the overall integrity of the coating film. This might be achieved by the addition of iron oxide red pigments to the coating formulation. The assumption is that iron oxide pigments show excellent binder/pigment bonding properties, resulting in an enhancement of the inherent barrier effect of the film. Zinc oxide is another example of a film reinforcer pigment. In this case, active anticorrosive pigments are used in combination with ZnO. Zinc oxide pigments may enhance the protection effect of the coating by acting as an ultraviolet absorber. In addition, zinc oxide has the ability to react with corrosive agents, thus making them harmless [35].

Table 9.2 contains a summary of the main important anticorrosive pigments including their composition and color [39].

Tab. 9.2: Composition and color of selected anticorrosive pigments [39].

Classification	Pigment name	Composition	Color
Cyanamide-containing	Zinc cyanamide	$ZnCN_2$	White-beige
	Lead cyanamide	$PbCN_2$	Lemon yellow
Chromate-containing	Basic zinc chromate	$ZnCrO_4 \cdot 4\,Zn(OH)_2$	Yellow
	Zinc chromate	$ZnCrO_4$	Lemon yellow
	Basic zinc potassium chromate	$4\,ZnO \cdot K_2O \cdot 4\,CrO_3 \cdot 3\,H_2O$	Lemon yellow
	Zinc potassium chromate	$KZn_2(CrO_4)_2OH$	Yellow
	Strontium chromate	$SrCrO_4$	Yellow
Molybdate-containing	Calcium zinc molybdate	$(Ca,Zn)MoO_4$	White
Phosphate-containing	Zinc phosphate	$Zn_3(PO_4)_2 \cdot 4\,H_2O$	White-beige
	Chromium phosphate	$CrPO_4 \cdot 3\,H_2O$	Green
	Aluminum zinc phosphate	$AlPO_4 \cdot Zn_3(PO_4)_2 \cdot x\,H_2O$	White
	Zinc calcium phosphate molybdate	$(Zn,Ca)_3(PO_4,MoO_4)_2 \cdot x\,H_2O$	White
Borate-containing	Barium metaborate	$BaO \cdot B_2O_3 \cdot H_2O$	White
	Zinc borophosphate	$ZnO \cdot x\,B_2O_3 \cdot y\,P_2O_5 \cdot 2\,H_2O$	White
Metallic	Zinc dust	Zn	Gray
	Lead powder	Pb	Gray
Oxidic	Lead(II,IV) oxide	Pb_3O_4	Orange-red
	Calcium plumbate	Ca_2PbO_4	Beige
	Zinc ferrite	$ZnO \cdot Fe_2O_3$	Gray
	Zinc oxide	ZnO	White
Platelet-like	Micaceous iron oxide	Fe_2O_3	Gray
	Talc	$Mg_9Si_4O_{10}(OH)_2$	White
	Muscovite mica	$KAl_2Si_3AlO_{10}(OH,F)_2$	Gray-white
	Zinc	Zn	Gray

9.2.2 Production and special properties of anticorrosive pigments

The broad diversity of anticorrosive pigments requires different production processes. Precipitation and solid state reactions as well as metallurgical processes are used, depending on the intended composition, pigment properties, and usage.

9.2.2.1 Lead-containing and chromate pigments

Anticorrosive pigments containing lead, especially red lead (Pb_3O_4), have already been used for corrosion inhibition for a long time. They have been proven to be outstanding anticorrosive substances. However, the increasing awareness in the toxicological properties of chemicals, occupational health, and environmental sustainability has led to a significant restriction in regard to the application of lead pigments.

As the most popular lead-based anticorrosive pigment, Pb_3O_4 can be produced industrially by oxidation of lead(II) oxide (PbO) at about 480 °C in an air stream [35].

$$3\,PbO + 1/2\,O_2 \rightarrow Pb_3O_4 \qquad (9.8)$$

Red lead is an oxidizing agent and acts as a cathodic passivator (reduction of Pb(IV) to Pb(II)). The main use of Pb_3O_4 before its replacement by lead-free substances was in linseed oil-based paints. Red lead is able to form soaps with the fatty acids in linseed oil. These lead soaps can contribute to the inhibition of rust formation. In addition, they can improve the mechanical properties of coatings, e.g., mechanical strength, water resistance, and adhesion to the metal substrate to be protected.

An anticorrosive pigment that contains lead and chromate together is the so-called basic lead silicochromate. It is characterized by a core-shell structure and was developed to replace Pb_3O_4. The core of the pigment consists of SiO_2. The anticorrosive, active part of the pigment is $PbCrO_4$ (shell), which is precipitated onto the inert SiO_2 particles. Basic lead silicochromate is losing its importance more and more due to the undesired content of lead and chromate [35].

The following lead-containing compositions have been used to inhibit corrosion in very specific and limited applications: lead(II) oxide, lead carbonate, lead cyanamide (mirror back coatings), basic lead silicate (electro-deposition primers), dibasic lead phosphite, tribasic lead phosphate silicate, basic lead silicate sulfate, and calcium plumbate [35, 36].

Chromate pigments have long been used for corrosion-inhibiting purposes. The anticorrosive action of these pigments is based on a certain amount of water soluble chromate. Chromate-containing pigments could successfully be used because most paint films are permeable to water. This permeability is the precondition for the partial dissolution of chromate pigments in the anticorrosive coating. A sufficient amount of chromate ions can migrate to the metal surface to initiate and sustain the formation of passive layers [36]. The atmospheric corrosion of iron and the inhibition effect of chromate pigments are schematically shown in Figure 9.4. The formation of chemically inactive Fe(OH)$_2$ · 2 CrOOH at the iron surface, coupled with the reduction of Cr(VI) to Cr(III), is the basis for the anticorrosive effect of chromate-containing pigments.

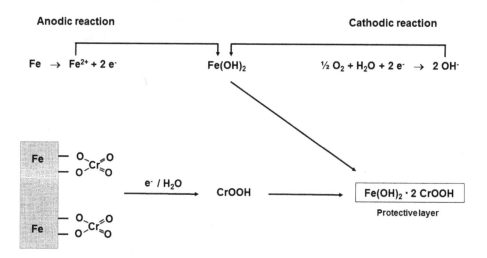

Fig. 9.4: Atmospheric corrosion of an iron surface and inhibitive effect of chromate pigments under formation of inert Fe(OH)$_2$ · 2 CrOOH.

The active surface of chromate pigments is important for their anticorrosive action. This means that particle shape and size, particle size distribution, and dispersibility of the pigments have to be adjusted in a suitable manner to achieve optimal corrosion inhibition.

Basic zinc chromate pigments (zinc tetrahydroxy chromate, ZnCrO$_4$ · 4 Zn(OH)$_2$) are characterized by a neutralizing and pH-stabilizing effect coming from the Zn(OH)$_2$ as a component of the pigment composition. This effect additionally contributes to the corrosion inhibition originating from the chromate. The pigments are produced by precipitation of the components as chromate and hydroxide followed by filtration, drying, and grinding. They are also obtained by the reaction of aqueous slurry of zinc oxide with potassium dichromate and sulfuric acid. Zinc tetrahydroxy chromates have mainly been used for the formulation and production of wash and shop primers, based on polyvinyl butyrale resins [35].

Zinc potassium chromate (KZn$_2$(CrO$_4$)$_2$OH), also known as zinc yellow, is produced by the reaction of aqueous zinc oxide or zinc hydroxide slurry with a solution of potassium chromate. Alternatively, the pigment can be synthesized by the addition of aqueous zinc salt solution to potassium chromate solution or vice versa. Filtration of the obtained precipitate, washing, drying, and grinding are the necessary subsequent steps.

Zinc chromate (ZnCrO$_4$) and strontium chromate (SrCrO$_4$) pigments are manufactured by precipitation from solutions of sodium dichromate and zinc chloride and strontium chloride, respectively, followed by filtration, washing, drying, and grinding. Strontium chromate has especially gained significance for primers used for inhibiting the corrosion of aluminum. This pigment is still in use despite its chromate content, because it is considered to be the most effective anticorrosive for application in aircraft and coil coating primers [35].

Barium chromate and calcium chromate were also used in the past as anticorrosive pigments, but only to a limited extent. The use of chromate-containing anticorrosive pigments has been reduced significantly over the years, due to the confirmed or suspected carcinogenic effects of chromates.

9.2.2.2 Phosphate pigments

The intention of the development and manufacture of phosphate pigments was to replace lead compounds and chromates in the field of corrosion protection. In particular, zinc-containing orthophosphates have gained great significance, and today belong to the most commonly used anticorrosive pigments. The main advantage of zinc orthophosphates is their versatile formulation, based on an extremely low solubility and reactivity compared to chromates and other anticorrosive pigments. Zinc phosphates can be used in a wide variety of corrosion-inhibiting binder systems.

Zinc orthophosphate pigments as white anticorrosive substances consist either of predominantly zinc phosphate dihydrate ($Zn_3(PO_4)_2 \cdot 2\,H_2O$), of a mixture of zinc phosphate dihydrate and zinc phosphate tetrahydrate ($Zn_3(PO_4)_2 \cdot 4\,H_2O$), or of zinc phosphate tetrahydrate. The specific corrosion-inhibiting properties of the various phosphate types in different application systems may differ.

Zinc orthophosphate pigments are produced on an industrial scale by the dissolution of zinc oxide and precipitation with orthophosphoric acid.

$$3\,ZnO + 2\,H_3PO_4 \rightarrow Zn_3(PO_4)_2 + 3\,H_2O \tag{9.9}$$

Filtration, washing, drying, and grinding are subsequent production steps [40–43].

Zinc orthophosphate pigments in anticorrosive systems perform through their ability to form adhesion and inhibitor complexes on the surface of the metal substrate. In addition, zinc phosphates have an appropriate electrochemical effectiveness, preferably in anodic areas, whereby small amounts of zinc phosphate hydrolyze in the presence of moisture. Zinc hydroxide and secondary phosphate ions may be generated under these conditions, which are able to form protective layers on the metal surface in anodic areas (Figure 9.5) [44, 45].

In addition, a reaction of zinc phosphates with inorganic ions or with carboxylic groups of the binder system under formation of basic complexes can also take place. Such complexes may react with metal ions to form so-called adhesion, cross-linking, and inhibitor complexes [44–46].

Hydrolytic reactions with the participation of zinc phosphate and the formation of related hydrolysis products are a prerequisite for the activation of the anticorrosive

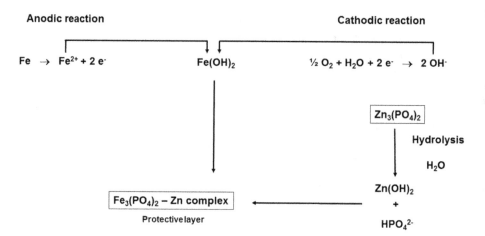

Anodic reaction

Cathodic reaction

$Fe \rightarrow Fe^{2+} + 2\,e^{-}$ $Fe(OH)_2$ $\frac{1}{2}\,O_2 + H_2O + 2\,e^{-} \rightarrow 2\,OH^{-}$

$Zn_3(PO_4)_2$

Hydrolysis

H_2O

$Zn(OH)_2$

$Fe_3(PO_4)_2$ – Zn complex   +

Protective layer HPO_4^{2-}

Fig. 9.5: Atmospheric corrosion of an iron surface and inhibitive effect of zinc phosphate pigments.

behavior of zinc orthophosphate pigments. These reactions depend on the permeability of the protective coating, which is influenced by the type of binder used and particularly by the pigment volume concentration. Thus, it can be concluded that the entire formulation (binder components, pigments, and fillers) has an important influence on the corrosion protection behavior of protective coatings containing zinc orthophosphate pigments [46].

Practical experience and theoretical considerations show that zinc phosphate is suitable for active anticorrosive pigments, but the outstanding protective effect of lead and chromate pigments can only be achieved in certain systems [47]. This was the reason behind the development of modified orthophosphate pigments with improved properties.

Modified orthophosphate pigments, meanwhile, include a broad variety of different compositions:

– zinc aluminum phosphates (with aluminum phosphate)
– basic zinc phosphates (partly with zinc molybdate, calcium molybdate, basic zinc borate, iron phosphate, calcium phosphate, potassium phosphate, barium phosphate, aluminum phosphate, zinc molybdate, partly treated with organic or inorganic compounds)
– zinc-free phosphates (substitution of zinc by calcium and/or magnesium)
– zinc phosphate silicates (partly with calcium silicate, strontium phosphate, barium phosphate, calcium carbonate, partly treated with organic compounds)
– zinc-free phosphate silicates (substitution of zinc by calcium, strontium and barium)

Phosphate borate combinations were developed to accelerate the readiness of the phosphate for hydrolysis and with it to activate the pigments for corrosion protection

[46]. The improved anticorrosive activity of zinc phosphate molybdate pigments is attributed to the inhibitive effect of water soluble molybdate ions [48].

Aluminum zinc phosphate pigments are produced by the coprecipitation of aluminum dihydrogen phosphate $(Al(H_2PO_4)_3)$ with zinc hydroxide, followed by a moderate thermal treatment. Pigments obtained in this way are characterized by phosphate content higher than that of standard zinc phosphate. The improved performance properties can be attributed to higher phosphate content. Modified phosphate pigments treated with organic compounds have the advantage of a closer bonding between pigment and binder and also between coating and metal substrate [35].

Phosphate silicate pigments belong to the so-called mixed phase or core pigments. Their effect in an anticorrosive formulation is characterized by fixing the active components onto the surface of calcium silicate (wollastonite) particles, which act as the core, and adjusting the pH value near neutral. Such synthesized pigments are universally applicable in corrosion-inhibiting systems [35].

Polyphosphates can also be used as anticorrosive pigments. The so-called modified polyphosphate pigments, especially, have found their place in the class of corrosion-inhibiting substances.

Polyphosphates are producible by the condensation of acidic orthophosphates, e.g., of $Al(H_2PO_4)_3$, at higher temperatures [35]:

$$Al(H_2PO_4)_3 \rightarrow AlH_2P_3O_{10} \cdot 2\,H_2O \tag{9.10}$$

Modified polyphosphates with relevance for anticorrosive pigments are based on acidic aluminum tripolyphosphate, in which aluminum is partially replaced by zinc, strontium, calcium, or magnesium. Phosphate can be substituted by silicate and chromate in some variations. There are also pigments treated with organic or inorganic compounds:

- zinc aluminum polyphosphates (partly with calcium silicate, strontium chromate, silicon dioxide, partly treated with organic compounds)
- zinc-free aluminum polyphosphates (partly with strontium phosphate, calcium silicate, magnesium phosphate, barium phosphate, partly treated with organic compounds)

There is also the intention to combine advantageous properties of different anticorrosive compounds to achieve synergistic effects within one product. Special synergistic effects can be generated by using selected phosphate-based pigments in combination with an organic corrosion inhibitor. Such combinations seem to have potential for the improvement of corrosion-inhibiting effects even in the initial phase of the exposure of a metal surface to atmospheric influences. As a result, improved long-term protection in specific anticorrosive applications may be possible [35].

An advantage of most modified orthophosphate and polyphosphate pigments is their compatibility with a wide range of binder systems. This is an additional reason why the development work for universal corrosion protection applications by using phosphate-based pigments is of further interest.

9.2.2.3 Other phosphorous-containing pigments

The continuous interest in improved corrosion protection has led to the development of further phosphorous-containing anticorrosive pigments. Zinc hydroxyphosphite (2 Zn $(OH)_2 \cdot ZnHPO_3 \cdot x$ ZnO, with x from 0 to 17) is one result of efforts in this respect. This pigment can be synthesized by the reaction of zinc oxide slurry with phosphorous acid in the presence of a zinc hydroxyphosphite complex promoter [49]. The effect of this white anticorrosive pigment is attributed to the ability of phosphite ions to inhibit anodic corrosion reactions by the formation of iron phosphites and phosphates. The corrosion protection effect is enhanced by the formation of zinc-containing soaps in oleoresinous binders. Such soaps are known for their corrosion-inhibiting potential [48].

Iron phosphide (Fe_2P) is another phosphorous-containing anticorrosive pigment. Industrially produced Fe_2P pigments contain traces of FeP and SiO_2. They are characterized by a metallic gray color and act as a conductivity enhancer in coating formulations. They were developed as a partial substitute for zinc dust in zinc-rich organic and inorganic anticorrosive coatings. The replacement of zinc may be up to 50% in a coating composition. The use of Fe_2P pigments can lead to an improvement in the weldability of the system [35].

9.2.2.4 Borate pigments

Borate based pigments are also used for corrosion protection. The alkaline character of many borates plays a major role in this application [36]. Alkaline borate pigments are favorably able to maintain a high pH value in the coating. They may also form soaps with acidic products derived from the coating film. Mainly calcium or barium soaps are formed depending on the pigment composition. Borates may also act as anodic passivators forming a protective layer on the metal substrate. They are effective particularly in the initial phase of corrosion protection [35].

Calcium borate silicate pigments are regarded as mixed phase or core pigments, comparable with phosphate silicates. The core consists of an alkaline earth silicate, e.g., calcium silicate. Industrially produced pigment grades differ primarily in the B_2O_3 content. They are mainly used in protective coatings based on solventborne alkyd resins [35].

Modified barium metaborates have also found application in anticorrosive coating systems. They are modified by silicon, which is used to reduce the solubility of the barium metaborate. Unmodified barium metaborates are soluble in water and other solvents and would not be suitable as pigments [48]. Such pigments can be produced by the precipitation of a mixture of barium sulfide (BaS) and sodium tetraborate ($Na_2B_4O_7 \cdot 5\ H_2O$) in the presence of sodium silicate (Na_2SiO_3) [50]. Modified barium metaborate pigments have a white appearance. They are mainly used in solventborne coatings [48].

9.2.2.5 Molybdate pigments

Molybdate-based anticorrosive pigments are characterized by the ability of their molybdate ions to pass into solution and to migrate to the metal surface. Thus, a protective layer on the metal is formed, which passivates the surface and prevents corrosion by this means. Molybdate pigments are also used in mixtures with phosphate pigments or inorganic fillers such as calcium carbonate [35]. Such mixtures are difficult to disperse in some application systems. To overcome this problem, micronized versions of such mixed pigments are available, which are easy to disperse.

9.2.2.6 Ion exchange pigments

Ion exchange anticorrosive pigments are based on slightly porous calcium ion-exchanged silica gel. The silica particles are characterized by a relatively high surface area. Ion exchange pigments have a different mode of action compared to other types of anticorrosive pigments [51, 52].

It is assumed that an ion exchange takes place when corrosion-causing ions enter the coating film and come into contact with the silica surface. The attacking ions are locked onto the silica, and calcium ions from the silica migrate to the metal surface. This form of ion exchange occurs continuously over a longer period. The released calcium ions form a protective layer with barrier properties at the interface between the metal and the coating film. This barrier layer prevents corrosion-relevant ions from making contact with the metal surface [51]. An application of ion exchange pigments is basically possible in coil coating primers, mostly in combination with other anticorrosive pigments [35].

9.2.2.7 Titanium dioxide-based pigments

Titanium dioxide-based pigments can also be used for corrosion protection purposes. In this case, the excellent hiding power of the white TiO_2 can be used together with suitable anticorrosive properties of the pigment. Titanium dioxide acts as a carrier material here. The surface of the TiO_2 particles is treated with phosphate-containing anticorrosive substances, e.g., with manganese aluminum phosphates. A precipitation process comparable to the procedure for the reduction of the photoactivity of TiO_2 pigments is used. An additional organic treatment provides good dispersing properties in both polar inorganic and organic media. The modified titanium dioxide-based pigments combine appropriate corrosion-inhibiting behavior and favorable application technical properties. They can be used in waterborne and solventborne anticorrosive coating systems [53, 54].

9.2.2.8 Inorganic organic hybrid pigments

The joint application of inorganic anticorrosive pigments and organic corrosion inhibitors can enhance the protection of a metal surface against corrosion. This has already been described for phosphate-containing pigments (Section 9.2.2.2). Other pigments following this concept are the so-called synergistic hybrid pigment-grade corrosion inhibitors. Examples of this include anticorrosive pigment compositions based on a stable unitary hybrid, which contains organic and inorganic solid phases. The inorganic phase consists of a phosphate, polyphosphate, phosphite, molybdate, or silicate of zinc, magnesium, calcium, strontium, iron, cerium, titanium, or zirconium. The organic phase may contain zinc or alkyl-ammonium salts of organic mercapto and thio compounds or derivatives thereof [55].

Other examples of inorganic organic hybrid pigments are anticorrosive compositions based on oxyaminophosphate salts of magnesium or of magnesium and calcium. It is supposed that the effect of these pigments in a protective coating is based on the generation of a magnesium oxide layer on the metal surface. In the case of a steel surface, such a layer acts most probably as a passivation layer. In the case of an aluminum surface, a barrier layer effect is assumed [35, 56, 57].

9.2.2.9 Zinc cyanamide pigments

Zinc cyanamide ($ZnCN_2$) can also be used for corrosion protection purposes. It was developed as a replacement for lead cyanamide in mirror coatings. The mode of action of cyanamides in anticorrosive coatings is explained by the formation of passivation layers [35].

Zinc cyanamide pigments are industrially produced by the reaction of calcium cyanamide with dissolved zinc salts or slurry of zinc oxide in water.

$$CaCN_2 + ZnCl_2 \rightarrow ZnCN_2 + CaCl_2 \tag{9.11}$$

The anticorrosive effectiveness of $ZnCN_2$ for application in mirror coatings is not as advantageous as that of $PbCN_2$. The use of zinc cyanamide anticorrosive pigments is, therefore, very limited.

9.2.2.10 Micaceous iron oxide pigments

Micaceous iron oxide (MIO, iron mica) has found a substantial usage as an anticorrosive barrier pigment for intermediate coats and topcoats on metals, preferably on steel surfaces. MIO is a special variety of naturally occurring α-Fe_2O_3, which crystallizes either in granular or more often lamellar crystals. The terms micaceous and mica are used because of the visible layer structure and the often-occurring platelet-like particle shape achieved by grinding this special form of hematite. This structure and the particle shape are outwardly similar to natural muscovite mica and its ground flat particles. Thus, MIO does not contain mica. It differs in particle size and shape from the well-known red iron oxide and the transparent iron oxide red pigments (Table 7.1).

MIO is characterized by a dark gray color shade with a metallic sheen. Thin MIO platelets appear red and translucent under an optical microscope with transmitted light. Thicker and granular particles, on the other hand, show a black shade. The mode of action of MIO in anticorrosive coatings is that of a physical barrier formed by overlapping of platelet-shaped particles. This special particle alignment results in the lengthening of the diffusion paths for moisture, gases, and corrosion stimulating ions through the coating film as shown in Figure 9.6 [58].

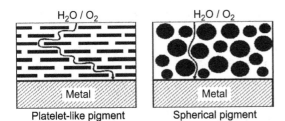

Fig. 9.6: Demonstration of the barrier effect: comparison of platelet-like and spherical pigment particles [57].

Other properties besides particle shape also have an influence on the anticorrosive effects of MIO pigments. A very high Fe_2O_3-content and a very low amount of water-soluble salts are also obviously important prerequisites for an effective protection behavior of MIO pigments [35].

MIO pigments are not only appreciated because of their anticorrosive properties but also due to their very good thermal and chemical stability, good hardness (abrasion resistance), and high electrical resistance. In addition, the pigments are recommended for the use in topcoats as protection means against ultraviolet degradation [48].

Synthetic MIO pigments produced by hydrothermal processes are also available. Such pigments do not find the same interest as natural pigments, primarily for economic reasons. Most of the MIO pigments on the market consist of platelet shaped particles to a large extent.

Aluminum and steel flakes, as well as phlogopite and muscovite platelets, are other lamellar materials used to a certain extent for corrosion protection based on the barrier effect.

9.2.2.11 Metallic zinc pigments

Metallic zinc in the form of dust is the only sacrificial anticorrosive pigment of significance. Metallic zinc pigments are offered as free-flowing powders with a bluish-gray color shade. The pigments consist of spherical particles. Zinc dust in former times was a by-product of zinc metal production. Since then, zinc pigments for the application in anticorrosive coatings have been specifically produced in various qualities for different binder systems.

The raw material for the manufacture of zinc pigments is purified zinc metal from different sources. This metal is evaporated, and the zinc vapor formed is condensed under controlled conditions under the exclusion of air. Sieving processes follow in order to obtain the required particle size fractions. Typical particle sizes of zinc dust are in the range of 2 to 8 μm.

The anticorrosive effectiveness of metallic zinc particles in protective coatings is understood as a combination of sealing effects and electrochemical reactions caused by the zinc. Zinc is highly reactive, and with atmospheric components, e.g., with water, oxygen, and carbon dioxide, it forms water-insoluble zinc salts, mainly basic zinc carbonates. Voids existent in the coating film are filled with the reaction products, which leads to a decrease of permeability and correspondingly, to an enhancement of the barrier properties [35, 59]. The high effectiveness of zinc is derived from its position in the electrochemical series of metals. In contact with a ferrous metal under the conditions of corrosion, zinc acts as sacrificial anode. The zinc anode protects the ferrous metal (cathode) and is thereby converted to Zn^{2+} ions. This protection mechanism can be efficient as long as zinc metal is available and able to go into solution.

Primers used for anticorrosive purposes typically contain zinc contents of more than 80% in the dry paint film. Zinc-containing coatings are used in large quantities for the protection of structural steel, e.g., for buildings, technical facilities, underwater steel constructions, and ship hulls. Zinc pigments can be used in organic and inorganic binder systems without problems [35, 59].

Flaky zinc particles are also used in coatings for corrosion protection. Anticorrosive pigments consisting of zinc flakes are mainly used in so-called microlayer corrosion protection systems. An example of the use of such protection systems is in threaded parts [35].

9.2.2.12 Graphene-based composite materials

Graphene, an allotrope of carbon, is a two-dimensional material that has potential as an anticorrosion filler because of its sheet-like geometry and high aspect ratio, which are favorable for the physical barrier effect. Furthermore, its hydrophobicity and exceptional impermeability to all gases and liquids are advantageous properties for corrosion resistance [60–62]. Consequently, graphene-based anticorrosive coatings have attracted attention in recent times. It was possible to show, for example, that graphene prepared by chemical vapor deposition (CVD) can effectively inhibit the oxidation of copper and nickel-copper alloys [63–66]. Thick organic coatings with added graphene fillers seem to be particularly suitable for long-term protection. The addition of titanium dioxide/graphene mixtures into an epoxy resin leads to encouraging protection results. The graphene flakes improve the corrosion resistance by blocking the micropore corrosion channels formed in epoxy resins due to solvent evaporation [67]. About 0.5 wt% of graphene filler incorporated into a polyaniline matrix significantly raises the corrosion protection efficiency [68]. It can be concluded that sheet-like graphene fillers, with a much higher aspect ratio than that of clays, can provide better physical barrier effects by forcing the corrosive agents to take a more torturous path through the coating. The optimized graphene loading varies from 0.1 to 8 wt%, depending on the choice of the matrix material, the dispersion uniformity of the filler as well as on the preparation and modification of the graphene [67, 68].

9.2.3 Specification of corrosion protection standards

Corrosion protection by protective coatings is specified in various national and international standards. DIN EN ISO 12,944 "Paints and varnishes – Corrosion protection of steel structures by protective paint systems" – describes the protection of steel by paint and varnish systems [69, 70]. The norm includes chapters on the classification of environments, design considerations, types of surface and surface preparations, protective paint systems, laboratory performance test methods, execution and supervision of paintwork as well as the development of specifications for new work and maintenance.

The norm contains information about relevant types of primers to be used for the different corrosion-inhibiting systems. Binder systems for primers according to DIN EN ISO 12,944 are alkyds, chlorinated rubbers, acrylics, polyvinyl chlorides, epoxies, ethyl silicates, and polyurethanes. Information on quality evaluation of a protective paint system by means of laboratory tests enables the user of anticorrosive compositions to select the most suitable test methods [69, 70].

9.2.4 Toxicology and occupational health

Changed requirements for anticorrosive pigments regarding toxicological and environmental aspects have led to a significant reduction in the use of lead-containing and chromate pigments for corrosion-inhibiting coatings.

In general, precautions must be respected and workplace concentration limits must be observed when handling a broad variety of anticorrosive pigments. Specific information on the toxicology of lead-containing and chromate pigments can be found in Section 4.5.4.

Lead-containing anticorrosive pigments like Pb_3O_4, zinc-containing pigments such as zinc chromate, zinc potassium chromate, and zinc phosphate as well as other chromate pigments like strontium chromate are defined as dangerous for the environment. They are classified as very toxic to aquatic organisms and may cause long-term adverse effects in aquatic environments. The pigments and the containers where they are kept must be disposed of as hazardous waste. Release into the environment must be avoided. Wastes containing these anticorrosive pigments cannot be recycled and must be brought to a special waste disposal site under proper control.

In addition, the mentioned anticorrosive pigments must be appropriately labeled. Special restrictions defined for carcinogenic and teratogenic substances and their corresponding preparations must be considered. Anticorrosive pigments containing lead or chromates are not permitted to be used by private consumers and must be labeled with the phrase "Only for industrial purposes" [35].

9.3 Conclusions

Magnetic and anticorrosive pigments are the two main representatives of inorganic functional pigments. They are characterized by their special magnetic and corrosion-inhibiting properties, while color and appearance play only a secondary role for these pigments. Further inorganic functional pigments include electrically conductive, IR-reflective, laser-marking, and UV-absorbing pigments.

Magnetic pigments are used in magnetic information storage systems, e.g., in audio cassettes, videotapes, floppy discs, hard discs, and computer tapes. They are also applied as toner pigments in photocopiers and laser printers. The compounds used for magnetic pigments are either ferrimagnetic or ferromagnetic. The pigment particles are dispersed for magnetic applications in appropriate binder systems and then applied to the storage system. Technically important magnetic pigments are γ-Fe_2O_3, Fe_3O_4, Co-containing iron oxides (Co-γ-Fe_2O_3, Co-Fe_3O_4), metallic iron, CrO_2, and $BaFe_{12}O_{19}$. The manufacturing methods for the pigments are diverse and oriented towards their specific compositions and physical properties.

Anticorrosive pigments are very important components in corrosion-inhibiting coatings used for the protection of metallic surfaces. The effectiveness of an anticorrosive system depends strongly on the interaction of the anticorrosive pigment with the binder. The selection of a specific anticorrosive pigment must, therefore, be tuned to the used binder system.

There are three types of action for anticorrosive pigments: chemical/electrochemical action (active pigments), physical action (barrier pigments), and electrochemical/chemical action (sacrificial pigments). The objective of the pigmentation in all cases is to prevent the contact of water, oxygen, and other corrosion-supporting components (ions from salts, acids, and hydroxides) with the metal surface to be protected. A variety of different compounds is used as the basis for anticorrosive pigments, e.g., lead-containing compounds, chromates, phosphates, borates, molybdates, titanium dioxide, zinc cyanamide, micaceous iron oxide, metallic zinc, and inorganic organic hybrid materials.

The production of anticorrosive pigments takes place using different routes, depending on the required pigment specification. Surface treatments may be used in some cases to improve the quality of the pigments based on the anticorrosive effectiveness and the compatibility with the application system.

Further efforts are necessary to develop an appropriate replacement for the lead-, zinc-, and chromate-containing anticorrosive pigments that are used less and less because of toxicological and environmental effects.

Study questions

9.1 What are the characteristics of functional inorganic pigments?
9.2 What are magnetic pigments, and where are they applied?
9.3 What are the main representatives of magnetic pigments?
9.4 Which types of magnetism exist, and which of these are relevant for magnetic pigments?
9.5 What are the essential characteristics of magnetic pigments?
9.6 What are anticorrosive pigments, and what does corrosion protection depend on?
9.7 Which are the main representatives of anticorrosive pigments?
9.8 How can the process of corrosion be described?
9.9 How can the main electrochemical reactions for the corrosion of a metal be summarized?
9.10 What are the three types of action for anticorrosive pigments in an application system?

Bibliography

[1] Pfaff G, in Winnacker-Küchler: Chemische Technik, vol. 7, Industrieprodukte, 5th edn. Dittmeyer R, Keim W, Kreysa G, Oberholz A, eds. Wiley-VCH, Weinheim, 2004, 375.
[2] Horiishi N, Pitzer U, in Industrial Inorganic Pigments, 3rd edn. Buxbaum G, Pfaff G, eds. Wiley-VCH Verlag, Weinheim, 2005, 195.
[3] Patent US 2,694,656 (Armour Research Foundation), 1947.
[4] Patent DE 10 61 760 (Bayer AG), 1957.
[5] Patent GB 765464 (EMI), 1953.
[6] Patent DE 12 04 644 (BASF), 1962.
[7] Patent US 4,202,871 (Sakai Chemical Industries), 1980.
[8] Corradi AR, Andress S, French J, Bottoni G, Candolfo D, Cecchetti A, Masoli F. IEEE Trans. Magn. 1984,20,33.
[9] Patent DE 12 66 997 (Bayer AG), 1959.
[10] Imaoka Y, Umeki S, Kubota Y, Tokuoka Y. IEEE Trans. Magn. 1978,14,649.
[11] Patent US 3,573,980 (3M), 1968.
[12] Patent DE 22 35 383 (Hitachi Maxell), 1972.
[13] Bate GJ. Appl. Phys. 1981,52,2447.
[14] Kishimoto M, Kitahata S, Amemiya M. IEEE Trans. Magn. 1986,22,732.
[15] Patent US 2,923,683 (DuPont), 1957.
[16] Patent DE 14 67 328 (Bayer AG), 1963.
[17] Patent US 3,034,988 (DuPont), 1958.
[18] Buxbaum G, Schwab E, in Industrial Inorganic Pigments, 3rd edn. Buxbaum G, Pfaff G, eds. Wiley-VCH Verlag, Weinheim, 2005, 199.
[19] Essig M, Müller MW, Schwab E. IEEE Trans. Magn. 1990,26,69.
[20] Horiishi N, in Industrial Inorganic Pigments, 3rd edn. Buxbaum G, Pfaff G eds. Wiley-VCH Verlag, Weinheim, 2005, 201.
[21] Hibst H. Angew. Chem. 1982,94,263.
[22] Stäblein H, in Ferromagnetic Materials, vol. 3. Wohlfarth EP ed. North Holland Publ, Amsterdam, Oxford, New York, Tokyo, 1982 441
[23] Patent EP 0 116 449 (Toda), 1984.
[24] Patent EP 0 164 251 (Toda), 1984.
[25] Patent EP 0 290 263 (Matsushita), 1987.

[26]   Patent DE 30 41 960 (Toshiba), 1979.

[27]   Patent DE 37 02 036 (BASF), 1987.

[28]   Fayling RE. IEEE Trans. Magn. 1979,15,1567.

[29]   Veitch RJ. IEEE Trans. Magn. 1990,26,1876.

[30]   Okazaki Y, Noda M, Hara K, Ogisu K. IEEE Trans. Magn. 1989,24,4057.

[31]   Speliotis DE. IEEE Trans. Magn. 1990,26,124.

[32]   Kathrein H, in Industrial Inorganic Pigments, 3rd edn. Buxbaum G, Pfaff G, eds. Wiley-VCH Verlag, Weinheim, 2005, 204.

[33]   Starkweather GK, in Printing Technologies for Images, Gray Scale, and Color, Proceedings of SPIE, vol. 1458. Dove D, Abe T, Heinzl J, eds. SPIE Digital Library, Bellingham 1991 120

[34]   Bielemann J. Lackadditive. Wiley-VCH Verlag, Weinheim, 1998, 309.

[35]   Krieg S, in Industrial Inorganic Pigments, 3rd edn. Buxbaum G, Pfaff G, eds. Wiley-VCH Verlag, Weinheim, 2005, 207.

[36]   Hare CH. Protective Coatings, Fundamentals of Chemistry and Composition. Technology Publishing Company, Pittsburgh, 1994, 331.

[37]   Funke W. Farbe + Lack. 1983,89,86.

[38]   Wienand H, Ostertag W. Farbe + Lack. 1982,88,183.

[39]   Pfaff G, in Winnacker-Küchler: Chemische Technik, vol. 7, Industrieprodukte, 5th edn. Dittmeyer R, Keim W, Kreysa G, Oberholz A, eds. Wiley-VCH, Weinheim, 2004, 378.

[40]   Meyer G. Farbe + Lack. 1962,68,315.

[41]   Meyer G. DEFAZET – Dtsch. Farben Z. 1966,20,8.

[42]   Meyer G. Farbe + Lack. 1967,73,529.

[43]   Ruf J. Werkst. Korros. 1969,20,861.

[44]   Brock T, Groeteklaes M, Mischke P. Lehrbuch der Lacktechnologie. Vincentz Verlag, Hannover, 1998, 155.

[45]   Ruf J. Korrosion – Schutz Durch Lacke + Pigmente. Verlag W. A. Colomb, Stuttgart, Berlin, 1972, 120.

[46]   Ruf J. Organischer Metallschutz. Vincentz Verlag. Hannover, 1993, 269.

[47]   Schik JP. Korrosionsschutz mit wässrigen Lacksystemen. Expert Verlag, Renningen-Malmsheim, 1997, 98.

[48]   Smith A. Inorganic Primer Pigments. Federation Series on Coating Technology, Philadelphia, 1988, 7.

[49]   Patent US 4,386,059 (NL Industries), 1983.

[50]   Patent US 3,060,049 (Buckman Labor Inc.), 1962.

[51]   Bravo CA, ABRAFATI Congress Proceed., Sao Paulo, 1995, 770.

[52]   Deya C, Blustein G, Del Amo B, Romagnoli R. Prog. Org. Coat. 2010,69,1.

[53]   Rentschler T, Winkler J. Farbe + Lack. 1999,105,101.

[54]   Küppers H-J, in Kittel – Lehrbuch der Lacke und Beschichtungen, vol. 5, 2nd edn. Spille J ed. S. Hirzel Verlag, Stuttgart, Leipzig, 2003, 188.

[55]   Patent US 6,139,610 (Wayne Pigment Corp.), 2000.

[56]   Du YJ, Damron M, Tang G, Zheng H, Chu C-J, Osborne JH. Progr. Org. Coat. 2001,41,226.

[57]   Sanchez C, Julian B, Belleville B, Popall M. J. Mater. Chem. 2005,15,3559.

[58]   Pokorny G. Eur. Coat. Conf. Proceed. Berlin, 2000, 135.

[59]   Jahn B, in Kittel – Lehrbuch der Lacke und Beschichtungen, vol. 5, 2nd edn. Spille J, ed. S. Hirzel Verlag, Stuttgart, Leipzig 2003 175

[60]   Bunch JS, Verbridge SS, Alden JS, Van der Zande AM, Parpia JM, Craighead HG, McEuen PL. Nano Lett. 2008,8,2458.

[61]   Berry V. Carbon. 2013,62,1.

[62]   Miao M, Nardelli MB, Wang Q, Liu Y. Phys. Chem. Chem. Phys. 2013,15,16132.

[63]   Chen S, Brown L, Levendorf M, Cai W, Ju SY, Edgeworth J. ACS Nano. 2011,5,1321.

[64]   Cho J, Gao L, Tian J, Cao H, Wu W, Yu Q. ACS Nano. 2011,5,3607.

[65]   Prasai D, Tuberquia JC, Harl RR, Jennings GK. ACS Nano. 2012,6,1102.

[66]   Krishnamurthy A, Gadhamshetty V, Mukherjee R, Chen Z. Carbon. 2013,56,45.

[67]   Yu Z, Di H, Ma Y, He Y, Liang L, Lv L, Ran X, Pan Y, Luo Z. Surf. Coat. Technol. 2015,276,471.

[68]   Chang CH, Huang TC, Peng CW, Yeh TC, Lu HI, Hung WI, Weng CJ, Yang TI, Yeh JM. Carbon. 2012,50,5044.

[69]   DIN-Taschenbuch 286/1: Korrosionsschutz durch Beschichtungen und Überzüge 1, 2nd edn. Beuth Verlag GmbH, Berlin, 2012.

[70]   DIN EN ISO 12944: Paints and varnishes – Corrosion protection of steel structures by protective paint systems, 2016.

10 Luminescent pigments

10.1 Fundamentals and properties

Luminescent materials or pigments, also called phosphors or luminophores, are synthetically generated crystalline compounds that absorb energy followed by the emission of light with lower energy
or in other words of longer wavelengths. The light emission often occurs in the visible spectral
range. The nature of luminescence is, therefore, different from blackbody radiation. External energy
must be applied to luminescent materials to enable them to generate light. Luminescence can have its
origin in very different kinds of excitation such as the photo or electroluminescence caused by X-rays,
cathode rays, UV radiation, or even visible light. It is based on electronically excited states in atoms and
molecules. The emission process is governed by quantum mechanical selection rules [1].

There is a distinction in luminescence between fluorescence and phosphorescence, going back to
the different energy transitions. Forbidden optical transitions are generally slower than allowed
ones. Emission originating from allowed optical transitions with decay times in the order of μs or
faster is defined as fluorescence. On the other hand, emission with longer decay times is called phosphorescence. The time in which the emission intensity decreases to 1/10 (for exponential decay) is
called the decay time. The occurrence of fluorescence or phosphorescence as well as the decay time,
depends on the structure and composition of a specific luminophore [1].

In general terms, fluorescence is the emission of light by a substance that has absorbed
light or other electromagnetic radiation. The emitted light typically has a longer wavelength and, therefore, lower photon energy than the absorbed radiation. For example,
the absorption of radiation can take place in the ultraviolet region of the electromagnetic spectrum (invisible to the human eye), while the emission of light happens in
the visible region. This gives the fluorescent substance a distinct color, which is only
visible when the substance is exposed to UV light. Fluorescent materials glow nearly
immediately after irradiation. The glowing stops practically at the same moment when
radiation is interrupted. Unlike fluorescence, a phosphorescent material does not immediately re-emit the light or other electromagnetic radiation it absorbs. A phosphorescent
material absorbs some of the radiation energy and re-emits it for a significantly longer
time after the radiation source is removed. The very complex and complicated processes of fluorescence and phosphorescence are shown schematically in Figure 10.1. An
electron excited to its singlet excited state S_2 may, after a short time between absorption
and emission (fluorescence lifetime), return immediately to the ground state, giving off
a photon via fluorescence (decay time). Sustained excitation may happen, which is followed by nonradiating transitions (intersystem crossing) to the triplet state T_1 that re-
laxes to the ground state by phosphorescence with much longer decay times. It should
be borne in mind that there are a large number of intermediate states in the singlet
states S_0, S_1, and S_3 as well as in the triplet state T_1 that play a role in the course of the
radiant and nonradiant transitions.

https://doi.org/10.1515/9783110743920-010

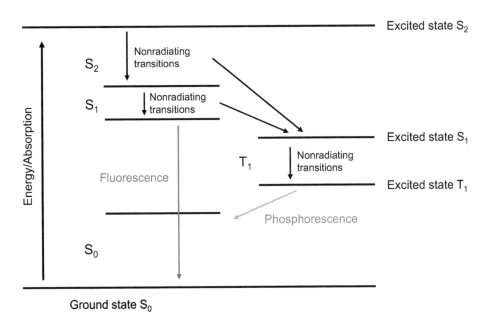

Fig. 10.1: Schematic illustration of the electron transition processes involved in fluorescence and phosphorescence.

Crystal lattices (host lattices) suitable for inorganic luminescent materials are based on colorless silicates, phosphates, sulfides, oxides, or halides, predominantly from alkaline earth metals or zinc. Activators (activator ions such as Mn^{2+}, Ag^+, Cu^+) are incorporated as emission centers in the crystal structure, partially in combination with sensitizers (sensitizing ions such as Sb^{3+}, Pb^{2+}, Ce^{3+}) in small amounts of $10^{-4}\%$ to $10^{-2}\%$. The activators are located on cation sites (e.g., Ag^+ or Cu^+ on Zn^{2+} positions in a wurtzite lattice). Electrical neutrality in the case of the incorporation of Ag^+ or Cu^+ in the ZnS lattice is preserved by the additional incorporation of Cl^- or Al^{3+} [2, 3].

Color and decay time of the emitted light depend in particular on the activators (transition metals or rare earth metals in various oxidation states), the lattice geometries and on influences of the crystal field of the host lattice.

Luminophores are widespread materials used in fluorescent lamps and tubes, LED lamps, displays, daylight paints, security features, safety signs, and modern medical equipment [4]. Organic luminescent materials have gained considerable interest in addition to inorganic luminescent materials. They are mainly used in the form of thin films and find broad application in organic light-emitting diodes (TV screens, computer monitors, smartphones, and others) [5].

10.2 Luminescence mechanisms

There are four effect mechanisms discussed for inorganic luminescent materials: center luminescence, charge–transfer luminescence, donor–acceptor pair luminescence, and long afterglow phosphorescence. Figure 10.2 shows schematically the structure of a luminescent material and its interaction with excitation radiation.

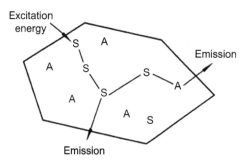

Fig. 10.2: Luminescent material containing activator ions A (ions with the ability to emit the desired radiation) and sensitizing ions S (ions on which excitation can take place) [1].

The energy, necessary to excite the luminescence, is absorbed either by the host lattice or by intentionally incorporated sensitizing ions. The excitation energy can be transferred through the crystal lattice by an energy transfer process. In most cases, the emission takes place at the activator ions. The emission color can usually be adjusted by choosing the proper activator and sensitizing ions, without changing the host lattice in which the ions are incorporated.

The mechanism of center luminescence is characterized by emission, which is generated at an optical center. This mechanism is distinguished from emission resulting from optical transitions between host lattice band states. A luminescence center can be an ion or a molecular ion complex. As a result of center luminescence, the so-called characteristic luminescence can be observed. Characteristic luminescence is defined in such a way that the emission could also occur at an ion in vacuum, i.e., when the optical transition involves electronic states of the ion only. Characteristic luminescence is shown by relatively sharp emission bands (spectral width typically a few nm), but also of broad bands that can have widths exceeding 50 nm. Such broad emission bands are seen when the character of the chemical bonding in the ground and the excited state differs significantly. Broad emission bands are observed for many optical transitions in the partly filled d-energy level of transition metal ions (d → d transitions) for transitions between the 5d-level and the 4 f-level of rare earth ions (d → f transitions) and also for emission on s^2 ions (these ions possess a "lone pair" of s electrons), like Tl^+, Pb^{2+}, or Sb^{3+}. Sharp emission bands are typical for optical transitions between electronic states having a chemical bonding character that is nearly the same for the ground and the excited state. There is a comparable situation for optical transitions between electronic states that only seldom participate in the chemical bonding (e.g., f → f transitions on rare earth ions) [1].

The nature of the bonding (covalent, ionic) and the symmetry of the site at which the emitting ion is incorporated play a very important role for optical processes involving electronic states that participate in chemical bonding. The ligand field theory is suitable for understanding and describing these processes. An example for a broad d → d emission band (in the green part of the spectrum) is the emission of Mn^{2+} in $BaMgAl_{10}O_{17}:Eu^{2+}$, Mn^{2+}. The weak blue emission band originates from a d → f optical transition on Eu^{2+}. This luminescent composition can be applied as a green phosphorescent material in very high-quality fluorescent lamps and also in plasma display panels. An example for a d → d emission consisting of a few relatively sharp bands is the emission of Mn^{4+} in $Mg_4GeO_{5.5}F:Mn^{4+}$. The emitting species is again a manganese ion, but its charge (and, therefore, its electronic configuration) is different. This composition can be used as a red phosphor primary in fluorescent lamps producing deep red colors [1].

Some green phosphors based on Y_2O_2S with Tb^{3+} as activator are suitable for applications in fluorescent lamps. $Y_2O_3:Eu^{3+}$ is the most important red phosphor with line emission at about 611 nm. The width and the position of the emission bands originating from the optical transitions within the f-energy level are almost independent of the chemical environment. The relative intensity of some separate bands, however, depends on the structure of the crystal lattice. Transitions on many rare earth ions are spin- and parity-forbidden. They are, therefore, rather slow. For a number of rare earth ions, however, broad emission bands also occur. The reason for this is the d → f emission, which is known for ions like Eu^{2+} or Ce^{3+}. Such transitions are allowed and, therefore, very fast. Consequently, several important industrial phosphors are based on rare earth ions [1].

The mechanism of charge–transfer luminescence is based on an optical transition that takes place between different kinds of orbitals or between electronic states of different ions. The width and position of the emission bands in this case also depend on the chemical environment. $CaWO_4$ as an example for charge–transfer luminescence has already been used for a long time. It displays luminescence originating from the WO_4 group. The transition can be explained by charge transfer from the oxygen ions to empty d-levels of tungsten. The chemical bonding situation changes very strongly during the transition, leading to a very broad emission spectrum. The incorporation of an activator in the $CaWO_4$ lattice is not necessary because this material operates by self-activation [1].

The mechanism of donor–acceptor pair luminescence occurs in some semiconducting materials doped with both donors and acceptors. The band model may be used for understanding this mechanism. Electrons from the valence band are excited and lifted into the conduction band where they are captured by ionized donors. The resulting electron holes in the valence band are caught by ionized acceptors. The emission comprises an electron transfer between the neutral donors and the neutral acceptors. As a result, ionized donors and acceptors are formed. The final state is Coulomb-stabilized. The wavelength of the emitted light depends on the distance between the donor and the acceptor in a donor–acceptor pair. The shorter this distance, the higher is the energy of the photon generated. In the crystal lattice of a luminescent material whose effect is based on the mechanism of donor–acceptor pair luminescence, many different donor-acceptor distances are possible. This leads to a relatively broad emission band. Donor–acceptor pair luminescence occurs in the blue- and green-emitting luminophores $ZnS:Ag^+$, Cl^-; $ZnS:Ag^+$, Al^{3+}; $ZnS:Cu^+$, Au^+, Al^{3+}; and $ZnS: Cu^+$, Al^{3+}. Among others, luminophores of this type can be used in color television picture tubes [1].

The mechanism of long afterglow phosphorescence is based on the phenomenon that optical excitation energy is stored in the crystal lattice by trapping photoexcited charge carriers. The composition $SrAl_2O_4:Eu^{2+}$, Dy^{3+} is an example for this mechanism. Eu^{2+} is oxidized to Eu^{3+} and Dy^{3+} is reduced to Dy^{2+} by optical excitation. Thermal excitation leads to the re-oxidation of Dy^{2+} to Dy^{3+}. The resulting free electron is captured by Eu^{3+}, and Eu^{2+} is formed again. This reduction step is accompanied by the delayed emission of light. The time delay for the emission is determined by the thermal excitation process for Dy^{2+}. Emission is generated by this material, which is still visible several hours after the excitation. Long afterglow phosphors may be used in exit signs for escape routes and other signs, watches, and safety features. $ZnS:Cu^+$ and $SrS:Bi^{3+}$ are other long afterglow materials [1].

10.3 Excitation mechanisms

There are three excitation mechanisms discussed for inorganic luminescent materials: optical excitation of luminescence, electroluminescence, and excitation with high-energy particles.

Optical excitation of luminescence occurs when the absorption of UV or visible light leads to the emission of light of a longer wavelength in the visible part of the electromagnetic spectrum. This kind of excitation is applied in fluorescent lamps and in phosphor-converted LEDs, in which phosphors are used to change the wavelength of the radiation emitted by the LED. Optical absorption can take place on activator ions or sensitizer ions, the so-called optical centers. Sensitizer ions are preferably used when the optical absorption of the activator ions is too weak to be suitable for practical applications, e.g., because the optical transition is forbidden. Energy transfer from the sensitizer ions to the activator ions is necessary in this case. Optical absorption may also happen in the host lattice itself (band absorption based on host lattice sensitization). Energy transfer from host lattice states to the activator ions is necessary. Sensitizer ions are involved too, in some cases. Material examples using optical excitation are $BaMgAl_{10}O_{17}:Eu^{2+}$ (blue emitting); $BaMgAl_{10}O_{17}:Eu^{2+}$, Mn^{2+} (green emitting); and $Ca_5(PO_4)_3$

(F, Cl):Sb^{3+}, Mn^{2+} (white emitting as a result of orange emission by Mn^{2+} and blue emission by Sb^{3+}). A further technically relevant sensitizer-activator pair is Ce^{3+},Tb^{3+}, e.g., in $GdMgB_5O_{10}$:Ce^{3+},Tb^{3+}. All green-emitting luminophores in high-quality fluorescent lamps contain this combination incorporated in a suitable crystal lattice [1].

Electroluminescence as an excitation mechanism for luminescent materials is divided into high-voltage electroluminescence and low-voltage electroluminescence. High-voltage electroluminescence is based on an electrical breakthrough in a semiconducting material. Typical voltages used are of the order of 100 V. Generation and acceleration of charge carriers take place in the host lattice. The charge carriers can excite the activator ions necessary for the generation of luminescence. Electroluminescent devices based on this mechanism can have a long lifetime, but their efficiency is only of the order of one percent. Consequently such devices are only used in applications where reliability is of high importance and efficiency plays only a minor role, e.g., for emergency and exit signs. ZnS:Mn^{2+}, ZnS:Cu^+, and SrS:Ce^{3+} are suitable luminescent materials for high-voltage electroluminescence [1].

Low-voltage electroluminescence is of high importance for the blue light-emitting diode (blue LED) and for organic electroluminescent devices. Efficient light-emitting structures using this mechanism can be realized that do not require either high or low pressure. Low-voltage electroluminescent devices may also be a way to overcome the cascade-based technique that is used in conventional luminescent devices. In fluorescent lamps, for instance, a discharge is initially generated, and the resulting invisible radiation is converted into visible light. A significant energy must be accepted here. In cathode ray tubes, as another example, first an electron beam is generated, which consists of electrons with high kinetic energy. These electrons subsequently impact the luminescent material, which finally leads to excitations where electrons in the conduction band are coupled to holes in the valence band (excitons). Excitons generated in this way are transferred to activator ions. As a result, the energy efficiency of the luminophores is limited to about 20% and of white light-emitting devices to less than 50% [6].

In low-voltage electroluminescent devices, on the other hand, the step leading to emission is the recombination of electrons in the conduction band with holes in the valence band. Only the band gap energy is required in this case to excite the luminescence. The color of the emitted light can be monitored to a certain extent by selecting the appropriate semiconductor. In summary, the generation of low-voltage electroluminescence may be very energy-efficient [1].

Luminescent materials for LEDs have to meet strict requirements. The Stokes Shift must be small, and the absorption has to be high. In addition, the materials must also operate efficiently at high temperatures; they have to be stable against radiation and they should not show a less-than-linear increase in output power with input power at high excitation densities (saturation) [1].

Excitation with high-energy particles is based on the effect that after the absorption of electrons or high-energy photons by the luminescent material electron–hole pairs are generated in the crystal lattice. The electron–hole pairs adjust their speed to the surrounding by collisions (thermalization), whereby band gap excitations can take place. The excitation is transmitted to an activator or sensitizer, after thermalization. As a result, light emission can be observed. A large number of electron–hole pairs are generated for each absorbed electron or high-energy phonon. Each electron–hole pair is able to initiate the emission of one photon on an activator ion [1].

10.4 Production of luminescent pigments

The production of luminescent materials or pigments takes place in the vast majority of cases via mixing and firing procedures. Homogeneous mixtures of starting materials are treated at temperatures between 1000 and 1500 °C in an oxidizing or reducing atmosphere. The powdery reactants are thoroughly mixed and ground in the dry or wet state and then calcined. Decisive for the calcination conditions are the crystal lattice of the luminophore to be formed and the activator. In many cases, a second or even a third calcination with intermediate cautious comminution steps is necessary. Overly intensive grinding leads to undesired lattice defects. Further thermal treatment for curing of such effects must be performed significantly below the synthesis temperature.

The production process can be facilitated and the crystallinity of the luminescent pigments formed can be improved by adding flux agents or melting salts. In this way, lower reaction temperatures and an optimization of the grain size of the luminophores are possible, knowing that the optimum particle size depends on the specific application. Suitable flux agents act by dissolving of at least one of the reactants. Two types of fluxes are used for the manufacture of luminescent pigments: a nonvolatile liquid that is usually a molten salt or a volatile liquid.

Melting salts used for the synthesis of luminescent pigments do not react with the starting materials. They also do not decompose or evaporate during the formation of the pigments. $Na_2B_4O_7$, $Na_4P_2O_7$, Na_2SiO_3, and Na_2MoO_4 are frequently used as melting salts.

Flux agents often react with the starting materials. They are used mostly in amounts of less than 10% by weight of the luminophore and undergo decomposition or evaporation during the formation of the phosphor. AlF_3, NaF, and NH_4Cl are suitable substances for flux agents.

Starting materials, such as carbonates, nitrates, or hydroxides, which decompose during calcination, can improve the reaction speed. The oxidic compounds formed during the thermal decomposition are much more reactive than the used starting materials.

Impurities of the starting materials should lie clearly below the concentration of the activators. They can act as discharge centers and, therefore, reduce the luminous efficiency of the luminophores. Particularly pure starting materials are achieved by precipitation reactions. For some compositions, coprecipitation can be used, whereby a reaction mixture is obtained, which contains the starting materials in the form of an intimate mixture.

The synthesis of a luminescent pigment is shown in the following using the example Y_2O_3:Eu^{3+}. The formation of this luminophore takes place via the precipitation of mixed oxalates starting from purified $Y(NO_3)_3$ and $Eu(NO_3)_3$ solutions. The filtered, washed, and at 600 °C pretreated precipitates are subsequently calcined for several hours at 1440 to 1500 °C, whereby the required crystal structure is formed [7]. This method can be used whenever suitable precipitates for each of the reaction partners exist.

Another process to obtain intimate reaction mixtures is spray drying. Small water droplets containing the reactants are transported in a gas stream and heated up. The water is evaporated very fast during heating, and an optimum reaction mixture for the calcination is obtained.

Some luminescent pigments can be synthesized using the polysulfide flux method, e.g., Y_2O_2S:Eu^{3+} and Gd_2O_2S:Pr^{3+} [8]. Here, oxides of the metals are mixed with excess sulfur and an alkali metal carbonate. During heating, the carbonate reacts under decomposition with sulfur to form a liquid polysulfide flux. This flux reacts with the metal oxides and oxysulfides are formed. Residues of the flux are removed by washing the reaction product with water.

In cases where the amounts of activator ions are very low, different routes must be used. Examples of this are specific ZnS-based luminescent pigments. The amounts of activator ions here can be in the order of 100 ppm. Tiny amounts of these ions can be precipitated on ZnS particles by preparing a suspension of ZnS in water, followed by the addition of a solution containing a solved activator salt. The precipitation of a composition containing the activator ions on the ZnS particles is accomplished by adding a suitable precipitating agent, e.g., $(NH_4)_2S$.

Some pigment compositions require a reducing atmosphere to incorporate activator ions in a nonpreferred low oxidation state, e.g., Eu^{2+}, or to prevent the oxidation of the host lattice, e.g., during the synthesis of ZnS-based pigments. The reaction can take place in such cases in a hydrogen/nitrogen mixture or in carbon monoxide [1].

10.5 Pigment properties and uses

Table 10.1 contains a summary of selected inorganic luminophores with compositions, activator ions, emission wavelengths, colors, and application fields. Luminescent pigments are applied in the form of thin layers and by means of luminophore/solvent suspensions, containing adhesive agents on a substrate [2].

The morphological properties of luminescent pigments depend strongly on their composition and the intended use. Typical particle sizes are in the range from 0.005 to 0.5 μm. The thickness of a pigment-containing layer on the inside of a television

Tab. 10.1: Composition and application fields of selected luminophors [2].

Activator	Composition of the luminophore	Emission (nm)	Color	Application field
Mn^{2+}	Zinc orthosilicate Zn_2SiO_4	525	Green	Fluorescent lamps, projection television tubes, plasma display panels
Mn^{2+}/Sb^{3+}	Calcium fluoro chloro apatite $Ca_5(PO_4)_3(Cl, F)$	480 and 580	White (blue and yellow)	Fluorescent lamps
Mn^{4+}	Magnesium fluoro germanate $Mg_2GeO_4 \cdot \frac{1}{2} MgO \cdot 0{,}5\ MgF_2$	710	Red	Mercury high-pressure lamps
Mn^{4+}	Magnesium arsenate $Mg_6As_2O_{11}$	700	Red	Fluorescent lamps
Sb^{3+}	(Sr, Ca) chloro bromo phosphate $(Sr, Ca)_5(PO_4)_3(Cl, F)$	480–500	Blue-green	Fluorescent lamps
Sn^{2+}	(Sr, Mg) orthophosphate $(Sr, Mg)_3(PO_4)_2$	630	Pink	Fluorescent lamps, mercury high-pressure lamps
Pb^{2+}/Ce^{3+}	Strontium thiogallate $SrGa_2S_4$	614	Orange-red	Flying spot scanners
Ce^{3+}	Calcium aluminum silicate $Ca_2Al_2SiO_7$ yttrium aluminate $Y_3Al_5O_{12}$	405 550	Blue Yellow	Flying spot scanners, fluorescent lamps
Eu^{2+}	Barium fluoride bromide $Ba(F, Br)$	440	Blue	X-ray detection
Eu^{3+}	Yttrium oxide Y_2O_3	625	Red	Fluorescent lamps, cathode ray tubes, projection television tubes, plasma display panels
Eu^{3+}	Yttrium vanadate YVO_4	630	Red	Plasma display panels
Tb^{3+}	Yttrium oxysulfide Y_2O_2S	525	Green	Cathode ray tubes, projection television tubes
Ag^+/Cl^-	Zinc sulfide ZnS	440	Blue	Cathode ray tubes, projection television tubes
Cu^+/Al^{3+}	Zinc sulfide ZnS	525	Green	Cathode ray tubes, projection television tubes

Tab. 10.1 (continued)

Activator	Composition of the luminophore	Emission (nm)	Color	Application field
Mn^{2+}	Zinc sulfide ZnS	585	Yellow	Radar tubes
Zn^{2+} (excess)	Zinc oxide ZnO	505	Green	Flying spot scanners
Without	Ca, Mg tungstate $CaWO_4$ $MgWO_4$	415 480	Blue-violet Blue-green	Fluorescent lamps

screen is in the order of 10 to 30 μm. The application of luminophores can take place in all common pigment-relevant application media and also in very special formulations used only for this type of material.

Cathode ray tube phosphors (CRT phosphors) emitting blue and red are in many cases coated with a daylight-absorbing pigment in order to enhance contrast in daylight viewing conditions. Regarding this, blue-emitting phosphors are coated with $CoAl_2O_4$ and red-emitting phosphors with α-Fe_2O_3. Coating of green-emitting phosphors with a green absorbing pigment is not common because of high sensitivity of the human eye to green light. Liquid crystal displays (LCDs) used in television, computer monitors, notebooks, digital cameras, and mobile telephones also require luminescent materials. All these devices need a backlight, which works either with thin or flat fluorescent lamps or with LEDs [1].

Fluorescent lamps work with mixtures of luminescent materials with different emission wavelengths to achieve a special color (tricolor phosphor mixtures). Even lamps generating white light with higher brightness than that of metal halide lamps are possible, based on this mixing principle. The resulting emission color is adapted to the sensitivity of the human eye. The problem with fluorescent lamps is the need for mercury in their function. This environmental disadvantage can be reduced by lowering the mercury content in fluorescent lamps. Mercury-free fluorescent lamps, based on a Xe/Ne-discharge may be an alternative here, but the efficiency of the lighting device is lower in this case. The use of vacuum UV radiation, however, could increase the efficiency of such mercury-free devices to the level of currently applied fluorescent lamps.

Some specific luminescent materials are suitable for X-ray and γ-ray detection. This phenomenon is utilized for computed tomography (CT) and position emission tomography (PET). X-rays in the case of CT or γ-rays in the case of PET are converted into visible light by these luminophores. Photomultipliers or photodiodes are used for the detection of the visible light generated in this way.

Luminescent materials used for X-ray detection can also be used in X-ray-intensifying screens. In this case, the X-rays are converted into visible photons, which subsequently irradiate a photographic film. This film is usually sandwiched between two luminophore sheets. The typical thicknesses of such luminescent layers are of the order of a few hundred micrometers [1].

Light-emitting diodes (LEDs) are another important application for luminescent materials. LED lamps generating white light with a low color temperature are of special interest. One way to produce such lamps is the combination of red, green, and blue LEDs in one LED device housing (RGB principle). A disadvantage of this method is a color shift at higher temperatures and aging of the LED. A more advantageous route is the use of the so-called phosphor principle. In this case a white LED can be produced by the combination of a blue-emitting LED with one or more phosphors. An example for the use of only one phosphor is the combination of the blue LED with the yellow emitting $(Y, Gd)_3Al_5O_{12}:Ce^{3+}$. Alternatively, the blue-emitting LED can be combined with a green (e.g., $CeMgAl_{11}O_{19}:Tb^{3+}$) and a red (e.g., $Y_2O_3:Eu^{3+}$) luminophore. Some nitridosilicate phases such as $LaSi_3N_5:Eu^{2+}$ (emission in the blue-green region) and $Ba_2Si_5N_8:Eu^{2+}$ (emission in the yellow-orange region) are interesting for their application as conversion phosphors in LEDs too [6, 9].

In most applications, luminescent materials operate at their physical limits. This refers to the absorption of the exciting radiation and its quantum efficiency (number of visible photons generated divided by the number of photons absorbed) with which the luminescence is generated. The quantum efficiency in plasma display panels, fluorescent lamps, and LEDs is almost 100%. The energy efficiency of the phosphors in cathode ray tubes is of the order of 20% [6]. In contrast, the average energy efficiency of luminescent devices is rather low (plasma display panels 1% to 2%, fluorescent lamps 15% to 25%, phosphor converted LED lamps 20%, cathode ray tubes 1% to 2%) [1].

There are two further groups of materials with possible potential for broader applications in solid-state lighting (SSL) and other applications: cascade phosphors and quantum dots.

Cascade phosphors are luminescent materials with more than one absorbed or emitted photon. A well-selected combination of energy levels, normally f-levels of lanthanide ions, is necessary to establish a cascade transition. $CaWO_4:Yb^{3+}, Er^{3+}$ is an example of cascade phosphors where the absorption of two photons with smaller energy (preferably in the infrared range of the spectrum) is succeeded by the emission of one photon with higher energy (in the visible range of the spectrum) [10]. Such behavior is called wavelength up-conversion. It is known for the summation of up to five photons. The efficiency of such cascade processes, with values <1%, is very low. Applications are, therefore, limited to areas where energy efficiency is not important.

The so-called wavelength down-conversion approach is the effect opposite to up-conversion for SSL. It uses a down-conversion material to generate visible light when excited by near-UV radiation or blue emission from InGaN LEDs [11]. A material suitable for down-conversion processes is $YF_3:Pr^{3+}$ [12, 13]. Two photons with lower energy are emitted per photon absorbed at a wavelength below 220 nm. Down-conversion phosphors are necessary to enable energy efficient mercury-free lamps using Xe/Ne discharge.

Quantum dots act on the basis of size-quantization effects that occur in semiconductors (e.g., InGaAs, CdSe, GaInP, or InP) with radii typically less than 10 nm. The spectral position of the emitted light can be shifted over a large spectrum, including the complete visible spectral range [14–20]. Adjustment of the emission is possible by simply varying the particle size. Tunable narrow band emission can be realized based on this size variation. Quantum dots, with diameters of 2 to 10 nm are small enough to enable transparent nonscattering luminescent layers. Based on this, electroluminescent displays and flexible light sources are possible to produce. Quantum dots may also be used as luminescent markers for optical imaging in medical devices [15, 21, 22].

Anti-counterfeiting of valuable documents, currency, and branded products is a further application field for luminescent materials. Security ink formulations based on lanthanide-doped luminescent nanomaterials or quantum dots are examples for the use of luminescent materials in anti-counterfeiting applications [23, 24]. Invisible full color luminescent images can be reproduced faithfully by printing luminescent inks containing lanthanide ions. The superposition of europium and terbium trisdipicolinate inks enables reproducing colors from red to orange, yellow, and green. The potential of this application as anti-counterfeiting feature is seen as high. Tagged documents can be easily authenticated using visualization under UV light. Counterfeiting of this document security feature is not possible without the appropriate luminescent inks, and without color reproduction software [23].

10.6 Toxicology and occupational health

Based on the criterion of composition, some luminescent pigments can be regarded as harmless and others as potentially hazardous. From the occupational health and safety perspective, operations with pigment dust require attention. Inhalation of the dust is most critical and, therefore, must be avoided by all available means.

Special care is necessary when the pigments contain larger amounts of zinc, cadmium, barium, or strontium. However, there is only a small risk during application due to the low solubility of most luminescent materials. Pigments with particle sizes in the nanometer range have to be handled very carefully. Measures must be taken here to minimize exposure to these materials for affected employees (Section 7.7).

10.7 Conclusions

Luminescent pigments (luminescent materials, luminophores, phosphors) as synthetically generated crystalline compositions absorb energy followed by emission of light
with lower energy, that is, with longer wavelengths. The light emission often occurs
in the visible spectral range. External energy is necessary to enable luminescent materials to generate light.

Luminescent pigments are divided into fluorescent and phosphorescent pigments.
This classification goes back to different energy transitions. Emission, based on allowed optical transitions, with decay times of the order of µs or faster is defined as
fluorescence. Emission with longer decay times is called phosphorescence. The occurrence of fluorescence or phosphorescence as well as the decay time depends on the
structure and composition of a specific luminophore.

There are four luminescence mechanisms discussed for inorganic luminescent
materials: center luminescence, charge–transfer luminescence, donor–acceptor pair
luminescence, and long afterglow phosphorescence. The emission of luminescent
light can have its origin in different excitation mechanisms, such as optical excitation
(UV radiation or even visible light), high-voltage or low-voltage electroluminescence
and excitation with high-energy particles (X-rays, γ-rays).

Inorganic luminescent pigments are used mainly in fluorescent lamps, cathode
ray tubes, projection television tubes, plasma display panels, LEDs, and for X-ray and
γ-ray detection. The pigment particles are dispersed for applications in specific binder
systems. They are applied in the form of thin layers and by means of luminophore/
solvent suspensions, containing adhesive agents, on a substrate.

Technically important inorganic luminescent pigments include a large variety of
compositions. It is possible to produce all relevant visible emission colors. Examples
of green are $ZnS:Cu^+$, Au^+, Al^{3+} and $Zn_2SiO_4:Mn^{2+}$, of yellow $Y_3Al_5O_{12}:Ce^{3+}$, of red Y_2O_2S:
Eu^{3+} and $Y_2O_3:Eu^{3+}$, of blue $ZnS:Ag^+$, Cl^- and $BaMgAl_{10}O_{17}:Eu^{2+}$, and of white $ZnS:Ag^+$ +
$(Zn, Cd)S:Ag^+$ or $Ca_5(PO_4)_3(F, Cl):Sb^{3+}$, Mn^{2+}.

The manufacturing methods for the pigments are diverse and oriented toward
specific compositions and physical properties.

? Study questions

10.1   What are the characteristics of luminescent materials?

10.2   What is the distinction between fluorescence and phosphorescence?

10.3   What are the four luminescence mechanisms discussed for inorganic luminescent materials?

10.4   Where are inorganic luminescent pigments mainly used?

10.5   Which inorganic luminescent pigments are technically important for green, yellow, red, blue, and
       white emission?

Bibliography

[1] Ronda C, in Industrial Inorganic Pigments, 3rd edn. Buxbaum G, Pfaff G, eds. Wiley-VCH Verlag, Weinheim, 2005. 269.

[2] Pfaff G, in Winnacker–Küchler: Chemische Technik, Prozesse und Produkte, vol. 7, Industrieprodukte, 5th edn. Dittmeyer R, Keim W, Kreysa G, Oberholz A, eds. Wiley-VCH, Weinheim, 2004. 373.

[3] Pfaff G, in Encyclopedia of Color, Dyes, Pigments, vol. 2, Pfaff G, ed. Walter de Gruyter GmbH, Berlin, Boston, 2022. 765.

[4] Feldmann C, Jüstel T, Ronda CR, Schmidt PJ. Adv. Funct. Mat. 2003,13,511.

[5] Kalyani NT, Dhoble SJ. Renew. Sust. Energ. Rev. 2012,16,2696.

[6] Robbins DJ. J. Electrochem. Soc. 1980,127,2694.

[7] Kotera Y, Higashi T, Sugai M, Ueno A. J. Lumin. 1984,31,709.

[8] Patent US 3,418,246 (M. R. Royce), 1968.

[9] Uheda K, Takizawa H, Endo T, Yamane H, Shimada M, Wang C-M, Mitomo M. J. Lumin. 2000,87-89,967.

[10] Auzel FCR. Ac. Sci. 1966,262,1016.

[11] McKittrick J, Shea-Rohwer LE. J. Am. Ceram. Soc. 2014,97,1327.

[12] Sommerdijk JL, Bril A, De Jager AW. J. Lumin. 1974,8,341.

[13] Piper WW, De Luca JA, Ham FD. J. Lumin. 1974,8,344.

[14] Colvin VL, Schlamp MC, Alivisatos AP. Nature. 1994,370,354.

[15] Empedocles SA, Neuhauser R, Shimizu K, Bawendi MG. Adv. Mater. 1999,11,1243.

[16] Talapin DV, Rogach AL, Kornowski A, Haase M, Weller H. Nano. Lett. 2001,4,207.

[17] Yoffe AD. Adv. Phys. 2001,50,1.

[18] Graham-Rowe D. Nat. Photonics. 2009,3,307.

[19] Coe-Sullivan S. Nat. Photonics. 2009,3,315.

[20] Bera D, Qian L, Tseng T-K, Holloway PH. Materials. 2010,3,2260.

[21] Shipway AN, Katz E, Willner I. Chem. Phys. Chem. 2000,1,18.

[22] Dujardin E, Mann S. Adv. Mater. 2002,14,775.

[23] Andres J, Hersch RD, Moser J-E, Chauvin A-S. Adv. Funct. Mater. 2014,24,5029.

[24] Kumar P, Singh S, Gupta BK. Nanoscale. 2016,8,14297.

11 Fillers

11.1 Fundamentals and properties

The term "filler" is used for solid materials in powder form that are irregular, acicular, fibrous or platy in shape and which are added to application media (plastics, coatings, paints, printing inks, paper, cosmetics, and drugs) to improve properties of the system or to lower the consumption of more expensive binder components. In the latter case, fillers are also called extenders. Fillers are typically applied in relatively large loadings. They are often used in combination with pigments to enhance properties of the material in which they are embedded [1–5].

Fillers are insoluble in the application medium and are not subject to shrinkage in the hardening process. They can improve coatings and other application systems with respect to density, porosity, and permeability for gases and water. The refractive index of fillers is usually below 1.7, although there is no fixed definition of this value. Naturally occurring white minerals are often used as fillers. For reasons of quality, synthetic products have established themselves for many uses, especially in the coatings and plastics industry. For coloristically driven pigment-related applications, only white or colorless fillers with particle sizes below 10 μm are used. The largest share of this is attributable to calcium carbonate. Table 11.1 contains a summary of important natural and synthetic fillers, including organic materials.

Tab. 11.1: Summary of natural and synthetic fillers.

Compound class	Examples
Natural fillers	
Oxides	Quartz, christobalite, diatomaceous earth
Carbonates	Chalk, calcite, limestone meal, dolomite
Sulfates	Barite, calcium sulfate
Silicates	Kaolin, talc, pyrophyllite, chlorite, mica (muscovite), hornblende, wollastonite, feldspars, slate powder, montmorillonite, nanoclays
Natural polymers	Cellulose fibers, wood flour and fibers, flax, cotton, sisal, starch
Synthetic fillers	
Oxides	Silicon dioxide (crystalline), silicic acid (amorphous), magnesium oxide, aluminum oxide
Carbonates	Calcium, magnesium, barium carbonates
Sulfates	Barium sulfate (blanc fixe), calcium sulfate
Silicates	Calcium, magnesium, sodium, aluminum silicates of various compositions
Hydroxides	Aluminum hydroxide, magnesium hydroxide
Metals	Boron, steel
Glass	Fibers, spheres, hollow spheres, flakes
Carbon, graphite	Carbon fibers, graphite fibers and flakes, carbon nanotubes, fullerenes, carbon black
Synthetic polymer fibers	Polyamide, polyester, aramids, polyvinyl alcohol

https://doi.org/10.1515/9783110743920-011

The chemical nature and physical properties of various inorganic and organic compounds used as fillers differ considerably. Most of them are rigid materials, immiscible with the matrix in both molten and solid states. Some fillers tend to agglomerate during storage or absorb humidity or additives of the application medium. This can be avoided by the use of a suitable surface treatment.

Reinforcing fillers for the enhancement of mechanical properties of application systems are characterized by a relatively high aspect ratio. Fillers consisting of spherical particles, therefore, show the lowest reinforcing capacity. Other intentions for the use of fillers, especially in plastics, are the modification of mechanical properties, enhancement of a flame retardant effect, modification of electrical and magnetic properties, modification of surface properties, and enhancement of manufacturability. Other functions of fillers include degradability enhancement, modification of barrier properties, anti-aging, bioactivity, radiation absorption, warpage minimization, optical clarity, or damping control.

Traditionally, the major effect of fillers in an application medium was seen in lowering the costs of materials by replacing the most expensive parts of the composition (e.g., polymers or pigments). Faster processing cycles for plastics as a result of increased thermal conductivity and fewer rejects in injection molding due to part warpage have been additional positive effects, already recognized long ago. Other polymer properties were also affected, e.g., the melt viscosity, which could be appreciably increased by the incorporation of fibrous materials. Another interesting effect observed for most inorganic fillers was the reduction of mold shrinkage and thermal expansion.

The following development work for reinforcing fillers was focused on process and material modifications, with the aim of increasing the aspect ratio of the particles, and to improve their compatibility and interfacial adhesion with the chemically dissimilar polymer matrix. Not only was the primary function of fillers as mechanical property modifiers enhanced by newly developed materials, but also additional functions, e.g., permeability to liquids and gases, can be reduced when specific flakes are used.

The performance of polymer composites containing functional fillers is affected by
- the morphological properties of the filler (particle shape, particle size and size distribution, aspect ratio, surface area, and porosity);
- the mechanical, chemical, thermal, optical, and electrical properties of the filler;
- the composition and crystal structure of the filler;
- interactions of the filler with the polymer at phase boundaries, which affect adhesion and stress transfer from the polymer to the filler particles;
- interfacial physical and chemical interactions related to the surface properties of the filler (surface tension and surface reactivity, which are relevant for the wetting and dispersion characteristics);

- the method of filler incorporation into the polymer melt and of dispersion of the filler particles;
- the distribution of the filler particles in the final polymer after all processing steps including control of the microstructure.

The most important natural fillers are calcium carbonate, magnesium carbonate, and some silicates. Natural calcium carbonate deposits are calcite, dolomite, and chalks. Chalks have a special status here because they consist of fine particle shell residues of small sea organisms. In view of their historical origins, a specific material with low abrasion resistance was formed by compaction. The comminution of this material by dry or wet grinding provides powders with small particle diameters. Due to the use of grinding steps during the production, the material obtained by this process is known as ground calcium carbonate (GCC).

Natural silicon-containing fillers are divided into pure silica and silicates. SiO_2 used for fillers occurs almost exclusively as quartz and is mined or gained from secondary deposits such as river and sea sands. Diatomaceous earth consists of 70 to 90% amorphous SiO_2. The deposits were developed from dead diatoms whose skeletons are built from silicic acid. Application fields for natural fillers are coatings, dispersion paints, sealants, and plastics. The characterization of fillers is performed using the methods described for pigments.

Demands for consistent quality and, preferably, finely divided fillers have led increasingly to synthetically generated filler materials. It is possible to produce carbonates of calcium, magnesium, and barium by precipitation as very pure and finely divided products. Such carbonate fillers have narrow particle size distributions and mean particle diameters down to about 50 nm. Precipitated magnesium and barium carbonates are only used in special products, while calcium carbonate (calcium carbonicum praecipitatum, precipitated calcium carbonate, and PCC) is applied as filler in dispersion and printing inks, coatings, and in the paper industry. On account of their high purity, PCC types can be used in pharmaceutical preparations and in the food sector. Calcium carbonates, equipped with an organic surface treatment, may improve shrinkage and flow behavior as well as the mechanical properties of plastics and rubber. Surface treatment consists mainly of fatty acids, e.g., of 0.5 to 1.5% stearic acid, to produce hydrophobic calcium carbonate with improved rheological and mechanical properties.

Synthetic barium sulfate fillers, known as "blanc fixe", are pure white, weather resistant, and lightfast products. Due to their low hardness, nearly no abrasion during the incorporation in polymer materials occurs. Barium sulfate acts as filler in coatings and plastics, and is suitable for the turbidity of transparent, colorless plastics, e.g., of polycarbonates. It is also used as an X-ray contrast agent. In addition to its refractive index, the particle size of barium sulfate plays a considerable role in the manufacture of translucent polymers.

Synthetically produced aluminum hydroxide, also known as alumina trihydrate $Al_2O_3 \cdot 3\ H_2O$ (ATH) or hydrated alumina, has a low hardness and a low solubility in water. It is characterized by a high degree of whiteness and a low density. These properties together with a positive zeta potential, which provide a good retention capacity, are the prerequisites for its application in paper finishing. Other applications are plastics, rubber materials, cable insulations, and several building materials. To enhance the overall performance of $Al(OH)_3$ fillers, a wide selection of chemical modifications of the surface can be carried out with surfactants, stearates, and organofunctional silanes. A suitable surface treatment on $Al(OH)_3$ filler particles can improve not only their processing but also chemical resistance, electrical performance, and mechanical and flame retarding properties.

Two synthetically produced silicon-containing materials are in use for filler applications. These are pure silica powders (SiO_2) and water-containing silica gels ($SiO_2 \cdot x\ H_2O$). The use of silica as a filler is often based on different requirements than those for carbonates and silicates. SiO_2 is characterized by its chemical inertness and high hardness. This comes into play where high abrasion resistance is required. Silicon dioxide types with very small particle sizes often take over rheological and optical functions. The following crystalline and amorphous silicon dioxide qualities are applied as fillers: quartz, cristobalite, diatomaceous earth, fumed silica, and precipitated silica.

11.2 Production of fillers

The processes used for the manufacture of natural and synthetic fillers are very different. Mechanical preparation processes are typically used for natural fillers, while synthetic fillers are produced mainly by precipitation reactions. Exceptions are glass fibers, spheres, hollow spheres, and flakes, which are produced directly from glass melt.

11.2.1 Natural fillers

Mining and processing technologies for natural fillers are dependent on the appropriate minerals, but the principal process steps necessary are nearly the same. They correspond to those used for the manufacture of natural pigments (see Section 2.1).

Representative of all natural fillers, the preparation of the calcium carbonates is described here – the most important class from a quantitative point of view. The process starts with the mining of calcite. The large calcite lumps obtained are sorted by quality and then pre-crushed in impact or jaw crushers. After this step, a dry separation or a washing procedure takes place in suitable plants. Washing includes wet separation using screws, hydrocyclones, or hydroseparators. The resulting granular raw material is ground in wet or dry form in pendulum roller mills, vibratory mills, or ball mills. The desired particle fraction for the different GCC types is obtained using suitable classification steps.

11.2.2 Synthetic fillers

The production of synthetic inorganic fillers is described here for four important examples: calcium carbonate ($CaCO_3$), barium sulfate ($BaSO_4$), aluminum hydroxide (Al$(OH)_3$), and silicon dioxide (SiO_2).

The production of synthetic calcium carbonate (PCC types) starts with the burning of natural limestone in an appropriate kiln at 900 to 1100 °C. Calcium oxide is formed, which is converted by using water to calcium hydroxide solution (lime water). Calcium carbonate is precipitated from this solution by injecting CO_2-containing exhaust gas from the burning kiln.

$$Ca(OH)_2 + CO_2 \rightarrow CaCO_3 + H_2O \tag{11.1}$$

Particle size and the form of the $CaCO_3$ precipitate are controllable by temperature and concentration of the reaction partners. Calcium carbonate is also formed as byproduct of ammonium nitrate fertilizer production. The reaction of calcium phosphate with nitric acid leads to calcium nitrate solutions, from which calcium carbonate is formed, in addition to ammonium nitrate by reaction with ammonia and carbon dioxide.

$$Ca(NO_3)_2 + 2\ NH_3 + CO_2 + H_2O \rightarrow CaCO_3 + 2\ NH_4NO_3 \tag{11.2}$$

Finely divided barium sulfate is only accessible by synthetic production routes. Raw materials for the synthesis are either barium sulfide or barium chloride. BaS is obtained by reductive roasting of barite ($BaSO_4$) or from $BaCl_2$, which is formed as byproduct of the synthesis of lithopone pigments. The production of BaS by roasting requires barite qualities with low silicate contents, because otherwise an undesired barium silicate is formed as a byproduct. BaS is dissolved in hot water, filtered, and then converted with sodium sulfate solution to $BaSO_4$ (blanc fixe) and dissolved sodium sulfide.

$$BaS + Na_2SO_4 \rightarrow BaSO_4 + Na_2S \tag{11.3}$$

Particle size and quality of the obtained blanc fixe depend strongly on the precipitation conditions. Finer grades are formed at higher precipitation temperatures and higher concentrations. The precipitates are washed with water and dried before they are ground to the desired particle size. The blanc fixe formation, starting from $BaCl_2$ and Na_2SO_4 solutions, leads to dough masses. These are commonly offered after washing as pastes with a solid content of about 70%.

Aluminum hydroxide produced using the Bayer process is too coarse grained and not suitable for filler applications. Finely divided products (particle sizes smaller than 0.5 µm) are synthesized, starting from sodium aluminate solutions. The adjustment of particle size is made by specific seed processes (precipitation from cooled and highly diluted solutions) or by neutralization of the aluminate brine with sulfuric acid, carbon dioxide, $NaHSO_4$, or $NaHCO_3$.

$$2\ NaAl(OH)_4 + CO_2\ \rightarrow\ 2\ Al(OH)_3 + Na_2CO_3 + H_2O \qquad (11.4)$$

X-ray amorphous aluminum hydroxides are possible to produce by precipitation from acidic aluminum salt solutions with ammonia, sodium hydroxide, or sodium aluminate solutions. The processing temperatures for all $Al(OH)_3$ grades must stay below the decomposition temperature at 205 °C.

Synthetic amorphous SiO_2 can be produced by a pyrogenic or thermal process (for pyrogenic or fumed silica) and by a wet chemical process (for precipitated or particulated silica). Fumed silica is synthesized by the flame pyrolysis of silicon tetrachloride in a hydrogen-oxygen atmosphere.

$$SiCl_4 + 2\ H_2 + O_2\ \rightarrow\ SiO_2 + 4\ HCl \qquad (11.5)$$

Hydrophilic-fumed silica, containing hydroxyl groups at the surface, is formed under these conditions. Hydrophobic-fumed silica is produced from hydrophilic-fumed silica by an appropriate surface treatment with silanes, siloxanes, or silazanes. The SiO_2 grades obtained are white, finely divided, and have a large surface area.

SiO_2 can also be synthesized by precipitation from a sodium silicate solution by the addition of sulfuric or hydrochloric acid. The precipitate is separated by filtration, washed, dried, and milled. Porosity, specific surface area, particle size and shape, density, and hardness are dependent on the process parameters (e.g., concentration, pH value, dosing rate, and temperature).

Synthetic amorphous silica gel is wet and chemically produced by the reaction of sodium silicate with sulfuric acid in aqueous solution.

$$Na_2SiO_3 + H_2SO_4 + x\ H_2O\ \rightarrow\ SiO_2 \cdot x\ H_2O + Na_2SO_4 + H_2O \qquad (11.6)$$

Precipitates obtained under these conditions contain about 75% water. They are filtered, washed, and then dried in a rotary kiln. The material is washed with hot alkaline water to reinforce the matrix, decrease shrinkage, and produce larger pores. The process for the synthesis of precipitated silica is closed by a milling step.

All fillers exhibit characteristic surface properties after their manufacture. These characteristics are dependent on the composition and on the granulometric characteristics. Decisive for many applications is the nature of the surface with respect to hydrophilicity or hydrophobicity. The surfaces of filler materials are hydrophilic in most cases. On the other hand, a hydrophilic surface is not beneficial for all applications. Therefore, the surface of certain fillers must be treated to obtain the necessary properties for the application. The surface treatment of fillers is done mostly using organic compounds. Fatty acids, e.g., stearic acid, are used in most cases for the treatment of carbonates, silicon dioxide, and silicates. The acid reacts with the surface of the filler to form thin layers of new compounds with the desired properties. As an example, the surface treatment of $CaCO_3$ with stearic acid can be demonstrated by the following equation [4]:

$$CaCO_3 + 2\ C_{17}H_{35}COOH \longrightarrow Ca(C_{17}H_{35}COO)_2 + CO_2 + H_2O \tag{11.7}$$

The reaction proceeds at temperatures above 80 °C. The amount of the fatty acid typically depends on the specific surface area of the filler particles to be treated [6, 7].

Organofunctional silicon and titanium compounds are also used for the surface treatment of fillers. Ester groups, silane, and titanate rests react in the presence of water with hydroxyl groups on the surface of the filler particles, and new chemical and physical bonds are formed. This leads to an effective functionalization of the filler surface. Properties such as the dispersion behavior in a distinct application medium can be steered in this way.

11.3 Filler properties and uses

Calcium carbonate fillers are used with a broad variety of particle sizes. Dry ground types are characterized by sizes from 5 to 20 μm (dry ground) and from 2 to 5 μm (wet ground). PCC types obtained by precipitation have particle sizes in the range from 0.01 to 0.35 μm. The corresponding values for the specific surface area can range from 1 to 30 m^2/g.

$CaCO_3$ fillers are mainly applied in polyvinyl chloride (PVC) and thermosets, primarily in fiberglass reinforced unsaturated polyesters (UP). The remaining quantity is used for the most part in polyethylene (PE), polypropylene (PP), ethyl-propyl-dien rubber (EPDM), polyurethane (PUR), polyamide (PA), and acrylonitrile butadiene-styrene (ABS).

The use of fillers in PVC is dominated by various $CaCO_3$ types (the estimated ratio of GCC types to PCC types is 7 to 1). Advantages of GCC are its high availability in different particle sizes and low production costs. GCC contributes to increased stiffness, strength in rigid formulations, low shrinkage, and improved dimensional stability of the PVC where it is used. GCC finds application in flexible PVC hose extrusion compounds, vinyl electrical wire insulation compounds, plastisols, organosols, or rigid PVC.

Typical for PCC are fine particle size and a narrow size distribution. It imparts specific properties to high value rigid PVC, e.g., improved mar resistance, increased surface gloss, reduced flex whitening, reduced plate-out, increased impact strength, and higher modulus. The addition of PCC improves the processing of rigid PVC and can, therefore, be regarded as a processing aid. The fine size of PCC is compatible with the PVC primary particles, which leads to an improvement of the dispersion of the formulation components.

The main application of GCC types in unsaturated polyesters is found in bulk molding compounds (BMC) and sheet molding compounds (SMC). Other uses are in hand lay-up and spray-up applications as well as in casting. The use of the GCC fillers improves the surface finish of the polymer by reduction of shrinkage and modification of the rheology in a positive manner.

Calcium carbonate fillers usually provide bulk molding compounds consisting of polyolefins with a higher ductility and mechanical stability. Applications in polyolefin parts are found in the automotive sector (automotive interior, exterior, under the hood), in household items, and in furniture (molded, filled polypropylene). The use of $CaCO_3$ fillers leads to increased stiffness and heat distortion temperature, and improves the dimensional stability of the polymer.

Calcium carbonate is also used in polyolefin film applications. Stiffness, dart impact resistance, and tear strength are improved for blown polyethylene films and biaxially oriented polypropylene films. $CaCO_3$ fillers in polyolefin films increase the surface roughness at low filler concentration and minimize "blocking" (tendency of adjacent film surfaces to stick together under pressure).

Barium sulfate fillers of the blanc fixe type are used in almost all applications, especially in paints, coatings, and plastics. They are resistant to water, acids, alkalis, and organic solvents as well as to light and industrial waste gases. They are also characterized by good dispersion properties, high gloss, and low binder demand. Most of the blanc fixe types have good matting properties. The mean particle sizes of $BaSO_4$ fillers are typically in the range from 0.7 to 4 µm. The surface of the particles is smooth and closed. Due to a high degree of whiteness, neutral white shades as well as color hues with high brilliancy are possible. Barium sulfate fillers have a relevant potential for the joint use with anticorrosive pigments in corrosion-resistant coatings.

$BaSO_4$ fillers can be used in a favorable manner in outdoor applications, e.g., in automotive and industrial coatings. They are particularly suitable for use in powder coatings. Other relevant applications are interior paints, antifouling ship paints, printing inks, and smoothing compounds.

Aluminum hydroxide is the largest volume flame retardant used in the world. More than 40% of all flame retardant chemicals consist of this filler material. Available ATH grades have particle sizes from 0.01 to 80 µm and specific surface areas from 1 to 40 m^2/g. The particle size chosen depends on the specific application purpose.

ATH has been used as a flame retardant since the 1960s. It combines flame retardant and smoke suppression properties. Applications for these fillers are transportation, construction, cast polymers, electrical and electronic devices, wires and cables, and leisure articles and appliances. ATH is used in a broad variety of polymers ranging from thermosets via sheet molding compounds and bulk molding compounds to wires, cables, and other thermoplastic applications (compounded with polyvinyl chloride, polyolefins or EPDM).

Aluminum hydroxide is very stable up to temperatures of about 205 °C. Above this temperature, $Al(OH)_3$ is endothermically decomposed to aluminum oxide and water.

$$2\ Al(OH)_3 \;\rightarrow\; Al_2O_3 + 3\ H_2O \tag{11.8}$$

The decomposition takes place slowly between 205 °C and 220 °C and becomes very fast above 220 °C. When used as a flame retardant, the water released from aluminum hydroxide during decomposition dilutes the gases that feed combustion. This dilution decreases the rate of polymer combustion. A vapor barrier is formed that prevents oxygen from reaching the flame. In addition, $Al(OH)_3$, with a heat capacity of 93.07 J/(mol · K), operates as a heat sink and absorbs a part of the combustion heat.

The loading level of ATH in a polymer is commonly very high and can exceed 60 wt%. As a consequence, the polymer properties are modified to a certain extent. An advantage for the use of aluminum hydroxide compared with some other flame retardant materials is its nontoxicity.

Silicon-containing fillers such as silica and silica gel are widely used in paints, coatings, and plastics. They act as surface modifiers with the aim to modify the appearance or to improve the performance characteristics of a polymer material. Properties such as anti-blocking, slipping, abrasion resistance, matting, and scratch resistance can be influenced positively by the presence of silica fillers. Silica is also used as white reinforcing filler in plastics and elastomers.

Diatomaceous earth or diatomite is a special silicon-containing naturally occurring material suitable for filler applications. Diatomite is an amorphous chalky sedimentary rock composed of skeletons of single-cell aquatic plants (diatomites). It is grown in a variety of different shapes and sizes, which range from 10 μm to 2 mm. Opal-like, water-containing amorphous silica is the basis of the skeletons, which are characterized by a high porosity in their structure. The calcination of diatomaceous earth for the use as filler leads to a loss of water of up to 40%. Commercially used diatomite fillers contain about 90 wt% SiO_2. Their porosity can take on values of up to 90%.

Fumed silica as an amorphous, white, and extremely fluffy synthetic filler is characterized by very small particle sizes, large surface area, high purity, and a refractive index of 1.46. The specific surface area of the fumed silica ranges from 100 to 400 m^2/g depending on the route and the parameters used for the synthesis. These high surface areas strongly affect the rheological characteristics, thixotropic and dispersion behavior, as well as their reinforcement properties. Higher surface areas are responsible for a more difficult dispersibility. They improve, however, the rheological control, the thixotropic behavior, and the reinforcement efficiency.

The structure of synthetic amorphous silica gel is based on a random arrangement of interlinked spherical silica particles. Diameters of silica gels are in the range from 2 to 10 nm, and specific surface areas range from 300 to 1000 m^2/g. The refractive index is 1.46, comparable with that of fumed silica. This value is very close to that of polymers such as polyethylene (1.52) and polypropylene (1.50), which are relevant for the use. Polymer films, containing fumed silica or silica gel as fillers, retain their high transparency and clarity due to a small difference in the refractive indices of the fillers and the polymer materials.

Silica fillers are used in high-quality plastic films mainly due to their appropriate antiblocking characteristics. Typical polymers used for such films are low density polyethylene (LDPE), polypropylene (PP), linear low density polyethylene (LLDPE), and polyethylene terephthalate (PET). PP and polyvinyl chloride (PVC) are preferably provided with synthetic silica fillers. Silica grades with larger particle sizes may be used in thicker polymer films. Synthetic silica fillers are significantly more efficient than natural types for blocking. They have, in addition, an improved adsorption capacity for PVC plasticizers due to a higher porosity.

The rheological properties of coatings can be improved by the use of synthetic silica fillers. Both hydrophilic and hydrophobic types are used for this purpose. They are also applied as flow control agents and anti-settling additives for pigments in coatings. In addition, the use of silica fillers may improve the scratch and abrasion resistance of transparent coatings. The application of these fillers does not negatively impact the transparency of the coatings because the difference of the refractive indices of the silica and of most automotive and industrial coatings is very small.

Silicates, mostly kaolin, a white porcelain clay, are mainly used as fillers in paper production. Kaolin makes the paper partially or completely opaque, whiter, and increases the bulk density. The filler also gives the paper a smoother surface as it fills the voids between the fibers. Paper can contain up to 30% filler, depending on the grade. Carbonates are often used as fillers, mostly chalk, but also sulfates, such as gypsum ($CaSO_4 \cdot 2\,H_2O$), or oxides, such as titanium dioxide (TiO_2). Barium sulfate is used as a filler for the production of the so-called barite paper. With the exception of titanium dioxide, fillers are less expensive than the pulp used. The use of fillers can, therefore, reduce the manufacturing costs of a paper. Today, the addition of fillers has hardly any noticeable effect on some of the mechanical properties of the paper. In particular, the tensile strength is not noticeably reduced, since in modern paper formulations, optimized additives and sizing agents are used to achieve a strongly resilient bond between the pulp and the filler. However, fillers serve to increase the degree of whiteness, influence the basis weight, and shape the paper absorbency, which are of great importance for print quality.

In the case of building materials and related materials, fillers are grouped under the term "aggregate".

Foodstuffs also contain a number of fillers (additives). These form a part of the volume of the food without contributing significantly to its usable energy content. This reduces the actual energy content per volume or per mass of the food. Fillers are thus used here to reduce the physiological calorific value of a food (e.g., in light products) and/or to increase the volume of a food (e.g., chewing gum). Some fillers additionally act as dietary fibers. The most important fillers in this application segment are cellulose (E 460), cross-linked carboxymethyl cellulose (E 468), and polydextrose (E 1,200). The use of fillers must be declared on the list of ingredients of the food. The functional class and the respective food additive or E number are mentioned. Different maximum quantity restrictions apply depending on the product and food additive.

In addition to the active substance and excipients, pharmaceutical products in the form of tablets also contain fillers, usually lactose, glucose, sucrose, starch, calcium sulfate, or microcrystalline cellulose, depending on the formulation.

11.4 Toxicology and occupational health

Based on the criterion of composition, most fillers can be regarded as harmless. From the occupational health and safety perspective, operation with filler dust requires special attention. Inhalation of the dust is most critical and must, therefore, be avoided.

The use of silica fillers especially requires caution. It is important in this case to differentiate between amorphous and crystalline forms of SiO_2. Crystalline silicon dioxide is classified as carcinogenic for humans. Occupational exposures to respirable microcrystalline silica (quartz, cristobalite) must be avoided to protect employees dealing with the material against lung cancer, silicosis, pulmonary tuberculosis, and airway diseases [8–11]. Unlike crystalline silicon dioxide, amorphous silica is not associated with these critical diseases. Natural silicates with larger amounts of crystalline SiO_2 must be handled similarly to pure crystalline silicon dioxide.

11.5 Conclusions

Fillers are powdered solid materials which are added to application media in order to improve properties of the system or to lower the consumption of more expensive binder components. Fillers that are used mainly for cost reduction reasons are also called extenders.

Fillers are often applied in combination with pigments to improve the properties of the medium in which they are embedded. The shape of filler particles can be irregular, acicular, fibrous, or platy. Relatively large loadings in the application system are typical for the use of fillers. Fillers, like pigments, are insoluble in the medium used. They have the ability to improve coatings and other application systems with regard to density, porosity, and permeability for gases and water. There is no clear differentiation between fillers and pigments with respect to the refractive index, but it is generally accepted that this index is below 1.7 for fillers and above this value for pigments.

Several naturally occurring white minerals, such as calcium carbonate, magnesium carbonate, silicon dioxide, and some silicates, are used as fillers. Demands for consistent quality and, preferably, finely divided fillers have led to the development of synthetic filler materials. These have especially established themselves in the coatings and plastics industry. Calcium carbonate, barium sulfate (blanc fixe), aluminum hydroxide, and silicon dioxide belong to the most important synthetic fillers. Calcium carbonate has the largest share of all fillers. The tendency of some fillers to agglomerate during storage or absorb humidity or additives of the application medium can be significantly reduced by the use of suitable surface treatment. Mining and processing technologies for natural fillers correspond to those used for the manufacture of natural pigments. Synthetic fillers are produced mainly by precipitation processes. The properties of synthetic fillers are controlled and adjusted by the process parameters chosen.

Main uses for fillers are found in plastics, coatings, paints, printing inks, paper, cosmetics, and drugs, but they can also be used in other application systems. Calcium carbonate is mainly used in plastics, and barium sulfate in automotive and industrial coatings, paints, and plastics. Aluminum hydroxide fillers are broadly used as flame retardant materials in many polymers. Silicon-containing fillers such as silica and silica gel are mainly applied in paints, coatings, and plastics.

Study questions

11.1	How are fillers defined?
11.2	Which natural and synthetic filler materials are technically important?
11.3	Which processing technologies are used for the production of natural and synthetic fillers?
11.4	What are the main application systems for fillers?

Bibliography

[1]   Pfaff G, in Winnacker–Küchler: Chemische Technik, Prozesse und Produkte, vol. 7, Industrieprodukte, 5th edn. Dittmeyer R, Keim W, Kreysa G, Oberholz A, eds. Wiley-VCH, Weinheim, 2004, 290.
[2]   Xanthos M, in Functional Fillers for Plastics, Xanthos M, ed. Wiley-VCH, Weinheim, 2005.
[3]   Wypych G. Handbook of Fillers, 3rd edn. ChemTec Publishing, Toronto, 2010.
[4]   Gysau D. Füllstoffe – Grundlagen und Anwendungen, Vincentz Verlag, Hannover, 2005.
[5]   Gysau D. Fillers for Paints, 2nd edn. Vincentz Verlag, Hannover, 2011.
[6]   Balard H, Papirer E. Progr. Org. Coat. 1993,22,1.
[7]   Kovacevic V, Lucic S, Hace D, Cerovecki Z. J. Adhes. Sci. Technol. 1996,10,1273.
[8]   Pandurangi RS, Seehra MS, Razzaboni BL, Bolsaitis P. Environ. Health Perspect. 1990,86,327.
[9]   Reuzel PGJ, Bruijntjes JP, Feron VJ, Woutersen RA. Food Chem. Toxicol. 1991,29,341.
[10] Merget R, Bauer T, Kupper HU, Philippou S, Bauer HD, Breitstadt R, Bruening T. Arch. Toxicol. 2002,15,625.
[11] O'Reilly KMA, Phipps RP, Thatcher TH, Graf BA, van Kirk J, Sime PJ. Am. J. Physiol. Lung Cell Mol. Physiol. 2005,288,1010.

12 Application systems for pigments

Automotive and industrial coatings, paints, plastics, printing inks, cosmetic formulations, and build-ing materials are the most important application systems for pigments. Paper, rubber, glass, porce-lain, glazes, and artists' colors are further media which are often colored with pigments and modified with fillers. Some of the application systems are only of relevance for inorganic pigments and fillers because organic pigments do not exhibit the necessary stability. Five important application systems will be considered in greater detail below.

12.1 Coatings

Depending on their compositions, coatings or paints can be divided into three groups: (1) solventborne, (2) waterborne, and (3) solvent-free (100% solid). Solventborne systems consist of resins, additives, and pigments that are dissolved or dispersed in organic solvents. Similarly, in waterborne paints, the ingredients are dispersed in water. In solvent-free compositions, the paints do not contain any solvent or water, and the ingredients are dispersed directly in the resin (powder coatings).

The final use of the coated product determines which type of coating is chosen. The so-called organic paints are often used in industrial and architectural applications, where appearance and function are of high importance. The so-called inorganic paints are chosen when protective properties are needed. Hybrid coatings are used to an increasing extent for specific industrial applications.

The properties of coating films are determined by the types of binders, pigments, and several additives (catalysts, driers, flow modifiers) used in the formulation. The nature of the substrate, substrate pretreatments, application methods, and conditions of film formation play additional roles in determining the end properties of a coating. The terms "coating" and "paint" are often used synonymously. In general, paints, var-nishes (transparent solutions), and lacquers (opaque or colored varnishes) are known as coatings [1, 2]. A coating can be defined as a dispersion that consists of one or more binders, volatile components, pigments, and additives [2, 3].

Drying of a liquid coating is mostly done by evaporative means or curing (cross-linking) using oxidative and thermal conditions or ultraviolet light. The binder (poly-mer or resin) is the component that forms the continuous film, adheres to the sub-strate, and holds the pigments and fillers in the solid film. The volatile component is the solvent that is used to adjust the viscosity of the formulation to ensure easy application.

Coatings occur in both organic and inorganic forms. Inorganic coatings are mainly applied for protective purposes, while organic ones are mostly used for decorative and functional applications [4, 5]. Organic coatings can be classified as either automotive or

https://doi.org/10.1515/9783110743920-012

industrial coatings. Industrial coatings may be subdivided into architectural (building, wall, and ceiling coatings), furniture, or coil coatings. Although organic and inorganic coatings may be used individually for industrial applications, for specific requirements, a combination of both systems is favored (hybrid coatings).

Coatings are usually applied as multilayered systems that are composed of both a primer and a topcoat. However, in some cases, for example, automotive coating systems, this can vary from four to six layers [2, 6]. Application can be performed manually, for example, with brushes or rollers, or by mechanical methods such as spraying, atomization by rotating disks or cones, dipping, pouring, rotating drums, and tumbling equipment, or automated application by rollers. Powder coatings are applied by electrostatic spraying or by dipping. Multicomponent coatings are applied with special multicomponent spraying equipment [7]. Each coating layer is applied to perform certain functions. The characteristics of each layer are influenced by the other layers of the system. The interactions between different layers and the interfacial phenomena play an important role in the overall performance of multicoat systems. Typical layer systems used for automotive and other special coatings consist of a substrate coated with a primer (for adhesion, durability, and anticorrosion), a basecoat (for opacity, color, flexibility, scratch resistance, diffusion chemical resistance, water uptake, and weather resistance), and a clearcoat (for light reflection, hardness, scratch resistance, surface slip, repellent properties, and erosion resistance). Pigment particles used to achieve a certain color tone or effect for the coating are located in a homogeneously dispersed manner in the basecoat.

Functional coatings or smart coatings are specific systems that possess an additional functionality besides the classical properties of a coating (i.e., decoration and protection) [2, 8]. This additional functionality may be diverse and depends upon the actual application of a coated substrate. Typical examples of functional coatings are self-cleaning and self-healing [9–12], easy-to-clean (antigraffiti) [13, 14], antifouling [15], soft feel [16], and antibacterial or antimicrobial [17–19]. Special pigments and fillers are used to achieve these functional properties. The term "smart coatings" goes back to a concept based on the assumption that coatings are able to sense their environment and to make a response to the source of stimulation [2].

Many coatings are based on organic polymers that serve as a backbone or matrix in film building. At the same time, these polymers have to integrate pigment and filler particles of differing morphology during the preparation of the coating. The surface to be coated can be inorganic (e.g., metals or metal alloys) or organic (e.g., plastics, wood, or composites). Distribution and orientation of the pigment particles in the final film affect its physical–chemical properties significantly. Another factor is the environment during application, film building, and treatment of the paint film (atmosphere, temperature, humidity, and downdraft speed). The most important properties of a coating are rheology (influence on the application properties), hiding power (important for extinction), pigment volume concentration (influence on stress, water uptake, water vapor permeability, and weather resistance), water vapor and water permeability (influence on the moisture balance of the substrate and thus the durability and protection of the substrate against weathering), adhesion to the substrate (particularly with alternating moist-dry, moist-hot, and moist-frosty conditions), and elasticity (in the case of crack covering coatings and wood coatings).

Coatings are produced from numerous components depending on the method of application, the desired properties, the substrate to be coated, and ecological and economic constraints. The components are classified as volatile or nonvolatile. Volatile coating components are organic solvents, water, and coalescing agents. Nonvolatile components include binders, resins, plasticizers, paint additives, pigments, and fillers. Chemical hardening of the binder can lead to volatile condensation products such as water, alcohols, and aldehydes, which are released into the atmosphere.

Most of the coating materials consist of solvents, binders, and pigments. The amounts of additives are comparatively small. Additives are responsible for an improved flow behavior, better wetting of the pigments, and catalytic acceleration of hardening. The binder as the most important component of a coating composition determines the application method, drying and hardening behavior, adhesion to the substrate, mechanical properties, chemical resistance, and resistance to weathering [3–7, 20].

Binders and resins are macromolecular products with a molecular mass between 500 and 30,000. Cellulose nitrate as well as polyacrylate and vinyl chloride copolymers are components of higher molecular weight products. Low-molecular-weight products contain alkyd resins, phenolic resins, polyisocyanates, and epoxy resins. Appropriate films are formed by the chemical hardening of these binders after application to the substrate. These films contain cross-linked macromolecules. Most binders are synthetic resins such as alkyd or epoxy resins. The paint industry also uses many synthetic hard resins, mainly based on cyclohexanone, acetophenone, or aldehydes [3–7, 20].

Plasticizers are organic liquids of high viscosity and low volatility. They lower the softening and film-forming temperatures of the binders and improve flow, flexibility, and adhesion properties. Examples of plasticizers are the esters of dicarboxylic acids.

Coating additives are auxiliary products added to coatings in small amounts, to improve specific film properties. The additive content is typically between 0.01% and 1%. Paint additives are used to prevent defects in the coating (e.g., foam bubbles, poor leveling, flocculation, and sedimentation) or to impart specific properties (e.g., better surface slip and UV stability) [21–24].

Additives are divided into the following categories: defoamers, wetting and dispersing additives (an important category for pigments in paints, in general), surface additives (an important category for pigments in paints, in general), rheology additives (an important category, e.g., for special effect pigments in paints), driers and catalysts, preservatives, and light stabilizers.

Wetting agents are one of the largest groups of coating additives. They belong to the surfactants and are added to coating compositions to aid wetting of the pigments by the binders and to prevent flocculation of the pigment particles, thus leading to the formation of a uniform, haze-free color and a consistently high luster of the coating film. Examples for wetting agents are sodium polyphosphates, and sodium and ammonium salts of low-molecular-weight polyacrylic acid. Wetting agents are materials with a typical polar–nonpolar surfactant structure. They reduce the surface tension of the binder solution to such an extent that the contact angle becomes zero, which is the requirement for good wetting.

Dispersing additives are stabilizing substances that are adsorbed onto the pigment surface via pigment-affine groups (anchor groups with a high affinity for the pigment surface) and establish repulsive forces between individual pigment particles [25]. Stabilization is achieved either via electrostatic charge repulsion or via steric hindrance due to molecular structures that project from the pigment surface into the binder solution [26–29]. Additive molecules can have relatively simple fatty-acid-based chemistry or more complex polymeric structures. In coatings with water-soluble resins, both mechanisms can be employed, but in practice steric stabilization is often preferred, especially in high-quality coating formulations. Good adsorption of the additive to the pigment surface is necessary for efficient stabilization. Wetting and dispersing additives can also solve flooding and floating problems.

The process of drying and film formation is very important for the quality of the final coating. As the paint dries on the substrate, a firmly bonded film is formed. The properties of this film are determined both by the substrate and its pretreatment (cleaning and degreasing), as well as by the composition of the coating and the application method used. Drying of the wet coating film takes place physically (evaporation of the organic solvents from solventborne paints, evaporation of water from waterborne paints, and cooling of the polymer melts for powder coatings) or chemically (reaction of low-molecular-weight products with other low- or medium-molecular-mass binder components by polymerization or cross-linking to form macromolecules). Film formation can be accelerated by drying at elevated temperatures (forced drying). In practice, the drying of coatings and paints does not take place by one method alone.

Several scientific fields are involved in the development of a coating, in its production, and in its application: inorganic chemistry (synthesis of inorganic pigments and fillers, and surface treatment of pigments and fillers), organic chemistry (synthesis of organic pigments and fillers, binders, and additives), metalorganic chemistry (pretreatment of substrates, surface treatment of pigments and fillers, adhesion promotion, and improvement of weather fastness), physical chemistry (interface and colloid properties, electrochemical aspects, and photochemical properties), material science (substrate development, nanomaterial aspects, and mechanical and functional properties), physics (mechanical properties, optical aspects, color theory, and structural colors), metallurgy (substrate properties, corrosion research, and metallic effect pigments), and experimental psychology (vision science and haptics). Coatings can, therefore, be regarded as an interdisciplinary field with the label "high tech". Some coating and paint systems and their applications were particularly developed in the past decades and have potential for further improvement: automotive coatings, dispersion paints (emulsion paints), powder coatings, coil coatings, plastic coatings, and radiation curing systems.

Most important for automotive coatings are durability of the exterior and interior finishes and the aesthetic characteristics. Each automotive company defines its color and appearance standards to meet customer expectations and to arouse interest in new colors and effects. The quality of automotive coatings can be judged among other things by the criteria protection against harsh environments, durability, and quality of appearance. In order to meet all requirements, only raw materials of highest quality are used. Effect coatings containing effect pigments have found more and more acceptance and are a particularly growing business segment [6, 30].

Dispersion paints or coatings are viscous coatings consisting of a chemical dispersion (usually an emulsion) of binders and solvents, pigments, and additives. In this general sense, the majority of liquid coatings (varnishes, paints, etc.) are dispersions. In colloquial terms, this usually refers to the approximately kilogram containers of white wall paint in oval buckets that can be found on pallets in any hardware store. Strictly speaking, these are synthetic resin dispersion paints. In addition, there are products on the market with a similar structure, but which use nonsynthetic and nonmineral oil ingredients as far as possible, and which are usually, somewhat vaguely, referred to as natural dispersion paints [2, 5].

Powder coatings contain 100% solid and are applied as dry powders that are formed to a film by heating. Typical components of a powder coating are binders (resins with hardeners and accelerators), pigments, additives, and fillers. Pigments used must fulfill certain requirements, such as thermal stability at the curing temperature, suppressed reaction with other components of the formulation, and stability to shear forces during extrusion and grinding. The solid binder melts during heating and binds the pigment [7, 31].

Coil coating is a process for coating flat approximately 2-m-wide steel or aluminum sheets on one or both sides. The resulting material is a composite of a metallic substrate and an organic coating. Common coating materials are paints and plastic films. Hot-dip galvanized steel or aluminum is used almost exclusively for the coil coating process. After production, the metal sheets are delivered to the coating plant in coiled condition. There they are uncoiled and fed through an accumulator or sheet extrusion system. In the next step, the sheets are freed from grease and oils that were applied to protect the surface after metal sheet production. This is done by alkaline cleaning. Subsequently, the sheets are chemically passivated by a pretreatment. A first coating layer, the primer, is then applied in a rolling process and baked at around 240 °C. A second coating layer, the topcoat, is deposited on the primer by a rolling procedure and again baked at around 240 °C. The coating layers are then typically passed through a curing process. The coating can additionally be protected by a laminating film. Finally, the metal sheets are again fed over an accumulator and rolled up into a coil. Sheets with coil coatings find a wide range of applications: facade construction (mostly for functional industrial buildings), household appliances (e.g., washing machines, microwave ovens, and refrigerators), lamp boxes, blinds, roofing systems, computer housings, metal furniture, and cassette systems for interior ceilings [32–34].

Plastic coating is the application of liquid polymers onto the surface of an object through dipping or immersion techniques. A thick plastic finish for protective and decorative purposes is formed by this process. The coated surface is equipped with additional resistance against scratches, wear, corrosion, and external influences. Plastic coatings are often seen in hand tools, handles, and grips. Other applications are shopping carts, baskets, forceps, covers, caps, and plugs. They are also applied because of their special heat and electrical insulating properties. Thus, they can serve as insulation layers in electronic components, for example, in wires, cables, and digital meter probes. The process of casting a plastic material onto a metal piece is called dip molding. The metal piece that is coated acts as the mold for the plastic or polymer coating which adheres at the surface. The surface to be coated is treated and preheated before it is dipped or immersed into the liquid polymer. The liquid polymer adheres to the metal surface and hardens upon cooling. The process of dip molding is related to dip coating, except that dip molding has an additional demolding or unloading step [35, 36].

Radiation curing is a technical process that uses electromagnetic radiation (mainly in the UV range) or ionizing radiation (mostly accelerated electrons) to induce a chemical chain reaction that leads from polyfunctional compounds to cross-linked polymers. UV curing of pigmented films is a challenging drying process and needs optimized parameters [37]. The main challenge is that the properties of pigments and photoinitiators in a coating layer compete when exposed to UV radiation. Using a suitable combination of photoinitiators, lamps, and curing conditions, the technical problems during curing can be overcome [38]. The raw materials for UV-curable coating systems are medium-molecular-weight resins. Main types of these resins are based on polyesters, polyepoxides, polyethers, polyurethanes, and epoxides. Driven by the technical advantages associated with a beneficial sustainability, pigment manufacturers develop specific product portfolios for radiation curing systems.

12.2 Plastics

Owing to the growing number of different plastics on the market, inorganic pigments and other colorants have to meet increasingly high requirements [39–43]. In making a sensible choice of pigments, the crucial factors besides the plastics are the processing method and the required fastness level. The intrinsic color of the plastic used is, in addition to the colorant, an important factor which has to be regarded for the production of the desired coloration. A very wide variety of additives such as antioxidants, UV absorbers, HALS stabilizers (hindered amine light stabilizers), plasticizers, slip agents, antislip agents, blowing agents, antistatic agents, flame retardants, and fillers are added to virtually all plastics.

Most plastics, for example, polyolefins and polystyrenes and their derivatives such as ABS (acrylonitrile butadiene styrene) and SAN (styrene acrylonitrile) are supplied by the manufacturers in a ready-to-use form. Most of the necessary additives are already incorporated in the polymer before it is sold. On the other hand, in the case of other polymers, for example, PVC, the end user has to add the additives and colorants. Fluid and high-speed mixers are suitable devices for the incorporation of additives. Gravity mixers or tumble mixers are also possible to use. The mixtures are homogenized on mixing rolls, kneaders, planetary extruders, or twin-screw kneaders and further processed.

The processing methods are very much governed by the desired end product. Injection molding, injection blow molding, extrusion into blow molding, calendaring, foaming, coating, as well as production of flat films, blown films, profiles, and fibers, are typical processes used. Injection molding machines or single-screw extruders do normally not have the shear forces necessary for the dispersion of the pigments in the polymer melt. This step is very often done by specialized producers of masterbatches. Specially equipped twin-screw extruders, kneaders, triple-roll mills, or bead mills in a suitable carrier material are used for this purpose. The pigment concentration of masterbatches and other preparations depends on the end product and the desired concentration in use. In preparations, the pigment is present in a well-dispersed form with the appropriate choice of polymer. Uniformly, homogeneously, and reproducibly colored plastics are, therefore, possible to produce under normal processing conditions [44].

Pigment preparations are offered in solid form as granules and powders or as pastes. They normally contain between 5% and 50% pigment. Such preparations are typically suitable for only a few plastics depending on the type of carrier material used. General-purpose preparations are possible to produce, but are of minor importance compared to the preparations tailored to specific plastics.

Good dispersion of the pigments in a preparation, and subsequently good distribution in the end product, requires an increase of the flowability of the preparation above that of the plastic to be colored. This can be achieved by using carrier materials with a low molecular weight wax, a copolymer, or an increased plasticizer content.

Color concentrates are of major importance for the coloration of plastics. These concentrates are known as color preparations that contain two or more colorants and produce a defined shade when mixed with a specific amount of the uncolored polymer. Pigment mixtures that have been mixed previously in the correct ratio to give desired shades are used for this purpose in some cases. High-speed mixers are mostly used to produce such mixtures.

Most inorganic pigments have larger particles than organic pigments. They cause significantly less dispersion problems compared with organic pigments. Mixtures with organic pigments are possible without any negative influence on the overall properties.

Plastic raw materials are offered and processed in powder or granulated form. Several processing methods have been introduced for the successful coloring with pigments or pigment preparations. Plastics in powder form are often premixed with pigment powders in fluid mixers. The pigments are dispersed by this method while the plastic is being processed.

Plastic granules are typically colored using granular pigment preparations or paste preparations, for example, pigment plasticizer pastes. Granular pigment preparations are advantageous for automatic metering by volumetric or gravimetric metering units and metering pumps. Such granules are, however, of only limited suitability for the coloring of plastics in powder form. The polymer–pigment distribution in the final plastic is often insufficient. In addition, there is a risk of separation particularly with pneumatic conveying.

Solvent fastness of the pigments in the plastic, particularly under processing conditions, is of high importance. Migration problems such as blooming and bleeding are based on the complete or partial "dissolving" of the pigment in the polymer at its processing temperature. Recrystallization is also attributable to pigments that have a certain "solubility" in the plastic. As a result, a change in transparency or opacity in transparent colorations and in the depth of shade in white reductions is observed.

The use of pigments in a number of plastics is restricted by high processing temperatures. The residence time (duration of thermal stress) is an important parameter here. Other factors are the pigment concentration as a function of heat resistance and the ratio of colored pigment to white pigment. The limit concentration, that is, the ratio of lowest possible amount of colored pigment to titanium dioxide, has to be determined in a test for thermal stability.

The lightfastness of a pigment depends on the polymer used. It can be low for a certain plastic after exposure to light. In other polymers, however, the pigments are almost not affected by light and fastness is high. Lightfastness in every case can only be stated for the entire pigmented system. Standards exist for determining the lightfastness of colored pigmented plastics in daylight and in xenon arc light, that is, in accelerated exposure equipment, and for determining the weathering fastness.

Inorganic pigments and other colorants applied in plastic articles used for packaging food or cosmetics must be not only migration fast and extraction resistant but also physiologically safe. Higher fastness requirements can only be met in many cases by more expensive pigments. A compromise between the fastness requirements and the price of a specific pigment often needs to be found.

Whether it is better to use organic or inorganic pigments in a particular coloration case depends on technical and economic considerations and also on regulations in individual countries. Organic pigments are preferred if a transparent coloration is desired and high color intensity, especially for thin-walled articles such as films and fibers, is needed. Organic pigments are also usually applied for multicolor printing on films. If high opacity, high light, and weathering fastness are required, inorganic pigments are the colorants of choice. The advantages of inorganic and high-quality organic pigments are often combined. The inorganic pigment with a lower color strength is usually present in excess in such cases in order to achieve the desired opacity, whereas the organic pigment used in smaller amounts primarily produces the color intensity of the pigment mixture and the brilliance of the shade.

12.3 Printing inks

Printing inks are liquid or pasty preparations consisting of binders, solvents, additives, and colorants. They are processed in high-speed printing machines [45–50]. Printing inks are similar to lacquers and paints, but typically dry faster and have a substantially smaller layer thickness after the application.

Printing inks are used in all printing processes such as letterpress printing, offset printing, flexographic printing, gravure printing, and screen printing. Sheet-fed as well as rotary printing presses are most common. Typical sample applications are books, newspapers, magazines, catalogues, packing materials, wallpapers, currency notes, stamps, and labels. Printing inks are delivered from the printing ink producers exclusively to printers and to specialized dealers and not to the private final consumers.

In order to print on flexible and rigid substrates for packaging and goods, it is necessary to adapt the printing ink to the structure of the substrate surface. Therefore, various ink systems with different formulations are produced and offered.

Industrial printing inks are highly transparent liquid systems which are specifically designed for the multicolor printing process (three-, four-, and seven-color printing process). The inks are selected in such a way that they can work effectively in the multicolor printing process with the chosen printing application. In order to achieve a maximum print quality, ink suppliers and printing houses work together according to the standardized processes. Defined standards include optical features of the three colors yellow, magenta (bluish red), and cyan (greenish blue) in the so-called color scales for several printing techniques. On the basis of the theory of colors, colored images are achieved with the multicolor printing process using these three standardized basic colors. The four-color printing process is supplemented by black and the rarely used seven-color printing process by orange, violet, and green.

Suitable binders for printing inks are mixtures of resins and solvents (all printing processes), transparent lacquers (gravure and flexographic printing), varnishes for pasty colors (offset printing), and compositions specifically designed for the needs of a specific printing process. Binders as components of printing inks are important for covering the pigments, the transfer of the ink during the printing process, the determination of the pigment particles in the printed layer after drying, and for protection of the printed ink film against chemical or mechanical influences.

Solvent-based ink systems contain solvents and nonsolvents. Problem-free and constant ink drying requires successive evaporation of the nonsolvents and the solvents. The latter are necessary for the adjustment of the viscosity/efflux time of the printing ink. Solvents which are used in physically drying inks for gravure, flexographic, screen, or tampon printing determine the drying of the inks by their evaporation time. Nonsolvents are also used to control the drying process of ink films. In addition, they act as retarders during the reduction of residual solvents. Solvent mixtures (cosolvents) are often used for the improvement of the solubility strength.

Additives for printing inks can be divided into waxes, plasticizers, driers, antioxidants, wetting and leveling agents, neutralizing agents, fungicides and bactericides, cross-linking agents, photoinitiators, defoamers, and complexing agents. Other additives used in printing inks include lubricants, thickeners, gelling agents, and preservatives.

Waxes are used to impart an improved slip, scuff, and block resistance to ink films. They consist of dispersions and emulsions of polyethylene, polypropylene, paraffin, vegetable waxes, or fatty acid amides that are typically dispersed in a solvent.

Plasticizers are applied for the improvement of elasticity by plasticizing the dried ink-film and for adhesive strength properties, especially on metal surfaces. The most common products are phthalic acid esters, often in combination with nitrocellulose in solvent applications.

Driers are soaps of cobalt, manganese, and other metals formed with organic acids such as linoleic or naphthenic acid. They catalyze the oxidation of drying oils and are, therefore, used in inks that dry by oxidation.

Antioxidants are based on oximes, substituted phenols, aromatic amines, and naphthol. They prevent or retard premature oxidation reactions and also skin formation and gelling in the ink container.

Wetting and leveling agents work on the basis of nonionic, anionic, cationic, or amphoteric compounds to pre-wet the surface of the substrate where the printing will take place. Effective interfacial substances based on silicones are used as leveling agents. Fluor additives are the most effective wetting agents.

Neutralizing agents in the printing ink react with acidic acrylates under formation of salts. Thus, water-soluble resin solutions, emulsions, or dispersions are formed. Ammonia, amines, and aminoalcohols are widely used for this purpose.

Fungicides and bactericides are preventive auxiliaries which are used for protection against the degeneration of the binder system in water-based formulations and against any odor which may form during storage.

Cross-linking agents are used to increase the resistance properties in two-component solvent or water-based ink formulations. Isocyanates are suitable agents in solvent-based printing inks and polyaziridines in water-based systems.

The purpose of photoinitiators is the acceleration of free-radical or cationic polymerization of acrylic or cycloaliphatic epoxidized oligomers and resins in UV inks in combination with reactive diluents. After polymerization, the photoinitiators should only form fragments of low residual odor.

Defoamers are based on interfacial substances like silicones, mineral oils, and polyglycols. Their purpose is the prevention of micro- and macro-foaming. Mixtures of defoamers are common. An ideal defoamer for all printing situations does not exist.

The purpose of complexing agents, mostly consisting of EDTA or tartrates, is the formation of insoluble calcium salts. EDTA works very well in buffer systems within a fountain solution.

The ink sequence during the printing process is chosen according to the printing forms and the associated printing inks. A suitable sequence is very important, especially in wet-on-wet printing, e.g., in two- or four-color offset machines. If required, the printing inks must be adjusted to a certain printing sequence. In practice, different sequences are used. Cyan, yellow, magenta, and black are typically printed successively. In the wet-on-wet printing process on two-color offset machines, the first printing process sequence might be black and yellow and the second one magenta and cyan. In the gravure and flexographic printing process, the printing sequence is always yellow, magenta, cyan, and black.

Printing is divided into four process types:

-   Planographics: The printing and nonprinting areas are on the same plane surface and the difference between them is maintained chemically or by physical properties; examples are offset printing, lithographic printing, collotype printing, and screenless printing.
-   Relief: The printing areas are on a plane surface and the nonprinting areas are below the surface; examples are flexographic printing and letterpress printing.
-   Intaglio: The nonprinting areas are on a plane surface and the printing areas are etched or engraved below the surface; examples are steel engraving and gravure printing.
-   Porous: The printing areas are on a fine mesh screen through which ink can penetrate, and the nonprinting areas are a stencil over the screen to block the flow of ink in those areas; examples are screen printing and stencil duplicator printing.

The thickness of the printed layer varies widely, depending on the printing method used. Dried layers have thicknesses from 1 to 3 μm (offset and flexographic printing) up to 15 μm (screen printing). Printing inks contain inorganic or organic pigments alone or jointly. The total pigment concentration in printing inks is typically in the range of 8–30%. Web offset inks for journals, catalogues, books, and advertising brochures contain 8–20%, sheet offset inks for books, journals, brochures, packages, calendars, and posters contain 10–30%, water-based inks for packaging (paper and cardboard), wallpapers, and wrapping paper contain 15–25%, and solvent-based inks for packaging (foils and decoration foils) contain 12–18% pigments.

12.4 Cosmetic formulations

Cosmetics are substances or products used to enhance or alter the appearance or fragrance of the human body. They are generally mixtures of chemical compounds: some derived from natural sources and some synthetic. Cosmetics include lipsticks, mascaras, eye shadows, foundations, rouge, skin cleansers and skin lotions, shampoo, hairstyling products, and perfumes. All these products are based on cosmetic formulations [51–54].

Two groups of color cosmetic products can be distinguished: those which are not so strongly affected by fashion styles (everyday products and products with a general beautifying action) and those highly affected by fashion and trends (fashion products and products that underline individuality). The first group of products serves to maintain even skin or enhance the expression of eyes, whereas the others convey the trendiness of the user and give color and optical effects.

There are internally and externally directed needs for the use of cosmetics, and of course for colored cosmetics as well. Internally directed needs are well-being, self-confidence, mood orientation, and protection. Those characteristics are mainly delivered by the formula (containing vitamins, protective oils, and UV filters), the texture (comfortable feeling, does not dry out the skin), and a beautiful casing and presentation of the product. Externally directed needs are beauty, attractiveness, femininity, variation of looks, and perfection. These characteristics are transported by a fashion-oriented brand, huge color assortment, product claims, and an appropriate packaging that transmits value and allows accurate application.

The production strategy for cosmetic products is determined by the technical requirements of the formulation, by the size of the batch produced, and the frequency of the production. Other factors are storage and filling capacity, quality control, and shelf life of intermediate products. Two different production approaches are used. In the first one, an intermediate product is made as a base and introduced in the production process together with other components of the formulation. The second approach works without such a base. All raw components are added consequently in the production process. The advantage of this method is that no change of production vessel is necessary, whereas the disadvantage is the necessarily longer time required for the process. Many pigment-containing emulsions are produced according to this method.

The manufacturing process starts usually with the preparation of a pigment dispersion. After adjustment of the color, the emulsification process is carried out. Alternatively, the manufacture of products with a wide color range starts with the preparation of a base. The dispersion method is used for all products containing large amounts of organic pigments, which are more difficult to disperse and grind than inorganic pigments. In order to achieve reproducible colors, all pigments have to be ground and dispersed very carefully and a reproducible procedure has to be established.

The color of the final cosmetic products is determined by
- the pigments and/or dyes in the formulation;
- the production process, especially the dispersion process of the pigments, which is decisive for reproducibility and the final shading of the product;
- the formula, knowing that there is a significant difference whether the formula is a liquid dispersion of the pigments in water, a wax based, or an emulsified dispersion (the same pigment composition looks different in different systems);
- the storage and the after treatment (they are important because color cosmetics are not sold in bulk size but in mg or ml; therefore, the bulk must be stored and then filled into the appropriate casings).

Important color cosmetic products are
- primers: in formulas to suit individual skin conditions;
- lip products: lipstick, lip gloss, lip liner, lip plumper, lip balm, lip stain, lip conditioner, lip primer, lip boosters, and lip butters;
- concealers: makeup that covers imperfections of the skin;
- foundations: used to smooth out the face and cover spots, acne, blemishes, or uneven skin coloration;
- face powders: used to set the foundation, giving it a matte finish, and to conceal small flaws or blemishes;
- rouge or blusher: cheek coloring to bring out the color in the cheeks and make the cheekbones appear more defined;
- contour powders and creams: used to define the face;
- bronzer: gives the skin a bit of color by adding a golden or bronze glow and highlighting the cheekbones, as well as being used for contouring;
- mascara: used to darken, lengthen, thicken, or draw attention to the eyelashes;
- eye shadows: pigmented powders/creams or substances used to accentuate the eye area, traditionally on, above and under the eyelids;
- eye liners: used to enhance and elongate the apparent size or depth of the eyes;
- eyebrow pencils, creams, waxes, gels, and powders: used to color, fill in, and define the brows;
- nail polish: used to color the fingernails and toenails;
- setting sprays, setting powders: used as the last step in the process of applying makeup, keeping the applied makeup intact for long periods.

Cosmetics can be liquid or cream emulsions, powders (both pressed and loose), dispersions, and anhydrous creams or sticks.

Pigments may also be used in classical body lotions or creams. Successful products on the market are those which use pearlescent or interference pigments to obtain a hue of color. As body products are normally applied to parts of the body that are covered by clothes, the effect cannot be seen, and even worse the color is transferred to the fabric. Therefore, color products for the body are mainly seasonal products and only applied to areas not covered by clothes.

Pigments are rarely applied in personal care cosmetics. Only pearlescent and interference pigments are the colorants, which find an application in this cosmetic segment. These pigments are used in many types of personal care applications to add color, luster, shimmer, and shine. Application examples are day creams, eye creams, anti-wrinkle creams, body lotions, toothpastes, hair and shower gels, bath salts, bar soaps, shaving gels, foot and body powders, deodorants, and antiperspirants [55, 56].

Fillers are often regarded as cheap ingredients that are added into a product just for cost reasons. In the case of cosmetic formulations, this does not apply. Cosmetic fillers enable homogeneous application of colors, support formulation stability, facilitate dispersion of pigments, and influence the texture of the finished products. They do not contribute primarily to the color of the formulation. They can, however, have an influence on optical effects and are used, for example, for wrinkle reduction. In addition, fillers help the formulator to adjust the portions of solids in colored compositions. They are used, therefore, to balance the solid phase of a formulation when a new color is developed. Most important fillers for cosmetic formulations are talc, kaolin, and mica. Other inorganic fillers used in cosmetics are silica, boron nitride, bismuth oxychloride, and barium sulfate [54].

12.5 Building materials

Building materials, which are also called construction materials, as well as coatings, plastics, printing inks, and cosmetic formulations belong to the most important application systems for colorants. Inorganic pigments are the only relevant colorants used in building materials [57]. The pigments have to meet high requirements with regard to color and stability in the application medium, especially in concretes. Several processes are used for coloring of cement and concrete.

Years of experience and tests on colored concrete products, exposed to different climatic conditions all over the world, have shown that particularly inorganic pigments based on metal oxides have the required fastness properties, which are necessary for the application in building materials. Pigments used in building materials must be stable enough to withstand the aggressive influence of strongly alkaline cement pastes. In addition, they must be lightfast, weather resistant, and insoluble in water. The oxidic pigments used in building materials cover all popular colors. Various red, brown, yellow, green, blue, and anthracite color shades as well as white and black are applied. Most important pigments for red, brown, and yellow colors are iron oxides such as iron oxide red, iron oxide yellow, and iron oxide black. Mixed metal oxide pigments based on the rutile structure have a certain importance in the yellow color segment. Green shades are achieved by the use of chromium oxide green, and blue shades are generated using cobalt blue. Titanium dioxide is used for white and carbon black for black tones. Typical pigment concentrations applied in concrete are 3–6% with regard to the content of cement. In special compositions, pigment quantities of up to 10% are possible.

The requirements regarding the processibility of pigments for building materials have changed in the course of time. In the past, mainly powders were used; nowadays, aqueous pigment slurries are applied. Advantages of slurries are reduced dust problems as well as easier handling and metering. However, care must be taken to ensure that the pigments in the slurries do not settle too much before final application. The supply of the pigments in the form of free-flowing dry pigment preparations is another possibility in the building industry. It allows easy emptying of silos, sacks, and bulk bags and avoids also the formation of dust.

Concrete produced with normal gray Portland cement is not suitable for bright colors. These are typically generated by the use of white cement. In cases of black coloring where black pigments are used, there is virtually no difference between concrete made by using white or gray cement. With yellow and blue pigments, this difference is very pronounced. The brighter and cleaner the desired shade, the greater the dependency on the type of cement.

The water–cement ratio has an enormous influence on the color of the concrete. Foam is formed during the process when the cement is incorporated in the water while stirring. The foam consists of many very small bubbles which scatter the light similar to white powder particles. The excess mixing water evaporates from the concrete and cavities remain in the form of pores. These scatter the incident light and make the concrete lighter in this way. The concrete appears the lighter, the lighter the water–cement ratio is.

Concretes produced under suitable conditions have an extremely long steadiness. The color of these concretes is stable as well when pigments such as iron oxides or chromium oxide green are used. Changes of the color shade, if any, are quite small. Changes of the shade can also be detected on unpigmented concrete. They can have various reasons and may be of temporary nature as in the case of efflorescence or of permanent nature as in the case of surface exposure of the aggregates contained in the concrete. The use of weather-resistant pigments for the coloring of building materials is, however, a mandatory prerequisite for a steady color under all weather conditions.

12.6 Conclusions

The application of inorganic pigments always takes place together with a suitable medium in which the pigment particles are incorporated. The pigments are embedded in the application systems to perform their tasks in regard to coloration or functionality. The same situation exists for organic pigments and fillers.

The most important application systems for pigments are automotive and industrial coatings, paints, plastics, printing inks, cosmetic formulations, and building materials. Paper, rubber, glass, porcelain, glazes, and artists' colors are further media which are often colored with pigments, modified with fillers, or equipped with functional properties. Some of the application systems are only of relevance for inorganic pigments and fillers because a high stability is required which is not achieved by organic pigments.

Each application system consists of different components which have to be adjusted to each other for a specific use. The pigments are just one component of a complex system which requires a systematic sequence of processing steps to produce the desired final product.

Well-dispersed pigments in the application medium are an important condition to obtain optimum results with respect to color and functionality. Pigment agglomerates and aggregates should be reduced as far as possible to achieve good dispersion behavior. The surface of the pigment particles should be suitable for effective wetting either by the solvents or other binder components of the application system.

The beautiful appearance or the effective functionality of a final product is always the result of a beneficial interaction of the pigment and the other components of the application medium. A close communication and cooperation of pigment manufacturers, producers of the different components of the application systems, and end product manufacturers is decisive for the achievement of useful and attractive final products.

Study questions

12.1 Which application systems for pigments are the most important?
12.2 In which systems can coatings and paints be divided, depending on their compositions?
12.3 What important components are used in coating systems besides pigments?
12.4 Which processing methods are used for pigmented plastics?
12.5 In which printing processes are pigmented printing inks used, and what are typical sample applications?
12.6 What are important color cosmetic products wherein inorganic pigments find their application?

Bibliography

[1] Koleske JV, in Encyclopedia of Analytical Chemistry, Meyers RA ed. John Wiley & Sons Ltd.,, Chichester, 2000.
[2] Maile FJ, in Encyclopedia of Color, Dyes, Pigments, vol. 1, Pfaff G ed. Walter de Gruyter GmbH, Berlin, Boston, 2022, 161.
[3] Lambourne R, in Paints and Surface Coatings, Theory and Practice, 2nd edn. Lambourne R, Strivens TA ed. Chem Tech Pub. Inc.,, 1999.
[4] Wilson AD, Nicholson JW, Prosser HJ. Surface Coatings, vol. 1, Elsevier Applied Science, London, 1987.

[5]   Wicks Jr ZW, Jones FN, Pappas SP, Wicks DA. Organic Coatings Science and Technology, 3rd edn. John Wiley & Sons, Hoboken, 2007.

[6]   Streitberger H-J, Dössel K-F ed. Automotive Paints and Coatings, 2nd edn. Wiley-VCH, Weinheim, 2008.

[7]   De Lange PG. Powder Coatings Chemistry and Technology, 2nd edn. Vincentz Network, Hannover, 2004.

[8]   Ghosh SK, in Functional Coatings, Ghosh SK ed. Wiley-VCH, Weinheim, 2006, 1.

[9]   Nun E, Oles M, Schleich B. Macromol Symp. 2002,187,677.

[10]   Parkin IP, Palgrave RG. J. Mater. Chem. 2005,15,1689.

[11]   Venkatesan AG, Kumar H, Nair S, Ramakrishna S. J. Mater. Chem. 2011,21,16304.

[12]   Prathapan R, Venkatesan AG, Nair SV, Nair S. J. Mater. Chem. A. 2014,2,14773.

[13]   Kuhr M, Bauer S, Rothhaar U, Wolff D. Thin Solid Films. 2003,442,107.

[14]   Lettieri M, Masieri M. Appl. Surf. Sci. 2014,288,466.

[15]   Perez M, Garcia M, Del Amo B, Blustein G, Stupak M. Surf Coat. Int. Part B – Coat. Trans. 2003,86,259.

[16]   Zhou LC, Koltisko B. JCT CoatingsTech. 2005,2,54.

[17]   Tiller JC, Liao CJ, Lewis K, Klibanov AM. Proc. Natl. Acad. Sci. USA. 2001,98,5981.

[18]   Johns K. Surf Coat. Int. Part B: Coat. Trans. 2003,87(B2),91.

[19]   Cloutier M, Mantovani D, Rosei F. Trends Biotechnol. 2015,33,637.

[20]   Stoye D, Freitag W ed. Paints, Coatings and Solvents, 2nd edn. Weinheim, Wiley-VCH, 1998.

[21]   Calbo LJ ed. Handbook of Coatings Additives, Marcel Dekker, New York, 1987.

[22]   Karsa DR ed. Additives for Water-based Coatings, The Royal Society of Chemistry, Cambridge, 1991.

[23]   Shore J ed. Colorants and Auxiliaries, in Auxiliaries, vol. 2, 2nd edn. Society of Dyers and Colourists, Bradford, 2002.

[24]   Bieleman J ed. Additives for Coatings, Wiley-VCH, Weinheim, 2002.

[25]   Bugnon P. Prog. Org. Coat. 1996,29,39.

[26]   Napper DH. Polymeric Stabilisation of Colloidal Dispersions, Academic Press, London, 1983.

[27]   Lyklema J. Fundamentals of Interface and Colloid Science, vol. 1, Fundamentals. Academic Press, London, 1991.

[28]   Lyklema J, Fundamentals of interface and colloid science, in Solid-Liquid Interfaces, vol. 2, Academic Press, London, 1995.

[29]   Lyklema J, Fundamentals of interface and colloid science, in Liquid-Fluid Interfaces, vol. 3, Academic Press, London, 2000.

[30]   Akafuah NK, Poozesk S, Salaimeh A, Patrick G, Lawler K, Saito K. MDPI Coat. 2016,6,24.

[31]   Du Z, Wen S, Wang J, Yin C, Yu D, Luo J. J. Mater. Sci. Chem Eng. 2016,4,54.

[32]   Sander J. Coil Coating, Vincentz Network, Hannover, 2013.

[33]   Jandel A-S, Meuthen B. Coil Coating – Bandbeschichtung: Verfahren, Produkte, Märkte, Springer Fachmedien, Wiesbaden, 2013.

[34]   Bianchi S, Broggi F. Key Eng. Mater. 2016,710,181.

[35]   Ryntz RA, Yaneff PV ed. Coatings of Polymers and Plastics, Marcel Dekker, Inc.,, New York, Basel, 2003.

[36]   Wilke G, Ortmeier J. Coatings for Plastics, Vincentz Network, Hannover, 2011.

[37]   Schwalm R. UV Coatings, Elsevier B.V., Amsterdam, 2007.

[38]   Stowe RW, Guarniere JW. JCT CoatingsTech. 2009,6,20.

[39]   Charvat RA ed. Coloring of Plastics, 2nd edn. John Wiley & Sons, Hoboken, 2004.

[40]   Müller A. Coloring of Plastics, Hauser Gardner, 2003.

[41]   Harris RM ed. Coloring Technology for Plastics, Plastics Design Library, 1999.

[42]   Christensen IN. Development in Colorants for Plastics (Rapra Review Report) 157, Rapra, 2003.

[43] Pfaff G, in Encyclopedia of Color, Dyes, Pigments, vol. 1, Pfaff G ed. Walter de Gruyter GmbH, Berlin, Boston, 2022, 287.

[44] Russel SD, in Coloring of Plastics, 2nd edn. Charvat RA ed. John Wiley & Sons, Hoboken, 2004, 268.

[45] Santos JD. Pigments for Inkmakers, Select, Industrial Training Association Ltd.,, London, 1989.

[46] Stiebner ED. Bruckmann's Handbuch der Drucktechnik, 5th edn. Bruckmann, München, 1992.

[47] Leach RH, Pierce RJ. The Printing Ink Manual, 5th edn. Kluwer Academic Publishers, Dordrecht, 1999.

[48] Bassemir RW, Bean A, in Kirk-Othmer Encyclopedia of Chemical Technology, John Wiley & Sons, Hoboken, 2001.

[49] Frank E, Rupp R, in Ullmann's Encyclopedia of Industrial Chemistry – Printing Inks, Wiley-VCH Verlag, Weinheim, 2006.

[50] Rathschlag T, in Encyclopedia of Color, Dyes, Pigments, vol. 1, Pfaff G ed. Walter de Gruyter GmbH, Berlin, Boston, 2022, 311.

[51] Schneider G, Gohla S, Schreiber J, Kaden W, Schönrock U, Schmidt-Lewerkühne H, Kuschel A, Petsitis X, Pape W, Ippen H, Diembeck W, in Ullmann's Encyclopedia of Industrial Chemistry – Skin Cosmetics, Wiley-VCH Verlag, Weinheim, 2002.

[52] Petsitis X, Kipper K. Dekorative Kosmetik Und Gesichtspflege, Wissenschaftliche Verlagsgesellschaft, Stuttgart, 2005.

[53] Handbook of Cosmetic Science and Technology, 3rd edn. Barel AO, Paye M, Maibach HI ed. Informa Healthcare USA, Inc.,, New York, 2009.

[54] Lanzendörfer-Yu G, in Encyclopedia of Color, Dyes, Pigments, vol. 1, Pfaff G ed. Walter de Gruyter GmbH, Berlin, Boston, 2022, 245.

[55] Thurn-Müller A, Hollenberg J. Liston J. Kontakte (Darmstadt). 1992,2,35.

[56] Pfaff G, Becker M. Household Personal Care Today. Monogr Suppl. Ser. "Colour Cosmet.", 2012,1,12.

[57] Pfaff G, in Encyclopedia of Color, Dyes, Pigments, vol. 1, Pfaff G ed. Walter de Gruyter GmbH, Berlin, Boston, 2022, 149.

Answers to the study questions

Chapter 1: Fundamentals, general aspects, color, application

1.1 Pigments are substances consisting of small particles that are practically insoluble in an application system and that are used as colorants because of their anticorrosive or magnetic properties. Dyes, like pigments, belong to the colorants, but they are practically completely soluble in the medium of application. The term "colorant" covers all colored compounds regardless of their origin and whether they are used for coloration or other purposes. Colorants are not only divided into pigments and dyes but also into natural and synthetic compounds.

1.2 Pigments are differentiated according to their chemical composition and with respect to their optical and technical properties. A fundamental distinction is that between inorganic and organic pigments. Inorganic pigments are classified into white (titanium dioxide, zinc sulfide including lithopone, and zinc oxide), colored (iron oxide hydroxide, iron oxide, chromium oxide, mixed metal oxides, ultramarine, iron blue, chromates, molybdates, bismuth vanadate, cadmium sulfide, cadmium selenide, cerium sulfide, oxonitrides, and yttrium indium manganese oxides), black (carbon black), and special pigments (effect pigments, transparent pigments, luminescent pigments, magnetic pigments, and anticorrosive pigments). Another classification for inorganic pigments is based on the chemical composition: oxides and oxide hydroxides, sulfides and selenides, chromates, molybdates, vanadates, ultramarine, hexacyanoferrate, and oxonitrides.

1.3 Fillers (extenders) are powdery substances that are like pigments practically insoluble in the application system. They are typically white and are used because of their chemical and physical properties. The distinction between pigments and fillers is made based on the specific application. Another criterion is the refractive index, which is for fillers usually below 1.7 and for pigments above this value. There is, however, no fixed definition of the value for the refractive index to distinguish both product classes.

1.4 Generally, inorganic pigments are more stable against light, weather, temperature, and chemicals than organic pigments. Another advantage of inorganic pigments is their lower manufacturing costs. Organic pigments need mostly multistage syntheses and more expensive raw materials for their production, which leads finally to higher prices for them in the market. They are, however, in various cases more color intensive and, therefore, more attractive than inorganic colored pigments.

1.5 The criteria for the choice of a pigment for a specific application are: color properties (color, tinting strength, lightening power, hiding power), general chemical and

https://doi.org/10.1515/9783110743920-013

physical properties (chemical composition, particle size, particle size distribution, density, moisture and salt content, content of water-soluble and acid-soluble matter, and hardness), stability properties (resistance against light, especially against UV light, heat, humidity, and chemicals, retention of gloss, and corrosion resistance), and behavior in binders and other application systems (dispersion properties, interaction and compatibility with binder components, and solidifying properties).

1.6 The optical properties of a pigmented system are determined by the pigment properties (particle size, particle size distribution, particle shape, refractive index, scattering coefficient, and absorption coefficient), the properties of the application system (refractive index and self-absorption), the pigment volume concentration in the pigmented system, and the hiding power of the pigments used (for white pigments in addition the tinting strength, for black pigments in addition the color strength).

1.7 Primary particles (individual particles) are particles recognizable as such by appropriate imaging techniques (optical microscopy and electron microscopy). Agglomerates are assemblies of primary particles and/or aggregates typically adhering to one another at corners and edges; the total surface area does not differ significantly from the sum of the surface areas of the individual particles and/or aggregates; agglomerates can be destroyed to a certain extent by shear forces. Aggregates are assemblies of primary particles typically adhering to one another side by side; the total surface area is smaller than the sum of the surface areas of the involved primary particles; aggregates can only be conditionally or not at all destroyed by strong shear forces. Flocculates are agglomerates occurring in suspensions or in liquid application media (paints, coatings, printing inks, nail lacquers, etc.), which can be destroyed already by weak shear forces.

1.8 Important analytical methods for the characterization of inorganic pigments are particle size measurement (sedimentation analysis, light scattering, and sieving analysis), X-ray diffraction, measurement of the specific surface area, spectroscopic methods (UV, VIS, IR, NMR, and MS), thermal analysis, and qualitative and quantitative chemical analyses.

1.9 Optical events which may occur when light strikes the surface of a pigmented system are light absorption by the pigment particles (in the case of a colored or black pigment), diffuse reflection of light at the pigment particles (scattering) or in specific cases (effect pigments) directed reflection, and passing of the light through the pigmented medium without further interaction.

1.10 The four theoretical considerations describing the fundamental relationships among the optical characteristics of pigments are Mie theory, the theory of multiple scattering, Kubelka–Munk theory, and colorimetry.

1.11 Color is the characteristic of human visual perception described through color categories, for example, blue, green, yellow, or red. The term "color" is used in general communication also for means that are used for the change of the color impression of objects (painting color and printing color).

1.12 There are six principles for the development of color, which are relevant for inorganic pigments: d–d transfers of electrons within transition metal polyhedrons, charge-transfer transitions of electrons from ligands to a central atom or ion, intervalence transitions of electrons between cations from one and the same element in different valence states, electron transitions based on radical ions in a solid, electron transitions from donor levels of "guest ions" (dopant ions of a transition metal) into the conduction band of the host lattice, and electron transitions from the valence band into the conduction band within a solid.

1.13 The main steps which have to be conducted simultaneously or successively during the process of pigment dispersion in an application system are wetting (removal of air from the surface of the pigment particles and formation of a solvate layer around the same), disintegration (destruction of agglomerates by energy input), and stabilization (maintenance of the dispersed state by generating repulsive forces between the pigment particles, e.g., by steric stabilization or by electrostatic repulsion of particles with identical loading).

1.14 Tinting strength, lightening power, scattering power, hiding power, and transparency are the most important parameters for the efficiency of pigments in paints, coatings, and other application media. An optimal pigment–binder interaction is a decisive factor for the application of inorganic pigments in the different media. Stability against light, weather, heat, and chemicals must be high enough to fulfill the requirements associated with the application of pigmented systems.

Chapter 2: General methods of manufacture

2.1 Mechanical processes used for the manufacture of inorganic pigments are crushing, grinding, classifying, blending, and conveying.

2.2 The objectives for calcination processes are an increase of the crystallite size, refinement of the crystal lattice, solid-state transformation, gas/solid reaction, and solid-state reaction. The most important process parameters are temperature, time, and heating rate.

2.3 Special preparation methods used for the synthesis of inorganic pigments are flux crystallization processes, hydrothermal reactions, and gas-phase reactions.

Chapter 3: White pigments

3.1 The optical action of white pigments is mainly based on nonselective light scattering. White pigments do not absorb light in the visible range. Their strong scattering power leads to high hiding power in applications. The scattering power is higher, the larger the difference between the refractive index of the white pigment and that of the surrounding medium.

3.2 Titanium atoms are surrounded octahedrally by six oxygen atoms in all three TiO_2 modifications and crystal lattices. TiO_6 octahedrons are, therefore, dominant structural units. The single octahedrons differ in their structural arrangement in the modifications. They are linked to one another in different ways via corners and edges. There are two common edges each between the TiO_6 octahedrons in rutile. The octahedrons in brookite are linked via three and in anatase via four common edges. The oxygen atoms in the three crystal lattices are surrounded by three titanium atoms in a trigonal arrangement. Rutile crystallizes in the tetragonal system, anatase as well.

3.3 The raw materials for the production of titanium dioxide pigments include natural products such as ilmenite, leucoxene, and rutile, and also synthetic materials such as titanium slag and synthetic rutile.

3.4 The sulfate process for the production of TiO_2 pigments consists of the following main steps: grinding of the raw materials, digestion, redigestion (dissolution and reduction), clarification, crystallization, hydrolysis, purification of the hydrolysate, doping of the hydrate, calcination, and grinding.

3.5 The chloride process for the production of TiO_2 pigments consists of the following main steps: chlorination, gas cooling, purification of $TiCl_4$, oxidation of $TiCl_4$, and recovery of TiO_2.

3.6 The surface treatment of the TiO_2 particles improves the weather resistance and lightfastness of the pigment itself and of the pigmented organic matrix (application system). It also improves the dispersibility of a certain pigment in the system chosen for the application. Surface treatment is the coating of the individual pigment particles with colorless inorganic and organic compounds of low solubility by depositing them onto the surface. The most common inorganic compounds used for the coating are metal oxides and oxide hydrates as well as silicates and phosphates of titanium, zirconium, and aluminum. Organic surface treatments may be either hydrophilic (alcohols, esters, ethers and their polymers, amines, and organic acids) or hydrophobic (silicones, organophosphates, and alkyl phthalates).

3.7 Other white pigments of technical importance are zinc oxide, zinc sulfide, and lithopone (pigment mixtures consisting of zinc sulfide and barium sulfate). Basic lead carbonate and antimony(III) oxide have lost their relevance as white pigments.

3.8 White pigments are applied in all pigment relevant systems, such as coatings, paints, plastics, artists' paints, cosmetics, printing inks, leather, building materials, paper, glass, and ceramics. As pigments without an absorption color, they are applied for whitening of surfaces and materials, and also for brightening of all colored pigments.

Chapter 4: Colored pigments

4.1 Colored pigments are inorganic or organic colorants that are insoluble in the application system, where they are incorporated for the purpose of coloration. Their optical action is based on the selective light absorption together with light scattering. White, black, gray, effect, luminescent, magnetic, and anticorrosive pigments do not belong to the colored pigments.

4.2 The colored pigments show the broadest variation ranging from blue (mixed metal oxides, ultramarine, iron blue, and yttrium indium manganese oxides) via green (chromium oxide and mixed metal oxides) and yellow (iron oxide hydroxide, mixed metal oxides, chromates, bismuth vanadate, and cadmium sulfide) up to red (iron oxide, cadmium selenide, molybdates, cerium sulfide, and oxonitrides).

4.3 Iron oxides and oxide hydroxides used for pigment purposes are α-FeOOH (goethite) as yellow pigment, α-Fe_2O_3 (hematite) as red pigment, Fe_3O_4 (magnetite) as black and magnetic pigment, γ-FeOOH (lepidocrocite) as magnetic pigment, and γ-Fe_2O_3 (maghemite) as magnetic pigment.

4.4 The production of synthetic iron oxide pigments takes place by solid-state processes (for red, black, and brown pigments), precipitation processes (for yellow, red, black, and brown pigments), and the Laux process (for yellow, red, and black pigments).

4.5 Chromium oxide pigments are produced starting from chromium(VI) compounds, mostly from alkali dichromates, which are reduced to chromium(III) oxide by sulfur or other reducing agents. Ammonium dichromate can directly be converted into Cr_2O_3 by thermal treatment.

4.6 The main representatives of rutile-type mixed metal oxide pigments are nickel rutile yellow with the composition $(Ti,Ni,Sb)O_2$ and chromium rutile yellow with the composition $(Ti,Cr,Sb)O_2$. Examples for spinel-type mixed metal oxide pigments are cobalt blue $(CoAl_2O_4)$, cobalt green (Co_2TiO_4), and spinel black $(CuCr_2O_4)$.

4.7 Ultramarine is a sulfur-containing sodium aluminosilicate. Its structure consists of a sodalite crystal lattice corresponding to the formula $Na_6Al_6Si_6O_{24}(NaS_n)$ with n-values from 2 to 4. There are free polysulfide radicals in the crystal structure. These radicals are normally very unstable, but incorporated as negatively charged polysulfide ions, for example, as S_3^- or S_4^-, in the aluminosilicate lattice together with sodium counterions, and they are stabilized.

4.8 Chromates as well as lead and cadmium compounds used for pigments do not exhibit acute toxicity. They are not irritating to the skin or mucous membranes. The pigments are compounds with low solubility. However, these pigments have to be handled carefully due to their components. Provisions have to be made and workplace concentration limits have to be kept when handling these pigments. The use of these pigments is controversial and no longer seen as optimal due to the principal possibility of solving at least small amounts of the heavy metals uncontrolled. The potential danger is seen that components of the pigments can escape into the cycle of nature unintendedly. Recycling of chromates as well as lead and cadmium compounds demands special care and advanced processing technologies.

Chapter 5: Black pigments (carbon black)

5.1 The most important inorganic black pigments are carbon black, iron oxide black, and spinel black. Carbon black pigments are by far of greatest importance from these blacks.

5.2 Carbon black contains structural motifs that are very similar to those of graphite, which is crystallizing in a layer structure of hexagonally bound C atoms. The tightly bound layers are connected among each other only by loose van der Waals forces. The carbon atoms within the layers of carbon black are arranged almost in the same way as in graphite. The carbon layers are nearly parallel oriented to each other. The relative position of these layers in carbon black is, however, arbitrary, unlike in the case of graphite. Thus, there is no order in the structural c direction for carbon black. Crystalline regions in carbon black are, therefore, only of small size, typically 1.5–2 nm in length and 1.2–1.5 nm in height, corresponding to four to five carbon layers.

5.3 Raw materials used for the manufacture of carbon black depend on the process. Furnace black process: aromatic oils on coal tar basis or mineral oil, and natural gas; gas black and channel black process: coal tar distillates; lamp black process: aromatic oils on coal tar basis or mineral oil; thermal black process: natural gas or mineral oils; acetylene black process: acetylene.

5.4 The main reaction for the formation of carbon black pigments is the pyrolysis of hydrocarbons:

$$C_nH_m + \text{energy} \rightarrow n\,C + m/2\;H_2$$

5.5 The main advantages of carbon black pigments compared with other black pigments are excellent hiding power, color stability, solvent resistance, acid and alkali resistance, and thermal stability.

5.6 Particle size, particle size distribution, surface quality, and structure determine the coloristic and application technical properties of the individual carbon black pigments. Oxidative aftertreatment is used in many cases to modify the surface of the pigments concerning the stability and the compatibility with the application system.

Chapter 6: Ceramic colors

6.1 Ceramic colors contain pigments with high temperature and chemical stability (ceramic pigments). They are mainly used in glazes and enamels for coloring of porcelain, earthenware bone china, and other ceramics. The intended application determines the selection of the ceramic color including the ceramic pigment. This pigment with its high thermal stability is the color giving component of a ceramic color.

6.2 Colored ceramic surfaces can be achieved by dissolution of metal oxides in the glaze or enamel themselves or by the use of ceramic colors. These colors are introduced in the ceramic materials before sintering or with the glaze. The use of ceramic colors is the most effective way to achieve reproducibly defined color tones for a glaze or enamel composition.

6.3 Technically important pigments which are components of ceramic colors are cadmium sulfide and cadmium sulfoselenides, metals such as gold, silver, platinum, and copper, diverse metal oxides, mixed metal oxides, and silicates, as well as zirconia and zircon-based compositions.

Chapter 7: Transparent pigments

7.1 Pigments show transparent behavior in application systems when the size of their particles is in the nanometer range below 100 nm. The smaller their particle sizes, the more transparent they are. Transparent pigments are also characterized by large specific surface areas. Most of the technically relevant pigments consist of particles, which are even smaller than 30 nm. They are classified as nanomaterials. Nanosized primary particles do not or only slightly scatter visible light, which is the reason for their transparency.

7.2 Important representatives of colorless transparent pigments are titanium dioxide and zinc oxide. They do not absorb light of the visible spectral range and can also be regarded as functional nanomaterials with special properties. Colored transparent pigment types such as iron oxides, cobalt blue, and iron blue are mainly characterized by their color properties. In some cases, however, they also have functional properties.

7.3 Transparent TiO_2 and ZnO pigments find their main applications as functional powders. They are characterized by strong UV absorption and are, therefore, applied as UV absorbers. As such, they are used for the protection of organic materials such as plastics and coatings. Moreover, they are used in a broad diversity of sunscreen products where they protect the human skin.

7.4 The production of transparent pigments takes place mostly using wet chemical or gas-phase reactions. Special attention should be paid to avoiding of aggregates and agglomerates. Their presence in the application medium would influence the effect of transparent pigments in a way that they would act similarly to nontransparent inorganic pigments.

Chapter 8: Effect pigments

8.1 Effect pigments are colorants that provide additional color effects, such as angular color dependence (luster, iridescence, and color travel) or texture to an application system. The term "luster pigments" is also often used for effect pigments because almost all effect pigments provide lustrous effects in their applications. Luster pigments consist of predominantly platelet-like particles, which readily align with a parallel orientation to the surface to which they are applied. A characteristic luster is generated arising from the reflection of incident light from the smooth surface of the pigment platelets.

8.2 Effect pigments are subdivided into metal effect pigments (metallic effect pigments) and special effect pigments including pearl luster pigments (pearlescent pigments, nacreous pigments) and interference pigments. They consist of μm-sized thin platelets that show strong lustrous effects when oriented in parallel alignment in application systems.

8.3 Metal effect pigments are luster pigments consisting of metallic flakes. After parallel orientation in their application medium, they show a metal-like luster by reflection of light at the surface of the platelet-like metal particles. The pigments are nontransparent for light and reflect the entire incident light in one direction. The pigment particles act similar to small mirrors and lead when orientated parallel in the application system to a reflecting metal luster (metallic effect).

8.4 Special effect pigments are in the very most cases synthetic pigments, characterized by high luster, brilliance, and iridescent colors known from optically thin films. The visual appearance has its origin in reflection and refraction of light at thin single and multiple layers. The pigments are either transparent, semitransparent, or light-absorbing platelet-shaped crystals or layer systems. They can consist of single crystals, and also of monolayer or multilayer structures in which the layers have different refractive indices and light absorption properties.

8.5 Metal effect pigments are subdivided into silver bronzes (pigments consisting of aluminum) and gold bronzes (pigments consisting of copper or copper/zinc alloys). There are also metallic pigments and powders besides the metal effect pigments, which are used for functional purposes in coatings, for example, for anticorrosion, IR reflectivity, heat resistance, and electrical conductivity. Metals used here include zinc, stainless steel, and silver.

8.6 Aluminum pigments are distinguished in the unregularly shaped "cornflake type", the lenticular "silver dollar type", and the "VMP type" (vacuum metallized pigment).

8.7 The production of aluminum pigments consists of the manufacture of aluminum grit (from aluminum ingots, which are melted and sprayed through a nozzle), grinding of the aluminum in ball mills in the presence of mineral spirit, and a lubricant (Hall process) under formation of flakes, sieving, classification, and blending with organic solvents (for aluminum pastes).

8.8 The most important substrate-free special effect pigments are natural fish silver, basic lead carbonate, bismuth oxychloride, and micaceous iron oxide.

8.9 Pigments based on the layer-substrate principle represent the dominant class of special effect pigments. The variation and combination possibilities of layer and substrate materials are very broad. Special effect pigments based on natural mica as substrate material are the most important representatives of layer-substrate pigments. Other substrates, such as alumina, silica, borosilicate glass, and synthetic mica, are used in the form of platelets as alternatives to natural mica.

8.10 The main steps for the production of metal oxide mica pigments are mica preparation (manufacture of mica platelets by grinding, followed by classification), coating of the mica platelets with hydrated metal oxide layers (e.g., hydrated TiO_2 and Fe_2O_3)

in an aqueous suspension, filtration, washing, drying, and calcination (formation of thin metal oxide layers on the mica platelets).

8.11 Functional pigments based on the layer-substrate principle consist typically of metal oxide layers on platelets of natural mica. The metal oxides used possess electrically conductive, magnetic, infrared reflecting, or other physical properties, and give rise to the functional behavior of the pigments. Technically important functional pigments belonging to this group are light-colored conductive pigments, solar heat-reflecting pigments, and pigments for the laser marking of polymers.

Chapter 9: Functional pigments

9.1 Functional inorganic pigments are characterized by special physical properties. Color and appearance do not play a role for this category of pigments. The two main representatives of inorganic functional pigments are magnetic and anticorrosive pigments. Further functional pigments include electrically conductive, IR reflective, laser marking, and UV-absorbing materials.

9.2 Magnetic pigments are finely divided powders with distinct magnetic properties. They are used in magnetic information storage systems, for example, in audio cassettes, videotapes, floppy disks, hard disks, and computer tapes. They are also applied as toner pigments in photocopiers and laser printers and find application in special printings, for example, in the printing of banknotes. The pigment particles are dispersed for magnetic applications in suitable binder systems and then applied to the storage system.

9.3 Main representatives of magnetic pigments are γ-Fe_2O_3, Fe_3O_4, Co-containing iron oxides (Co-γ-Fe_2O_3 and Co-Fe_3O_4), metallic iron, CrO_2, and $BaFe_{12}O_{19}$.

9.4 There are five types of magnetism: diamagnetism, paramagnetism, ferromagnetism, ferrimagnetism, and antiferromagnetism. From these, only ferromagnetism and ferrimagnetism are relevant for magnetic pigments.

9.5 The essential characteristics of magnetic pigments are saturation magnetization, remanence, and coercive field strength, as well as particle shape and size. The coercive field strength determines the area of application. High coercivities are prerequisites for high storage densities. The coercive field is the magnetic field required to demagnetize the material. The saturation magnetization is a specific constant for the magnetic material. The magnetic properties of a pigment are typically determined by measurement of hysteresis curves on the powder or on magnetic tapes.

9.6 Anticorrosive pigments play an important role in the protection of metallic materials against corrosion. Damage arising from corrosion of metals is enormous. The performance of a protective coating can significantly be improved by the addition of anticorrosive pigments. The corrosion protection of a coating depends particularly on the effective interaction of the anticorrosive pigment used and the binder system chosen.

9.7 The main representatives of anticorrosive pigments are Pb_3O_4, chromates, phosphates, phosphites, borates, molybdates, zinc cyanamide, metallic zinc, and micaceous iron oxide. Other important anticorrosive materials are ion exchange pigments, titanium dioxide-based pigments, and inorganic–organic hybrid pigments.

9.8 Corrosion is a process in which a solid, especially a metal, is eaten away by chemical or electrochemical reactions resulting from exposure of the material to weathering, moisture, chemicals, or other agents in the environment into which it is placed. Corrosion can also be described as an electrochemically driven process of energy exchange. The assumption is that the metals, which were originally found in nature in the form of metal compounds in ores, are reconverted to their ores. Iron and other corroding metals are thermodynamically not stable. They, therefore, tend to oxidation under formation of compounds, for example, oxides, sulfides, and silicates, in which they have a stable oxidation state. In summary, corrosion can be regarded as the natural tendency of metals to achieve their most stable form.

9.9 The main electrochemical reactions for the corrosion of a metal can be summarized as follows (e.g., corrosion of iron):

Anodic reaction: $\qquad\qquad\qquad Fe \rightarrow Fe^{2+} + 2\,e^-$

Cathodic reaction: $\quad 1/_2\,O_2 + H_2O + 2\,e^- \rightarrow 2\,OH^-$

Summary: $\qquad\qquad Fe + 1/_2\,O_2 + H_2O \rightarrow Fe(OH)_2$

9.10 The three ways of action for anticorrosive pigments in an application system are chemical/electrochemical action (active pigments), physical action (barrier pigments), and electrochemical/chemical action (sacrificial pigments). The objective of the pigmentation in all cases is to prevent the contact of water, oxygen, and other corrosion-supporting components (ions from salts, acids, and hydroxides) with the metal surface to be protected.

Chapter 10: Luminescent pigments

10.1 Luminescent materials and pigments are synthetically generated crystalline compounds, which absorb energy followed by emission of light with lower energy and longer wavelengths. The light emission often occurs in the visible spectral range. The nature of luminescence is, therefore, different from blackbody radiation. External energy has to be applied to luminescent materials to enable them to generate light. Luminescence can have its origin in very different kinds of excitation such as photo- or electroluminescence caused by X-rays, cathode rays, UV radiation, or even visible light. It is based on electronically excited states in atoms and molecules. The emission process is governed by quantum mechanical selection rules.

10.2 The distinction between fluorescence and phosphorescence goes back to different energy transitions. Emission based on allowed optical transitions, with decay times in the order of µs or faster, is defined as fluorescence. Emission with longer decay times is called phosphorescence. The occurrence of fluorescence or phosphorescence, as well as the decay time, depends on the structure and composition of a specific luminophore.

10.3 The four luminescence mechanisms discussed for inorganic luminescent materials are center luminescence, charge-transfer luminescence, donor–acceptor pair luminescence, and long-afterglow phosphorescence. The emission of luminescent light can have its origin in different excitation mechanisms such as optical excitation (UV radiation or even visible light), high-voltage or low-voltage electroluminescence, and excitation with high-energy particles (X-rays and γ-rays).

10.4 Inorganic luminescent pigments are used mainly in fluorescent lamps, cathode ray tubes, projection television tubes, plasma display panels, LEDs, and for X-ray and γ-ray detection. The pigment particles are dispersed for the applications in specific binder systems.

10.5 All relevant visible emission colors are possible to produce with inorganic luminescent pigments. Examples for green are $ZnS:Cu^+,Au^+,Al^{3+}$ and $Zn_2SiO_4:Mn^{2+}$, for yellow $Y_3Al_5O_{12}:Ce^{3+}$, for red $Y_2O_2S:Eu^{3+}$ and $Y_2O_3:Eu^{3+}$, for blue $ZnS:Ag^+,Cl^-$ and $BaMgAl_{10}O_{17}:Eu^{2+}$, and for white $ZnS:Ag^+ + (Zn,Cd)S:Ag^+$ or $Ca_5(PO_4)_3(F,Cl):Sb^{3+},Mn^{2+}$.

Chapter 11: Fillers

11.1 The term "filler" defines powdered materials that are irregular, acicular, fibrous, or platy in shape and which are added to application media to improve properties of the system or to lower the consumption of more expensive binder components. In the latter case, fillers are also called extenders. Fillers are typically applied in relatively large loadings. They are often used in combination with pigments to better properties of the material, in which they are embedded.

11.2 Several naturally occurring white minerals, such as calcium carbonate, magnesium carbonate, silicon dioxide, and some silicates, are used as fillers. Important synthetic fillers are calcium carbonate, barium sulfate (blanc fixe), aluminum hydroxide, and silicon dioxide. Calcium carbonate has the largest share of all fillers.

11.3 Mining and processing technologies for natural fillers are similar to those used for the manufacture of natural pigments. Synthetic fillers are produced mainly by precipitation techniques. The properties of synthetic fillers are controlled and adjusted by the specific process parameters used.

11.4 The main application systems for filler materials are coatings, paints, and plastics. However, fillers can also be used in all other application systems. Calcium carbonate is mainly used in plastics, and barium sulfate in automotive and industrial coatings, paints, and plastics. Aluminum hydroxide fillers are broadly used as flame-retardant materials in many polymers. Silicon-containing fillers such as silica and silica gel are mainly applied in paints, coatings, and plastics.

Chapter 12: Application systems for pigments

12.1. The most important application systems for pigments are automotive and industrial coatings, paints, plastics, printing inks, cosmetic formulations, and building materials. Paper, rubber, glass, porcelain, glazes, and artists' colors are further media that are often colored with pigments and modified with fillers. Some of the application systems are only of relevance for inorganic pigments and fillers because organic pigments do not exhibit the necessary stability.

12.2 Depending on their compositions, coatings, respectively, paints can be divided into three groups: (1) solventborne, (2) waterborne, and (3) solvent-free (100% solid). Solventborne systems consist of resins, additives, and pigments that are dissolved or dispersed in organic solvents. Similarly, in waterborne paints, the ingredients are dispersed in water. In solvent-free compositions, the paints do not contain any solvent or water, and the ingredients are dispersed directly in the resin.

12.3 Important components used in coatings besides pigments are binders, resins, plasticizers, defoamers, wetting, and dispersing additives.

12.4 Processing methods used for pigmented plastics are injection molding, injection blow molding, extrusion into blow molding, calendaring, foaming, coating, and production of flat films, blown films, profiles, and fibers.

12.5 Pigmented printing inks are used in all printing processes such as letterpress printing, offset printing, flexographic printing, gravure printing, or screen printing. Sheet-fed as well as rotary printing presses are most common. Typical sample applications are books, newspapers, magazines, catalogues, packing materials, wallpapers, cash notes, stamps, and labels.

12.6 Important color cosmetic products wherein inorganic pigments find their application are primers, lip products, concealers, foundations, face powders, rouge, contour powders, mascara, eye shadows, eye liners, nail polish, setting sprays, and setting powders.

Index

Printed in the USA
CPSIA information can be obtained
at www.ICGtesting.com
LVHW080157140624
783111LV00003B/343